Transfer and Storage of Energy by Molecules

Volume 2
Vibrational Energy

Transfer and Storage of Energy by Molecules

(A Multi-volume Treatise)

Volume 1 : ELECTRONIC ENERGY

Volume 2 : VIBRATIONAL ENERGY

Volume 3 : ROTATIONAL ENERGY

Additional Volumes to Follow

Transfer and Storage of Energy by Molecules

Volume 2
Vibrational Energy

Edited by

George M. Burnett *Professor of Physical Chemistry, University of Aberdeen*

Alastair M. North *Professor of Physical Chemistry, University of Strathclyde, Glasgow*

WILEY — INTERSCIENCE

A division of John Wiley & Sons Ltd

London — New York — Sydney — Toronto

Library of Congress catalog number 77–78048

SBN 471 12431 1

Printed in Great Britain by J. W. Arrowsmith Ltd., Bristol 3

Preface

This is the second volume in the series presenting authoritative articles on molecular energy transfer processes. Volume 1 covered electronic energy, for which the quantities involved are frequently rather large. By contrast the energies involved in vibrational excitation are only slightly greater than thermal energies at room temperature, and consequently vibrational excitation can be brought about by a variety of methods, excitation by thermal means being very important. It is the purpose of this volume to review the current theories of vibrational energy transfer, to cover the background of the experimental methods currently used to study the phenomenon, and to discuss the experimental results obtained to date.

Two techniques—ultrasonic relaxation spectroscopy and the shock tube—are of outstanding importance in this field, and consequently are afforded considerable coverage. Although shock tube techniques are mentioned in the two opening chapters reviewing the theory and experimental results of vibrational energy transfer in gases, detailed coverage of this powerful tool is provided in a separate article.

In rather the same way the use of ultrasonic relaxation spectroscopy to study liquid systems usually involves a slightly different approach than when gases are involved. Consequently, although the technique is mentioned in the chapters dealing with the gas phase, a separate chapter on liquid systems is included. Indeed the ever-widening application of ultrasonic and related relaxation techniques to condensed phases is one of the current 'growth points' in physical chemistry.

A complete understanding of the chemical significance of vibrationally excited species is fundamental to the discipline of chemical kinetics, and this inter-relationship is emphasized in the chapter on collisional excitation and chemical reaction.

The division of energy transfer processes into electronic energy and vibrational energy is somewhat arbitrary, since so often both forms of energy are involved simultaneously. However it is hoped that the division has not resulted in any important omissions, and that a coverage of the combined field is provided by Volumes 1 and 2 together.

Contributing Authors

BORRELL, P.

Senior Lecturer in Chemistry, University of Keele, Staffordshire.

ORVILLE-THOMAS, W. J.

Professor of Physical Chemistry, University of Salford, Lancashire.

PRITCHARD, H. O.

Professor of Chemistry, Centre for Research in Experimental Space Science, York University, Toronto 12, Canada.

STRETTON, J. L.

Head of Chemistry Department, Central Electricity Generating Board, South West Region Scientific Services Department, Portishead, Bristol.

WYN-JONES, E.

Lecturer in Chemistry, University of Salford, Lancashire.

Glossary of Symbols

Figures in parentheses indicate the chapter where a particular usage is to be found.

A	empirical constant descriptive of relaxation strength (4)
A	amplitude of oscillation (2)
$\bar{A}$	average movement of molecules
$A^2 \Sigma^+$	spectroscopic molecular state
a_1	sound speed in a gas .
a_4	sound speed in driver gas
a	$h\nu/kT$
B	the second virial coefficient of a gas (1)
B	empirical constant descriptive of classical sound attenuation (4)
b	energy function
C	Sutherland constant
C_p	specific heat at constant pressure
$C_{p\infty}$	specific heat at constant pressure at very high frequencies
$C_{p_{eff}}$	effective specific heat
ΔC_p	relaxation part of specific heat
C_v	specific heat at constant volume
c	velocity of light
D_i	relaxation constant
d	distance between an atom and the nearest atom of a molecule
d_0	half the internuclear distance in a simple harmonic oscillator
E	instantaneous value of the relaxation energy

ΔE	energy converted from translation to vibration
$E(T)$	value of the relaxation energy at the equilibrium temperature, T
e	specific internal energy
F	transient driving force on an oscillator during collision
F_1	relaxation parameter
f	sound frequency
f_c	relaxation frequency
g_i	degeneracy of ith state
H	Hamiltonian of an unperturbed oscillator, enthalpy
h	enthalpy of a gas
I	moment of inertia
J	flux between states, angular momentum quantum number (1), (2) relaxation parameter (ultrasonic, in a liquid) (4)
K	coefficient of thermal conductivity (1)
$\mathbf{K}$	modulus of a system (4) equilibrium constant for a two-state system
K_c	compressional modulus
K_s	sheer modulus
K'	real part of the modulus
K''	imaginary part of the modulus
K_r	relaxing part of the modulus
$k'_{01} k'_{10}$	bimolecular rate constants for vibrational excitation (de-excitation)
k_i	wave number
L_f	mean free path
L, L'	subsidiary collision functions
L_{ik}	phenomenological coefficients
M	mass number
M_A	mole fraction of A
N	number of molecules per c.c., number of quantum states in an oscillator
$\tilde{N}$	number of collisions per second
n	number of atoms in a molecule
	population of nth state
P	transition probability
P	probability of change of vibrational energy in a collision
p	pressure in a liquid due to an ultrasonic wave (4) gas pressure (3)

P^{i-j}	probability of vibrational excitation
P^{1-0}	probability of transfer of vibrational energy per collision (Napier Probability)
p_c'', p_c'''	final pressures after a shock
Q	normal coordinate of vibration
q_{ij}	inelastic cross-section
r	separation of molecules (1), (2), (3), relaxation strength (4)
r	the distance between two molecules
S	stress in a system, entropy
s	compression (1), distance travelled by one molecule in the repulsive field of another (2), strain in a system (4)
T'	instantaneous vibrational temperature
T_{tr}	translational temperature
$\overline{T}$	mean translational temperature
U	complex velocity, potential energy
U_0	complex velocity at very low frequencies
u	particle velocity (1), relative velocity at collision (2)
$u_1 u_2$	gas flow velocities in a shock tube
V	potential energy of a system
V_{ii}	quantum mechanical matrix element
V_q	total number of vibrational quanta
v	vibrational quantum number (3), sound velocity in a liquid (4)
W	measured velocity of a sound in a gas, real phase velocity of a wave
W_0	velocity of sound at very low frequencies
W_1	shock wave velocity
W_2	particle velocity behind a shock wave
W_{ij}	matrix element from the oscillator time-dependent wave equation
w_1	shock velocity relative to observer
w_2	gas velocity relative to observer
X_i	thermodynamic driving force
$X^2\pi$	spectroscopic molecular state
x	Navier–Stokes coefficient
χ_d	length of driver section in a shock tube, rate coefficient
Z (Napier Number)	total number of collisions per second undergone by a molecule

Z_{rot}	number of collisions required to reduce a perturbation of the rotational energy to $1/e$ of its original value
α	absorption coefficient, exponent in repulsive potential, thermal average approach velocity (1), (2)
α'	exponent in repulsive potential
α_{cl}	coefficient of classical absorption
β	generalized relaxation time, adiabatic compressibility of a liquid (4)
β_T	isothermal compressibility
γ	ratio of specific heat at constant pressure to specific heat at constant volume, subsidiary vibration frequency parameter (1), (2)
γ_0	ratio of C_p to C_v at very low frequencies
γ_{eff}	effective ratio of C_p to C_v
γ_∞	ratio of C_p to C_v at very high frequencies
$\tilde{\gamma}$	ratio of C_p to C_v in an ideal gas
γ_4	heat capacity of driver gas
δ	phase of oscillation (2), cross-section for deactivation (3)
ε	relaxation strength, Lennard–Jones parameter
ε_a	vibrational energy of oscillator in a th state
η	coefficient of viscosity
θ	angle between line of approach and line of centres (1), (2), characteristic temperature of vibration (equals hv/k) (3), coefficient of thermal expansion (4), scattering angle in Brillouin scattering (4)
$\kappa_{S_{eff}}$	effective adiabatic compressibility
κ_{S_0}	adiabatic compressibility at very low frequencies
λ	wavelength of sound, displacement of oscillator from equilibrium position (1), (2)
λ_e	wavelength of incident light in Brillouin scattering
μ	reduced mass of a collision complex (2), absorption per wavelength (4)
μ_m	maximum value of attenuation per wavelength
v	frequency of radiation, oscillator frequency
ξ	dimensionless energy term
ρ	density
$\bar{\rho}$	mean density
τ_{pS}	adiabatic relaxation time at constant pressure

τ_{pT}	isothermal relaxation time at constant pressure
τ_R	radiative lifetime
τ_{VS}	adiabatic relaxation time at constant volume
τ_{VT}	isothermal relaxation time at constant volume
Φ	amplitude factor of a plane wave
ϕ	any disturbance in a plane harmonic damped wave
χ	the state of oscillation of a molecule
Ψ	time-dependent wave function for an oscillator
ψ	wave function of ith state
Ω	reduced mass of contact atoms in collision
ω	angular frequency

Contents

1   The Measurement of Vibrational Relaxation Times in Gases
1.1   Introduction   1
1.2   Acoustic methods   6
    1.2.1   The propagation of sound in a relaxing gas   .   7
    1.2.2   Acoustic instruments   27

2   Vibrational Energy Exchange at Molecular Collisions
2.1   Introduction   58
2.2   Calculation of the transition probabilities   .   .   .   60
    2.2.1   One-dimensional collision between an atom and a
        diatomic molecule   62
    2.2.2   Three-dimensional collision   69
    2.2.3   The intermolecular potential   72
    2.2.4   Completion of the quantum-mechanical calcula-
        tion of the inelastic cross-section: the SSH–
        Tanczos formula   75
    2.2.5   Vibrational and steric factors   79
    2.2.6   Discussion of the methods of calculation   .   .   81
2.3   Calculation of relaxation times from transition proba-
    bilities   89
    2.3.1   The SSH–Tanczos method of rate equations   .   90
    2.3.2   The Bauer method using irreversible thermo-
        dynamics   96
    2.3.3   Schuler's investigations of relaxation   .   .   99
    2.3.4   The calculation of transition probabilities from
        experimental results: series and parallel excitation   100
2.4   Review of experimental relaxation times and com-
    parison with theoretical ones   103
    2.4.1   Pure gases and mixtures with only one relaxing
        component   103
    2.4.2   Binary mixtures containing two relaxing com-
        ponents   140

2.5 General discussion 164
  2.5.1 The effect of rotation 164
  2.5.2 Conclusion 168

3 The Measurement of Vibrational Relaxation Rates by Shock Techniques
3.1 Introduction 180
  3.1.1 Scope of the chapter 180
  3.1.2 Some terminology and conventions: Napier time, vibrational states and levels, relaxation . . 181
3.2 Methods for studying vibrational relaxation . . . 185
  3.2.1 The simple shock tube 186
  3.2.2 Ancillary shock methods: expansions . . 188
  3.2.3 Ultrasonic methods 189
  3.2.4 The spectrophone 190
  3.2.5 Flash photolysis 191
  3.2.6 Infrared fluorescence 192
  3.2.7 Ultraviolet fluorescence 192
  3.2.8 A Raman method 192
  3.2.9 Summary 193
3.3 Elementary shock-tube theory 193
  3.3.1 The wave patter in a simple shock tube . . 193
  3.3.2 Coordinate systems and nomenclature . . 195
  3.3.3 Conservation equations: the Rankine–Hugoniot relations 196
  3.3.4 The Hugoniot relation 197
  3.3.5 Solution of the conservation equations for ideal gases 201
  3.3.6 Solution of the conservation equations for gases with active vibrations 205
  3.3.7 Compression of the observation time . . 206
  3.3.8 Shock experiments in real gases . . . 209
  3.3.9 Miscellaneous equations for calculation of shock-tube parameters 210
  3.3.10 Reflected shock 211
3.4 Experimental methods in shock tubes 212
  3.4.1 Construction 212
  3.4.2 Velocity measurements 212
  3.4.3 Techniques for relaxation measurements . . 213

3.4.4   Interferometry   213
3.4.5   Schlieren and other densitometry techniques; the
        theory of density changes   214
3.4.6   Line and band-reversal techniques   .   .   .   216
3.4.7   Infrared fluorescence   219
3.4.8   Ultraviolet fluorescence and absorption   .   .   221
3.5   The adequacy of models used in the interpretation of
      Napier Times   222
3.5.1   Introduction   222
3.5.2   Two-state model   223
3.5.3   Landau–Teller model for an harmonic oscillator .   224
3.5.4   Mixtures   226
3.5.5   Absolute calculations of Napier times   .   .   228
3.5.6   Empirical suggestions   229
3.5.7   The applicability of the two-state and Landau–
        Teller models   230
3.5.8   Higher vibrational levels   232
3.5.9   Temperature dependence   235
3.5.10 Mixtures   240
3.6   Results for individual molecules   241
3.6.1   Table of Napier times   241
3.6.2   Hydrogen   243
3.6.3   Deuterium   244
3.6.4   Nitrogen   245
3.6.5   Oxygen   247
3.6.6   Air   249
3.6.7   Carbon monoxide   249
3.6.8   Nitric oxide   252
3.6.9   Hydrogen halides   254
3.6.10 Chlorine   255
3.6.11 Cyanide radical   255
3.6.12 Nitrous oxide   255
3.6.13 Carbon dioxide   256
3.6.14 Methane   258
3.7   Some conclusions   258

4   Molecular Relaxation Processes in Liquids
4.1   General considerations   265
4.1.1   Introduction   265

4.1.2   Energy of liquids   267
4.1.3   Relaxational coupling   268
4.1.4   The interaction of acoustic waves with liquids   .   269
4.2   Experimental methods   282
4.2.1   Introduction   282
4.2.2   Resonance and resonance-reverberation methods .   284
4.2.3   Tuning fork   285
4.2.4   Acoustic streaming   286
4.2.5   Pulse methods   288
4.2.6   Optical method   290
4.2.7   Radiation pressure   291
4.2.8   Ultrasonic interferometer   292
4.2.9   Comparison measurements   294
4.2.10 Brillouin spectroscopy   295
4.3   Analysis of experimental data   296
4.3.1   Calculation of thermodynamic data   .   .   .   296
4.3.2   Calculation of kinetic parameters   .   .   .   302
4.3.3   Numerical example   304
4.4   Rotational isomerism   314
4.4.1   Substituted ethanes   314
4.4.2   Substituted cyclohexanes   329
4.4.3   Tertiary amines   334
4.4.4   Unsaturated molecules   338
4.5   Carboxylic acids   345
4.6   Vibrational relaxation   349
4.6.1   Introduction   349
4.6.2   Basic theory   349
4.6.3   Specific systems   353
4.6.4   General discussion   356

5   Collisional Excitation and Chemical Reaction
5.1   Introduction   368
5.2   The accumulation of vibrational energy   .   .   .   368
5.3   Dissociation of diatomic molecules diluted in a large
      excess of an inert gas   374
5.4   Dissociation of undiluted diatomic gases   .   .   .   377
5.5   Dissociation of polyatomic molecules   378
5.6   Bimolecular reactions   385
5.7   Conservation of spin in chemical reactions   .   .   .   385

Contents

Author Index 391

Subject Index 401

1

The Measurement of Vibrational
Relaxation Times in Gases*

J. L. Stretton

1.1 INTRODUCTION

The specific heat of a substance is defined as the rate of change of its energy with temperature. If a sound wave is passed through a gas it produces a sinusoidal variation of the temperature for each elemental volume of the gas. At low frequencies all the energy changes with the temperature, local thermodynamic equilibrium is maintained and the specific heat at constant volume (C_V), calculated from the measured velocity (W) using the well-known formula for an ideal gas,

$$W^2 = \gamma \frac{RT}{M} = \frac{RT}{M}\left(1 + \frac{R}{C_V}\right) \qquad (1.1)$$

is in good agreement with the values obtained by calorimetric and spectroscopic methods. At higher, usually ultrasonic, frequencies not all the energy changes with the fast-varying temperature so the specific heat is lower, and consequently the velocity is higher than at low frequencies. Such a variation of velocity with frequency is termed dispersion.

This effect was first observed by Pierce[1], working with carbon dioxide. He found the velocity increased with frequency over the range 20 kc/s to 200 kc/s, and then became constant again. Herzfeld and Rice[2] explained this by suggesting that energy could be exchanged between the translational and vibrational degrees of freedom of the gas molecules only at a finite rate, and so, after an abrupt change in the translational energy, there is a delay before the vibrational and translational energies are in equilibrium again. The translational energy is continuously varying in a sound wave and,

* The author would like to thank the Central Electricity Generating Board for support during the period when the majority of his research in this field was carried out.

1

when its period for a cycle is comparable with this delay, energy equilibrium is not attained and the vibrational degrees of freedom are not contributing fully to the specific heat. The changes in the vibrational and translational temperatures are out of phase, the former lagging behind the latter with a smaller amplitude. The smaller the amplitude the less the specific heat contribution and, at the high frequency end of the dispersion zone, Pierce found that the apparent specific heat calculated from the velocity had fallen to that expected for translational and rotational motion only.

Because of this lag, energy is absorbed as the sound wave travels through the gas and so its amplitude decreases. This can be demonstrated by considering the simple system of an ideal gas enclosed in a cylinder by a frictionless piston and subjected to alternate adiabatic compressions and dilations, represented for simplicity by a step function. If there is no lag the pressure (p) and volume (V) vary together giving the p–V diagram as shown in Figure 1.1a. If however the vibrational energy lags behind the translational the situation is as in Figure 1.1b. The pressure depends on

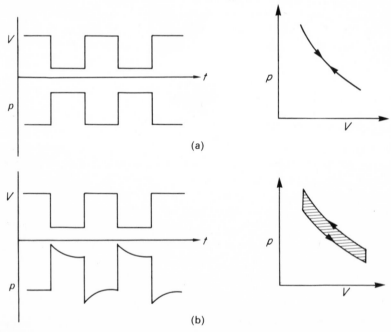

Figure 1.1 (a) Square wave and p–V diagram without relaxation; (b) square wave and p–V diagram with relaxation

the translational energy and the initial rise (or fall) exceeds the equilibrium value; then, as the energy flows into the vibrational modes, it drops to this value. Because of the lag there is now a net absorption of the energy by the gas during the cycle. This is represented by the shaded area of the p–V diagram in Figure 1.1b. The velocity and absorption vary with frequency as shown in Figure 1.2, the absorption reaching a peak near the middle of the dispersion zone. As can be seen from Figure 1.1 the gas is less compressible at the higher frequencies.

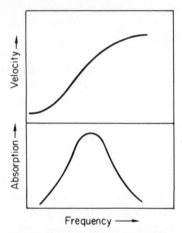

Frequency ⟶

Figure 1.2 Dispersion and absorption curves for a relaxation zone

When a physical quantity fails to follow rapid changes in another one, with which it is normally in equilibrium, it is said to be relaxed. With the carbon dioxide in Pierce's experiment it is the vibrational energy which has relaxed. Relaxation does not produce anything which would not have occurred anyway in a static process, but may prevent something that would have then occurred. The lag inherent in the relaxation process is characterized by the relaxation time (β) defined, following the form suggested by Landau and Teller[3], by the general equation,

$$-\frac{dE}{dt} = \frac{1}{\beta}\left(E - E(T)\right) \tag{1.2}$$

E is the instantaneous value of the relaxation energy and $E(T)$ is the value which it would have at the equilibrium temperature T. Thus the rate of attainment of equilibrium is proportional to the extent of the deviation from it. Equation (1.2) integrates to give an exponential time function as is shown for the relaxation in Figure 1.1b; it is also a linear equation and this is very important for the theory of relaxation as has been shown by Meixner[4]. The general application of the relaxation equation has been discussed by Herman and Shuler[5]. The relaxation time can be calculated from either dispersion or absorption measurements as will be seen in Section 1.2.1.

Why can energy exchange only occur between vibrational and translational modes at a finite rate? It is because energy exchange can only take place at intermolecular collisions (neglecting changes involving radiation which are usually of minor importance[6,7]), and such collisions are generally inefficient for this purpose because the vibrational energy is quantized and only changes when the vibrational modes are sufficiently perturbed. This requires kinetic energies several times greater than the average thermal values and the transfer of a vibrational quantum needs, on average, from ten to several thousand collisions for polyatomic molecules; diatomic molecules have larger vibrational quanta and can require as many as 10^{10} collisions at room temperature, though this figure drops sharply with increasing temperature. At collisions the vibrations are perturbed by the intermolecular repulsion forces and this is fully discussed in Chapter 2, where the theoretical calculation of the number of collisions required for vibrational energy exchange is described. This chapter is solely concerned with the measurement of vibrational relaxation times in gases; a critical comparison of theory and experiment is carried out in the second half of Chapter 2. It will be assumed that there is neither chemical reaction nor ionization in the gases discussed and only energy exchange with the low-lying vibrational states is considered.

Vibrational relaxation times depend not only upon the probability of a collision causing exchange of translational and vibrational energy, but also upon the molecular collision rate, which is of course pressure dependent. Experimentally the relaxation times are found to vary as the inverse of the pressure showing that the exchange of energy occurs at bimolecular collisions. Advantage is taken of this when investigating a dispersion zone as it is usually easier to vary the pressure (p) rather than the sound frequency (f), and the velocity (or absorption) measurements are plotted against the effective frequency (f/p). By convention p is in atmospheres and relaxation times are quoted for a pressure of one atmosphere. Vibrational relaxation is an example of *thermal* relaxation, which is the failure of an energy mode to follow rapid temperature fluctuations.

Other forms of energy, such as rotational, can also be relaxed but this is outside the scope of this chapter, except where it affects the measurement of vibrational relaxation times. The translational and rotational modes of the gas molecules (the external degrees of freedom) will generally be assumed to be in equilibrium with each other while the vibrational modes (the internal degrees of freedom) are relaxing.

The majority of experimental measurements of vibrational relaxation times have been made acoustically, either by measuring the sonic wavelength in the gas in order to determine the velocity or by measuring the absorption. For polyatomic gases at one atmosphere pressure the relaxation times can be determined using ultrasonic frequencies up to a few Mc/s*, the range being extended by increasing the pressure for very long relaxation times or decreasing it for short ones. Polyatomic molecules generally have sufficient specific heat for there to be a conveniently measurable amount of dispersion; for diatomic molecules with very long relaxation times and minute specific heats reverberation methods have been developed during the last decade for measuring their absorption. The big disadvantage of acoustical methods is that they cannot be used over a very wide temperature range because there is a lack of suitable means for generating ultrasonic waves at temperatures above about 300°C. This is unfortunate because the temperature dependence of a relaxation time provides an important test of the theory.

The best apparatus for carrying out measurements over a wide temperature range is the shock tube. This essentially consists of a long tube with two compartments separated by a thin diaphragm; one compartment is filled to a high pressure with the driver gas, the other with the test gas at a low pressure. On rupturing the diaphragm the expansion of the driver gas sends a shock wave through the test gas; the shock wave is a sharp discontinuity only a few mean free molecular path lengths thick and it produces a nearly instantaneous rise in pressure and translational temperature, the latter increasing by as much as several thousand degrees. The internal degrees of freedom then increase their temperatures and the relaxation times can be determined from say the density profile behind the shock wave. The measurement of vibrational relaxation times using shock waves is the subject of Chapter 3 by Dr P. Borrell.

Besides the shock tube there is one other aerodynamic method for measuring vibrational relaxation times and that is the impact tube. The rest of the methods use light to excite the vibrational modes whose

* Frequencies are also expressed on the SI system in Hz (1 Hz $\equiv$ 1 c/s).

relaxation is then followed by various methods such as absorption spectroscopy or the measurement of resonance fluorescence. If the incoming radiation is chopped then the vibrational energy is also converted to translational energy in bursts which give rise to a periodic change of pressure, i.e. a sound wave. This is the optic-acoustic effect which is the principle behind the measurement of vibrational relaxation times by the spectrophone.

Before the various experimental techniques are described one general but very important point must be made. It is essential that the samples of gases used are pure. It is useless reporting results on impure gases or on a mixture of uncertain composition. Consider what is admittedly an extreme example. Oxygen at room temperature has a relaxation time of about $\frac{1}{50}$ of a second; addition of only 0·1 mole % of water vapour shortens this by a factor of forty! It is thus vital that gases should be thoroughly dried; powerful desiccants such as phosphorous pentoxide, if applicable, should be used and good high-vacuum techniques are required for the gas handling; if possible the apparatus should be baked out to remove absorbed water vapour. Small quantities of substances whose molecules are small and polar can, if present in the gas being investigated, markedly shorten the vibrational relaxation time. Gases investigated should be analysed by any available techniques such as mass spectrometry or infrared absorption. As a general rule impurities shorten the relaxation time and so in comparing different experimental results the longer ones are usually to be preferred. If gas mixtures are being investigated their composition should be checked by measuring their molecular weight, especially if acoustic measurements are being made, when the average molecular weight enters into the calculations (Equation 1.1). The sensitive gas-density microbalance as developed by Lambert and Phillips[8] is a convenient method and has the required accuracy, one part in a 1000.

There is a lack of comprehensive reviews of the apparatus used for measuring vibrational relaxation times largely because of the rapid development in the last few years. Cottrell and McCoubrey[9] have produced the most comprehensive review and Callear[10] has recently described the optical techniques. Short histories of the subject's development have been written by Kneser[11] and Herzfeld[12].

1.2 ACOUSTIC METHODS

Before the experimental methods are described and discussed it is necessary first to consider the theory behind them. In particular several corrections have usually to be applied to the experimental measurements; then comes the determination of the relaxation time from the corrected values. The general theory for this is given below; for more detailed accounts Herzfeld and Litovitz[13], Cottrell and McCoubrey[9] or Bauer[14] should be consulted.

1.2.1   The Propagation of Sound in a Relaxing Gas

1.2.1.1   *General Equations*

In a progressive sound wave there are alternating compressions and dilations accompanied by changes in temperature and average particle velocity. The perturbations of these properties are considered to be small so that their products can be neglected and the equations of change are thus linearized; the importance of this is discussed by Meixner[4]. It is convenient to express harmonic variables in complex exponential form; because of the linearity, the real (physical) part can be recovered at the end. If ϕ is any disturbance in a plane, harmonic, damped wave travelling along the x axis with real phase velocity W, then ϕ is given by,

$$\phi = \Phi \exp[i\omega(t - x/W) - \alpha x] \tag{1.3}$$

where $\omega \, (= 2\pi f)$ is the angular frequency. α is the absorption coefficient, i.e. in travelling a distance $1/\alpha$ the wave's *amplitude* is decreased by a factor e. $\phi \, (x, t)$ can represent any variable such as average particle velocity (u), temperature (T), pressure (p) or density (ρ). An additional variable is the compression (s)† defined by:

$$s = (\rho - \bar{\rho})/\bar{\rho} = d\rho/\bar{\rho} \tag{1.4}$$

where $\bar{\rho}$ is the mean density. The behaviour of the gas is described by five fundamental equations:

  (i) the equation of continuity,
 (ii) the equation of motion,
(iii) the equation of state,
 (iv) the enthalpy equation and
  (v) the relaxation equation.

The continuity equation expresses the fact that the rate of change of mass, contained in a small cross-section of the fluid of thickness dx,

---

† Also called the 'condensation' by Herzfeld and Litovitz[13], p. 29.

must equal the net rate of flow of mass into the element:

$$\bar{\rho}\,dx\left(\frac{\partial s}{\partial t}\right) + \bar{\rho}\left(\frac{\partial u}{\partial x}\right)dx = 0$$

$$\therefore \quad \left(\frac{\partial s}{\partial t}\right) + \left(\frac{\partial u}{\partial x}\right) = 0 \tag{1.5}$$

Newton's second law of motion, applied to this situation, states that the nett force per unit area acting on an element in the fluid is $(\partial p/\partial x)\,dx$ and equals the rate of change of momentum of the element given by $\bar{\rho}(\partial u/\partial t)\,dx$. Following Coulson[15] this gives, as a consequence of the linearization (Herzfeld and Litovitz[13], Section 3),

$$\bar{\rho}\left(\frac{\partial u}{\partial t}\right) = -\left(\frac{\partial p}{\partial x}\right) = -\frac{\partial}{\partial x}\left(\frac{\partial p}{\partial s}\right)s \tag{1.6}$$

Equation (1.6) neglects the effect of a viscous restoring force. Substitution of Equation (1.3) into Equation (1.5) gives,

$$s = \frac{1}{W}(1 - i\alpha W/\omega)u \tag{1.7}$$

then using Equations (1.3) and (1.7), Equation (1.6) can be written as a dispersion equation,

$$W^2 = \left(1 - \frac{i\alpha W}{\omega}\right)^2 \frac{1}{\bar{\rho}}\left(\frac{\partial p}{\partial s}\right) \tag{1.8}$$

$(\partial p/\partial s)$ must now be evaluated. It depends not only upon the relaxation equation but also upon the equation of state. Since it is convenient to discuss relaxation in terms of an ideal gas, equations must be derived for correcting velocity measurements for the non-ideality shown by real gases.

1.2.1.2   *The Correction for Real Gases*

It can be shown experimentally that in normal sound propagation when there is no relaxation, the sound velocity is given by the adiabatic formula of Laplace (Equation 1.1). A relaxation process is, while adiabatic, not isentropic, because it is irreversible. However, if the wave's amplitude is small its propagation can be considered to be isentropic to a first order of approximation (see Herzfeld and Litovitz[13], p. 166, and Bauer[14], Section 1.4.4, for fuller discussions). Then taking the real part of Equation

(1.8) for constant (f/p),

$$W^2 = \frac{1}{\bar{\rho}}\left(\frac{\partial p}{\partial s}\right)_{\text{S}} \tag{1.9}$$

Substituting for ∂s using Equation (1.4) and $\bar{\rho} = M/\bar{V}$ (M is molecular weight and V the molar volume) gives,

$$W^2 = -\frac{\bar{V}^2}{M}\left(\frac{\partial p}{\partial V}\right)_{\text{S}} \tag{1.10}$$

This can be converted to a more convenient form using Reech's Theorem[16], $(\partial p/\partial V)_{\text{S}} = \gamma(\partial p/\partial V)_{\text{T}}$, to give,

$$W^2 = -\frac{\bar{V}^2\gamma}{M}\left(\frac{\partial p}{\partial V}\right)_{\text{T}} \tag{1.11}$$

In order to be able to evaluate Equation (1.11) an equation of state for a real gas is required; the virial expansion is usually used as far as the second term,

$$pV = RT(1 + B/V + ...)$$

this can be rearranged, neglecting second order terms in B, to give:

$$pV = RT + Bp \tag{1.12}$$

Substituting Equation (1.12) into (1.11) gives, after rearranging,

$$W^2 = \frac{\gamma}{M}\frac{(RT + Bp)^2}{RT} \tag{1.13a}$$

a simpler form, which is often used, is

$$W^2 = \frac{\gamma}{M}(RT + 2Bp) \tag{1.13b}$$

This involves appreciable error if $|B| > 500$ cc/mole. For a perfect gas $B = 0$ and equation (1.13) reduces to Laplace's formula,

$$\tilde{W}^2 = \tilde{\gamma}RT/M = \frac{RT}{M}\left(1 + \frac{R}{\tilde{C}_{\text{V}}}\right) \tag{1.14}$$

where the tildes denote ideal gas quantities.

Keesom and van Itterbeek[17] have worked out the correction for non-ideality when dispersion is *not* occurring. It can be derived using the standard thermodynamic relations (e.g. see Roberts and Miller[18], Chapter VI, 3):

$$C_p = \tilde{C}_p - pT \cdot \frac{d^2B}{dT^2} \tag{1.15}$$

$$C_V = \tilde{C}_V - p\left(\frac{2dB}{dT} + T\frac{d^2B}{dT^2}\right) \tag{1.16}$$

and $$\tilde{C}_p = R + \tilde{C}_V \tag{1.17}$$

Division of Equation (1.15) by Equation (1.16), neglecting all remainders which are products of differentials, and use of Equation (1.17) gives the ratio $(\gamma/\tilde{\gamma})$. This is substituted into Equation (1.13b) and again neglecting small terms one obtains,

$$\left(\frac{W}{\tilde{W}}\right)^2 = 1 + \frac{2p}{RT}\left(B + \frac{RT}{\tilde{C}_V} \cdot \frac{dB}{dT} + \frac{RT^2}{2\tilde{C}_V(\tilde{C}_V + R)} \cdot \frac{d^2B}{dT^2}\right) \tag{1.18}$$

Equation (1.18) can be used as a measure of the second virial coefficient of a gas in a region where dispersion is not occurring; $(W/\tilde{W})^2$ is plotted against p at various temperatures. An example of this is the recent work of Cottrell, Macfarlane and Read[19].

In regions where the gas is dispersing, C_V is dependent upon (f/p) and Equation (1.18) cannot be used. Subtracting Equation (1.16) from (1.15) and using Equation (1.17) gives:

$$\tilde{C}_V = \frac{R}{\gamma - 1} + p\left(\frac{2\gamma}{\gamma - 1} \cdot \frac{dB}{dT} + T \cdot \frac{d^2B}{dT^2}\right) \tag{1.19}$$

Thus using Equations (1.13) and (1.19), $\tilde{C}_V$ can be derived from W the measured velocity. $\tilde{W}$ is then obtained from $\tilde{C}_V$ using Equation (1.14). It is important that the pressure-dependent correction for non-ideality should be properly applied; if possible the virial data should be checked by measuring $\tilde{C}_V$ in a region where there is no dispersion and comparing it with the value calculated from spectroscopic data. The correction to W can be as high as 5% for polyatomic gases at one atmosphere which is comparable with the effect on W due to the dispersion. From now on it will be assumed that all quantities have been corrected for non-ideality of the gas and so the tildes can be dropped.

1.2.1.3   The Effective Specific Heat of a Relaxing Gas

Now return to the evaluation of $(\partial p/\partial s)$ for a relaxing gas with a single vibrational relaxation time; let the static $(f/p = 0)$ specific heat at constant volume be C_{V_0}. It is contributed to by the external degrees of freedom (translation and rotation) and by the vibrations; the static contribution of the latter to C_{V_0} is denoted by C. The assumption is now made that the equation of state of the gas depends only on its translational temperature (T_{tr}), as was first done by Herzfeld and Rice[2]. Then for an ideal gas whose equation of state can be expressed as $p = \rho RT/M$, one can write:

$$\frac{\mathrm{d}p}{p} = \frac{\mathrm{d}T_{tr}}{T} + \frac{\mathrm{d}\rho}{\rho}$$

or

$$(p - \bar{p})/\bar{p} = s + (T_{tr} - \bar{T})/\bar{T} \tag{1.20}$$

for small deviations, with the bar denoting the mean value. The enthalpy equation for one mole of an ideal gas is simply $C_p \, \mathrm{d}T = V \, \mathrm{d}p$; however for this case of a relaxing gas C_p is considered in two parts, the internal part depending upon T', the instantaneous vibrational temperature, and the external upon T_{tr}. Thus:

$$(C_p - C)(T_{tr} - \bar{T}) + C(T' - \bar{T}) = V(p - \bar{p}) = R\bar{T}(p - \bar{p})/\bar{p} \tag{1.21}$$

this neglects the effect of heat conduction.

The last of the five equations governing the behaviour of the gas is the relaxation equation. Before applying this, one needs to examine more closely the example given in the Introduction of the gas enclosed in a cylinder by a frictionless piston. Following Kneser[11], consider just one inward stroke of the piston producing a stepwise increase of T_{tr}, after which T' relaxes towards T_{tr}. If the cylinder is immersed in a heat bath at T_{tr}, the relaxation will be isothermal (T_{tr} constant), as shown in Figure 1.3a, with both pressure and volume remaining constant as well so that the relaxation time is denoted by τ_{pT} ($\equiv \tau_{VT}$). Without a heat bath relaxation proceeds adiabatically (S constant) and there are two further possibilities, the process is either isobaric (p constant) or isochoric (V constant). In both cases the energy for the internal modes is drawn from the external ones so that T' and T_{tr} relax towards each other as shown in Figures 1.3b and 1.3c. Thus compared with the isothermal case less energy has to be transferred to the internal modes and $\Delta T'$ is smaller, hence the

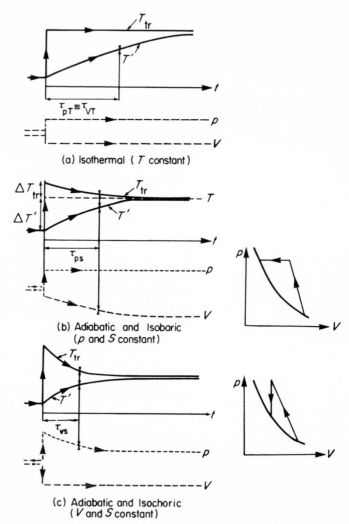

(a) Isothermal (T constant)

(b) Adiabatic and Isobaric
(p and S constant)

(c) Adiabatic and Isochoric
(V and S constant)

Figure 1.3   Relaxation under isothermal and adiabatic conditions

two relaxation times τ_{pS} and τ_{VS} are both shorter than τ_{pT}. Further because $C_p > C_V$, τ_{VS} is the shortest of the three relaxation times and Meixner has proved the general result:

$$\tau_{pT} \geqslant \tau_{pS} \geqslant \tau_{VS} \tag{1.22}$$

Equation (1.1) is the relaxation equation as suggested by Landau and Teller, β is the isothermal relaxation time τ_{pT}. This equation is an energy relaxation equation and for small deviations from the mean value $\overline{T}$ one can substitute $dE = C\,dT'$ and so forth to give the temperature relaxation equation,

$$\frac{dT'}{dt} = -\frac{1}{\tau_{pT}}(T' - T_{tr}) \tag{1.23}$$

In a sound wave the relaxation is adiabatic not isothermal and, following Kneser[11] (p. 139), T_{tr} is best developed in terms of p and T',

$$T_{tr} = \overline{T} + \left(\frac{\partial T_{tr}}{\partial T'}\right)_{pS}(T' - \overline{T}) + \left(\frac{\partial T_{tr}}{\partial p}\right)_{ST'}(p - \bar{p}) \tag{1.24}$$

If T_{tr} is varied adiabatically and isobarically then T' will vary in proportion to the internal and external specific heats as shown in Figure 1.3b; therefore $(\partial T_{tr}/\partial T')_{pS} = -C/(C_p - C)$. Herzfeld and Litovitz[13] (Section 4) show that generally $(\partial T/\partial p)_S = TV\beta/C_p$; β is the coefficient of thermal expansion which equals $1/T$ for an ideal gas. If T' is held constant the gas behaves as though it has no internal degrees of freedom and thus:

$$\left(\frac{\partial T_{tr}}{\partial p}\right)_{ST'} = \frac{\overline{V}}{(C_p - C)} = \frac{R\overline{T}}{\bar{p}(C_p - C)} \tag{1.25}$$

Substituting then Equation (1.24) becomes,

$$\frac{dT'}{dt} = -\frac{1}{\tau_{pT}} \cdot \frac{C_p}{(C_p - C)}\left[T' - \overline{T} - \frac{R\overline{T}}{\bar{p}C_p}(p - \bar{p})\right] \tag{1.26}$$

In a sound wave $(p - \bar{p})$ varies sinusoidally about $\bar{p}$; according to equation (1.26) $(T' - \overline{T})$ does the same even though a phase shift is involved. With simple harmonic motion the changes are proportional to $e^{i\omega t}$ so that $(dT'/dt) = i\omega(T' - \overline{T})$. Equation (1.26) then becomes,

$$[C_p + i\omega\tau_{pT}(C_p - C)](T' - \overline{T}) = R\overline{T}(p - \bar{p})/\bar{p} \tag{1.27}$$

Substitution for $(T' - \overline{T})$ using Equation (1.21) and rearrangement gives:

$$\left(C_p - \frac{Ci\omega\tau_{pT}}{1 + i\omega\tau_{pT}}\right)(T_{tr} - \overline{T}) = \frac{R\overline{T}}{p}(p - \bar{p}) \tag{1.28}$$

Comparison of Equation (1.21) with (1.28) shows that the gas behaves as though it had an effective specific heat $C_{p\,eff}$ given by:

$$C_{p\,eff} = C_p - \frac{Ci\omega\tau_{pT}}{1+i\omega\tau_{pT}} \tag{1.29}$$

If $\omega\tau_{pT} \ll 1$ then $C_{p\,eff} = C_p$, and for $\omega\tau_{pT} \gg 1$, $C_{p\,eff} = C_p - C$; when $\omega\tau_{pT} \sim 1$ there is a phase lag between $(T_{tr} - \bar{T})$ and $(T' - \bar{T})$. Using Equations (1.28) and (1.29) one can now eliminate $(T_{tr} - \bar{T})$ from Equation (1.20) and obtain the effect of the relaxation on the compression;

$$s = \left(\frac{p-\bar{p}}{\bar{p}}\right)(1 - R/C_{p\,eff}) \tag{1.30}$$

Use of the equation of state then gives:

$$\frac{ds}{dp} = \bar{\rho}\frac{R\bar{T}}{M}\left(1 - \frac{R}{C_{p\,eff}}\right)^{-1} \tag{1.31}$$

At low frequencies well below the dispersion zone (denoted henceforth by a zero subscript) the sound velocity is given by $W_0^2 = \gamma_0 RT/M$; substitution of this and Equation (1.31) into Equation (1.8) results in:

$$\gamma_0\left(1 - \frac{R}{C_{p\,eff}}\right) = \left(\frac{W_0}{W}\right)^2\left(1 - \frac{i\alpha W}{\omega}\right)^2 \tag{1.32}$$

One now defines the complex velocity (U) as $U = W(1 - i\alpha W/\omega)^{-1}$ and, since $U_0 = W_0$, Equation (1.32) can be written generally as:

$$\left(\frac{U_0}{U}\right)^2 = \frac{\gamma_0}{\gamma_{eff}} = \frac{C_{po}}{C_{V_0}}\cdot\frac{C_{V\,eff}}{C_{p\,eff}} \tag{1.33}$$

Separating the real and imaginary parts of Equation (1.33) gives the dispersion and absorption equations:

$$\left(\frac{W}{W_0}\right)^2 = 1 + \frac{R\cdot C\cdot C_{V_\infty}\cdot\tau_{pT}^2\cdot\omega^2}{(C_{V_0}+R)(C_{V_0}^2+C_{V_\infty}^2\cdot\tau_{pT}^2\cdot\omega^2)} \tag{1.34}$$

and

$$\alpha\lambda = \frac{\pi\omega RC\tau_{pT}}{(R+C_{V_0})C_{V_0}+\omega^2\tau_{pT}^2(R+C_{V_\infty})C_{V_\infty}} \tag{1.35}$$

The infinity subscript denotes quantities measured at high frequencies well above the dispersion zone when all the vibrational energy has relaxed; λ is the sound's wavelength. In deriving Equation (1.34) the small terms in $(\alpha W/\omega)^2$ were neglected (see Herzfeld and Litovitz[13], p. 64). It should be noted that whereas the absorption is a first-order effect, the dispersion is second order. Specimen curves for dispersion, absorption and $C_{V\text{eff}}$ are shown in Figure 1.4.

The ways of expressing the absorption and dispersion equations are legion. They can be expressed in terms of each of the three relaxation times, which are related in terms of the low and high frequency specific heats:

$$\tau_{\text{pS}} = \tau_{\text{pT}}(C_{\text{p}\infty}/C_{\text{p0}}) \quad \text{and} \quad \tau_{\text{VS}} = \tau_{\text{pT}}(C_{\text{V}\infty}/C_{\text{V0}}) \tag{1.36}$$

These relations can be proved simply for the case of the cylinder and piston (Figure 1.3); let $\Delta T_{\text{tr}} = T_{\text{tr}} - T$ and $\Delta T' = T - T'$ as shown in Figure 1.3b then Equation (1.23) becomes:

$$\frac{d\Delta T'}{dt} = -\frac{1}{\tau_{\text{pT}}}(\Delta T_{\text{tr}} + \Delta T')$$

Since the process is adiabatic $(\Delta T_{\text{tr}}/\Delta T') = (C/C_{\text{p}\infty})$; substituting this and integrating gives,

$$\Delta T' = \Delta T'_{(t=0)} \exp\left[-\frac{t}{\tau_{\text{pT}}} \cdot \frac{C_{\text{p0}}}{C_{\text{p}\infty}}\right] = \Delta T'_{(t=0)} \exp\left[-\frac{t}{\tau_{\text{pS}}}\right]$$

A similar argument leads to the expression for τ_{VS}; for the general proofs see Bauer[14], Section IVB. The differences between these relaxation times depend upon the magnitude of the relaxing specific heat and can be quite small if this is not large.

τ_{pS} could have been conveniently substituted for τ_{pT} in Equation (1.26) and then the equations subsequent to this derived in terms of an effective adiabatic compressibility $\kappa_{\text{S eff}}$ which is given by,

$$\kappa_{\text{S eff}} = \frac{1}{V}\left(\frac{\partial V}{\partial p}\right)_{\text{S}} = \left(\frac{\partial s}{\partial p}\right)_{\text{S}} = \kappa_{\text{T}}/\gamma_{\text{eff}} \tag{1.37}$$

The isothermal compressibility equals $1/p$ for an ideal gas and is independent of the frequency. Corresponding to Equation (1.29) it can be shown[11] that:

$$\frac{\kappa_{\text{S eff}}}{\kappa_{\text{S0}}} = \frac{1+(1-\varepsilon)i\omega\tau_{\text{pS}}}{1+i\omega\tau_{\text{pS}}} \tag{1.38}$$

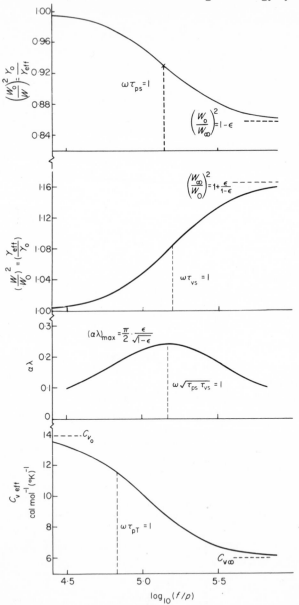

Figure 1.4 Specimen dispersion and absorption curves. Calculated assuming: $C_{V_\infty} = 3R$, $C_{V_0} = 7R$, $\tau_{VS} = 1\ \mu s$, $M = 75$ and $T = 300°K$

where ε is the relaxation strength,

$$\varepsilon = 1 - \frac{\gamma_0}{\gamma_\infty} = \frac{RC}{C_{V_0} \cdot C_{p_\infty}} \tag{1.39}$$

ε is thus a measure of the amount of absorption or dispersion and depends on the magnitude of the relaxing energy C. The dispersion equation is simply defined in terms of τ_{pS} and ε as:

$$\left(\frac{W_0}{W}\right)^2 = 1 - \varepsilon \cdot \frac{\omega^2 \tau_{pS}^2}{1 + \omega^2 \tau_{pS}^2} \tag{1.40a}$$

This is also shown in Figure 1.4 and the inflection point occurs when $\omega\tau_{pS} = 1$. Similar relations can be derived for the other curves and relaxation times and a full discussion is given by Bauer[14] (Section IVD). A comprehensive treatment is also given by Herzfeld and Litovitz in Chapter II of their book[13]; they derive for the inverse of Equation (1.40a) (their Equation 13–5);

$$\left(\frac{W}{W_0}\right)^2 = 1 + \frac{\varepsilon}{1-\varepsilon} \cdot \frac{\omega^2 \tau_{VS}^2}{1 + \omega^2 \tau_{VS}^2} \tag{1.40b}$$

This differs slightly from one given by Bauer[14] on his p. 70 because, as he points out, the derivations are only approximate involving as they do neglect of terms in $(\alpha W/\omega)^2$.

Confusion has been rife in the past because of the failure to adopt a standard nomenclature for the different relaxation times; the thermodynamic notation (Equation 1.36) is strongly recommended because the subscripts make it self-evident which relaxation time is meant. Table 1.1 summarizes the various notations which have been used. Some of the correspondences are not exact because of the approximations mentioned in the previous paragraph.

1.2.1.4 Obtaining Vibrational Relaxation Times from Experimental Results

Besides correcting experimental results for non-ideality of the gas as described in Section 1.2.1.2 allowances have also to be made for relaxation processes other than those due to the transfer of vibrational energy. In deriving Equation (1.6), the equation of motion, the effect of viscosity was neglected, and in Equation (1.21), the enthalpy equation, the effect of thermal conductivity was also omitted. The viscous drag opposes the setting up of pressure differences in the gas and the thermal conductivity

Table 1.1
Relaxation time nomenclature

Authors	Isothermal relaxation times	Adiabatic relaxation times		
Bauer[14], Kneser[11], Davies and Lamb[20], Lamb[21]	$\tau_T = \tau_{pT} = \tau_{VT}$	τ_{pS}	τ_{VS}	$\sqrt{(\tau_{pS} \cdot \tau_{VS})}$
Herzfeld and Litovitz[13], p. 82	τ	τ'	τ''	τ'''
Cottrell and McCoubrey[9]	τ	—	—	—
Lambert[22]	β	—	—	—
Tanczos[23], Stretton[24]	—	—	τ	—
Relations		$\tau_{pS} = \tau_{pT}(C_{p\infty}/C_{p0})$	$\tau_{VS} = \tau_{pT}(C_{V\infty}/C_{V0})$	$\sqrt{(\tau_{pS} \cdot \tau_{VS})} = \dfrac{W_0}{W_\infty} \cdot \tau_{pS} = \dfrac{W_\infty}{W_0} \cdot \tau_{VS}$
Remarks		At Halfpoint of $(W_0/W)^2 - 1$, $\omega \cdot \tau_{pS} = 1$	At Halfpoint of $(W_0/W)^2 - 1$, $\omega \cdot \tau_{VS} = 1$	At maximum of $\alpha\lambda$, $\omega\sqrt{(\tau_{pS} \cdot \tau_{VS})} = 1$

of the gas tends to equilibrate the temperature gradients produced by the sound wave. These two effects were first allowed for by Stokes[25] and Kirchhoff[26]; both produce absorption which increases quadratically with the frequency[2]. Their incorporation in the equation for sound propagation has been discussed by Markham, Beyer and Lindsay[27], by Herzfeld and Litovitz[13], and by Cottrell and McCoubrey[9] (Section 2.4); it can be shown that the absorption is given by,

$$\alpha_{cl} \cdot \lambda_0 = \frac{2\pi^2}{\gamma_0}\left[\frac{4}{3}\eta + (\gamma_0 - 1)\frac{KM}{C_{p0}}\right]\frac{f}{p} \tag{1.41}$$

where η is the coefficient of viscosity and K the thermal conductivity. This absorption is called 'Classical Absorption', and measurements made at values of (f/p) below $10^{7\cdot5}$ c/s atm can be corrected by simply subtracting the classical absorption calculated according to Equation (1.41).

Consider two examples: the absorption in sulphur hexafluoride at room temperature reaches a peak at $\log_{10}(f/p) = 5{\cdot}85$ where $\alpha\lambda = 0{\cdot}31$ for the vibrational relaxation[28], the classical absorption is only 1% of this and so the correction only affects half of the absorption curve, see Figure 1.5(a); with deutero-acetylene (C_2D_2) the correction is more important because at the higher values of (f/p) the classical absorption is greater and the absorption due to the vibrational relaxation is less than with SF_6 as shown in Figure 1.5(b), for which the data are taken from Jones[29].

For values of (f/p) exceeding 10^8 the classical absorption has increased to the point where it is usually greater than the absorption due to vibrational relaxation. As usual there is dispersion corresponding to the absorption; it is called translational dispersion and allowance has to be made for it at values of (f/p) exceeding about 3×10^7 c/s atm. The absorption can be considered as being due to the relaxation of the translational energy, the relaxation time being comparable with the mean time between two collisions by the same molecule ($\sim 10^{-10}$ s). When the sound frequency is comparable with the molecular collision rate the gas can no longer be regarded as a continuum and the absorption is extremely high; this makes the experimental measurements, particularly of the sound velocity, difficult and tends to obscure other relaxation processes which have very short relaxation times. Then clearly correct allowance for the classical absorption or translational dispersion is important. Greenspan[30] has shown how correction formulae can be obtained from the hydrodynamical 'Navier–Stokes' equation for polyatomic molecules,

$$\alpha_{cl} \cdot \lambda_0 = 2\pi \left\{ \frac{-1 + \sqrt{(1+x^2)}}{2(1+x^2)} \right\} \tag{1.42}$$

$$\frac{W_0}{W} = \left\{ \frac{1 + \sqrt{(1+x^2)}}{2(1+x^2)} \right\} \tag{1.43}$$

where $x = 2{\cdot}6\pi f\eta/p$, and the absorptions are no longer quite additive[30]. These equations should be used for values of $(f/p) > 10^{7\cdot5}$ c/s atm. An example of the use of Equation (1.42) is the absorption curve for carbon tetrachloride measured by Hinsch[31], Figure 1.6. Even the Greenspan treatment is only an approximation and a better approximation, due to Burnett[32], is more involved but fits the experimental results for the monatomic gases and has been used by Holmes, Jones and Pusat[33,34] for determining rotational relaxation times and the vibrational relaxation

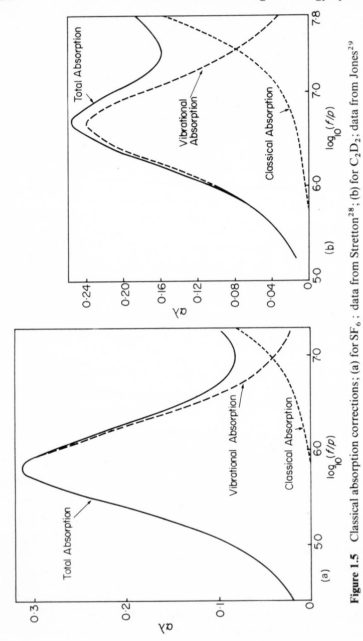

Figure 1.5 Classical absorption corrections; (a) for SF_6; data from Stretton[28]; (b) for C_2D_2; data from Jones[29]

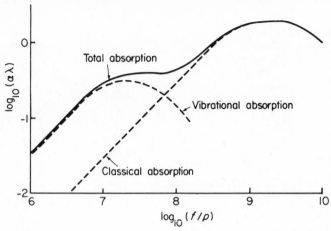

Figure 1.6   Absorption in carbon tetrachloride; data from Hinsch[31]

times of hydrocarbons. For full discussions of the propagation of sound at very high values of (f/p) see Greenspan[35] and Truesdell[36] †.

The relaxation of other forms of energy can also interfere with the determination of vibrational relaxation times. Rotational energy usually has very short relaxation times‡ and Z_{rot}, the number of collisions required to reduce a perturbation of the rotational energy to $1/e$ of its original value, is usually less than ten. Correction can be made by adding an extra term to the classical absorption using the approximate formula of Herzfeld and Litovitz[13], p. 236,

$$\alpha_{cl+rot} = \alpha_{cl}(1 + 0.067 \, Z_{rot}) \tag{1.44}$$

Another form of energy which can relax is the electronic; the one known example of this is provided by nitric oxide (NO) which has another electronic state only 0·02 eV above the ground one due to spin splitting. Bauer and Sahm[37] have measured the sound absorption in NO as a function of (f/p) and their results, shown in Figure 1.7, provide a good illustration of how the various relaxation processes can be separated;

† Note added in proof: for a further theory of classical absorption see the recent paper by Morgan and Kern[129]. Results are obtained similar to the Burnett theory just about up to the absorption peak.

‡ Rotational relaxation is discussed in Volume 3 of this series.

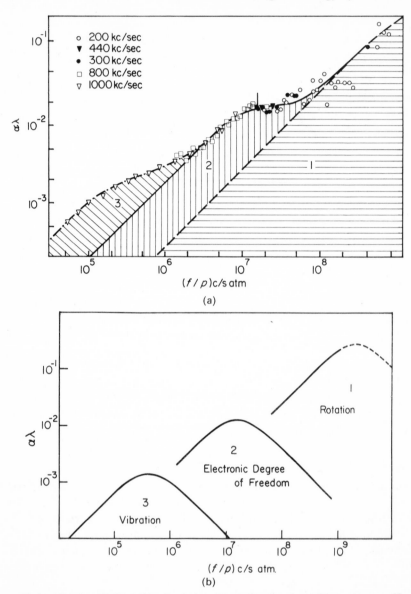

Figure 1.7 Sound absorption in nitric oxide (NO) taken from Bauer and Sahm[37]; (a) experimental points with the classical absorption subtracted; (b) individual contributions by the various degrees of freedom

the plotted points are the measured absorption minus the classical. Note how at low (f/p) all the absorption lines are parallel and proportional to (f/p), then successively each reaches a peak and declines. At low (f/p), $(\omega\tau_{pT} \ll 1)$, Equation (1.35) reduces to:

$$\alpha\lambda = \frac{\pi\omega R C \tau_{pT}}{C_{V_0}(R + C_{V_0})} \tag{1.45}$$

This equation can be used to determine an approximate value for τ_{pT} at effective frequencies well below the relaxation zone provided that due allowance can be made for the other absorptions.

The stage has now been reached where the experimental absorption or dispersion results have been corrected for relaxation processes other than those due to the vibrational energy; generally more corrections have to be applied to the absorption measurements because absorption is a first-order effect. Plots of the results against (f/p) can be compared with theoretical curves calculated using say Equations (1.34) and (1.35); the latter are obtained as functions of $\omega\tau_{pT}$ and are independent of the value of τ_{pT}; then, by superimposing the theoretical curve on top of the experimental one and sliding along the abscissa until the experimental points fit the theoretical curve, one can deduce the effective frequency (f/p) at which $\omega\tau_{pT} = 1$, and hence τ_{pT}, from either the dispersion or absorption results. The height of the absorption curve depends upon C, the amount of vibrational specific heat relaxing, and it can be shown (e.g. see Bauer[14], p. 69) that the maximum occurs at $\omega^{-1} = \sqrt{(\tau_{pS} . \tau_{VS})}$ and is given in terms of the relaxation strength ε by,

$$(\alpha\lambda)_{max} = \frac{\pi}{2} \frac{\varepsilon}{\sqrt{(1-\varepsilon)}} \tag{1.46}$$

Thus by measuring the height of the absorption peak or the extent of the dispersion, one can determine the amount of vibrational energy relaxing providing that the absorption is measured in the region of the peak or, with dispersion measurements, that the velocity is measured at both ends of the zone. This can be inconvenient if one of these is beyond the effective frequency range of the apparatus, but there is another way which was devised by Cole and Cole[38] in their investigations of dielectric relaxation. The real and imaginary parts of Equation (1.38) can be separated to give[14],

$$\frac{\kappa_{S\,eff}}{\kappa_{S_0}} = \frac{W_0{}^2}{W^2} - i\frac{\alpha\lambda}{\pi}\cdot\frac{W_0{}^2}{W^2}$$ (1.47)

The Cole–Cole plot is a graph of the imaginary part of Equation (1.47) against the real part, i.e. it is an Argand diagram. The result is a semicircle of diameter ε with a minimum at $\omega = 1/\tau_{pS}$. For such a plot one must have absorption and dispersion measurements made simultaneously at the same value of (f/p); an example is shown in Figure 1.8, it is for C_2H_2 and the data are taken from Jones[29]. An alternative but equivalent Cole–Cole plot has been discussed by Johnson[39] who has also suggested a straight-line plot†.

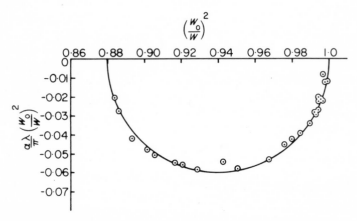

Figure 1.8   Cole–Cole plot for acetylene; data from Jones[29]

If one obtains a semicircular Cole–Cole plot it means that the vibrational energy has all relaxed with a single relaxation time, and such a plot is a sensitive test for this. It has been found that the majority of pure gases have single relaxation times. This is true even for polyatomic molecules which have several vibrational modes, the usual explanation being that all the modes relax *via* the one with the lowest frequency whose energy exchange with translation provides a rate-determining step for the relaxation of all the vibrational energy; this is discussed fully in Chapter 2.

† Note added in proof: analysis of results in terms of straight lines has recently been discussed by Holmes and Stott[127].

There are three well-characterized examples of multiple relaxation: sulphur dioxide (SO_2)[40], methylene chloride (CH_2Cl_2)[41] and ethane (C_2H_6)[42,43]. When this occurs one has to assume a mechanism for the relaxation of the vibrational energy in order to determine the relaxation times from the experimental results and it is better that theoretical dispersion or absorption curves, obtained as described in Chapter 2, be compared direct with the experimental measurements.

In mixtures, multiple relaxation is commoner and two examples are shown: In Figure 1.9 for the absorption measurements made by Bauer

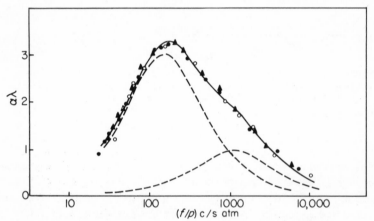

Figure 1.9 Sound absorption in oxygen–carbon monoxide mixture containing 80% O_2. Solid line denotes total absorption, dotted ones the contributing parts; $T = 365°K$ (from Bauer and Roesler[44])

and Roesler[44] on an oxygen and carbon monoxide mixture and in Figure 1.10 for the dispersion measurements of Lambert, Edwards, Pemberton and Stretton[45] on a mixture of CF_4 and C_2F_4. It is general that with multiple relaxation the dispersion curve is less steep and the absorption peak lower and broader compared with a single relaxation process involving all the specific heat. If there are two relaxation times then as their ratio increases the dispersion curve becomes two distinct steps and the absorption curve splits into two separate peaks: then each relaxation zone can be treated separately using the equations derived above for a single relaxation time.

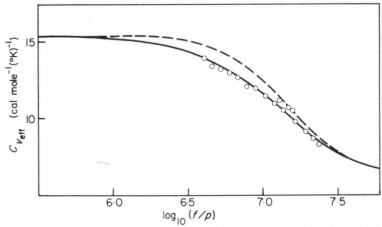

Figure 1.10 Dispersion curve for mixture of carbon tetrafluoride (CF_4) with 60·2 mole % tetrafluoroethylene (C_2F_4). Solid line is the theoretical curve for double relaxation process, dotted one is for single relaxation process involving total vibrational energy of both components (from Lambert, Edwards, Pemberton and Stretton[45])

With mixtures, properties such as molecular weight are averaged according to the mole fractions and a virial correction has to be calculated for the mixture[45]. Also another contribution to the classical absorption can arise due to the density and temperature gradients produced by the sound wave causing 'unmixing' of components which have different diffusion coefficients. In practice account has only to be taken of this if one of the components is hydrogen or helium. A recent example is furnished by the work of Bauer and Liska[46] on carbon dioxide and helium mixtures: the theory of this effect is given by Herzfeld and Litovitz[13], Section 41.

Finally a few words about accuracy: absorption measurements need to be accurate to a few per cent; if they are quoted as $\alpha\lambda$, then λ only need be known to a similar accuracy. However, since dispersion is a second-order effect and over a whole zone the velocity (and wavelength) might only change by a few per cent it is necessary that velocity measurements should, if possible, be accurate to 0·1 %. Inaccurate measurement of the sonic velocity and insufficient control of the temperature will lead to experimental scatter, whilst use of the incorrect virial coefficient or molecular weight in calculating the results will result in a systematic error or bias for the experimental dispersion curve. The absorption curve will have a bias if the correct allowances are not made for the absorptions due to effects other than vibrational relaxation.

1.2.2   Acoustic Instruments

The majority of acoustically determined vibrational-relaxation times have been measured with the acoustic interferometer and based solely upon dispersion measurements. The interferometer uses only one transducer and can cover the middle part of the effective frequency range from about $10^{4\cdot5}$ to $10^{7\cdot5}$ c/s atm. With nearly all the acoustic techniques the effective frequency is varied by altering the gas pressure rather than the ultrasonic frequency which is either fixed or restricted to a comparatively narrow range. The interferometer is now being superseded by apparatus which use two transducers, one acting as a receiver the other as a transmitter; this allows one to measure both absorption and dispersion and such instruments can operate up to the highest values of the effective frequency range by using transducers suitable for operating at very low pressures ($\sim 10^{-4}$ atm).

Molecules with only high vibrational frequencies have small specific heats and usually long relaxation times. Thus in the low part of the effective frequency range, i.e. below about 10^4 c/s atm, the relaxing vibrational specific heats are very small and the dispersion is hardly detectable so that only measurements of the absorption are usually made. Included in the methods for the lower part of the effective frequency range is the tube method which has proved to be very useful for making measurements with corrosive gases. All the above instruments use purely acoustical techniques: however one method, which has recently been revived, uses the ultrasonic waves travelling through the gas as a diffraction grating and the sonic wavelength is measured optically. In the following review emphasis will be placed upon describing the newer instruments.

1.2.2.1   *The Acoustic Interferometer*

The essentials of an interferometer are a device for converting electrical oscillations into mechanical ones, i.e. a transducer, and a movable reflector. The transducer transmits continuous waves at the reflector through the test gas and these usually interfere destructively with those reflected back. However, if the separation between the transducer and reflector is an integral number of half wavelengths, standing waves are set up: these positions are detected so that one determines the wavelength of sound in the gas from which, knowing the frequency, one can calculate the sonic velocity.

Certain crystals have the property that, if an electric field is applied across them, they change their length very slightly; whether they expand or contract depends upon the direction of the field and thus an electric oscillation can be converted into a mechanical one, i.e. a sound wave, and the reverse process is also possible. Such crystals are said to be piezoelectric and are used as the interferometer transducer; for discussions of piezoelectric crystals see Mason[47] and Berlincourt, Curran and Jaffe[48]. The crystal most widely used so far is quartz though it is likely to be superseded by lead zirconate. The crystal vibrates longitudinally and its two end faces are usually coated with gold to form electrodes; the electrode facing the reflector is earthed in order to reduce capacity effects.

The apparatus which contains the crystal and movable reflector is of fairly simple design; the one described recently by Blythe[49] and Blythe, Cottrell and Day[50] is typical and is shown in Figure 1.11. It is mainly constructed out of stainless steel and is immersed in a thermostat, the temperature being maintained constant to within 0·01°C. The acoustic chamber should be as large as possible in order to reduce wall effects and eliminate transverse modes; on the other hand if it has a big volume it will need a large quantity of a perhaps rare gas in order to fill it. The chamber in Figure 1.11 has an internal diameter of 3·8 cm and a volume of about 60 cm³. The reflector can travel 2 cm (equivalent to about six wavelengths in air at 100 kc/s) and the movement is transmitted by an Invar rod which is itself driven by an electric motor. One thus has to have a movable seal; metal bellows, as shown in Figure 1.11, are the most popular though glands made with 'O' rings have been used[51]. Although bellows can be a little difficult to evacuate and slightly change the volume of the apparatus, and hence the gas pressure, during a run they do give a good, grease-free vacuum-tight seal; if the volume change is appreciable it can be counteracted[52] by arranging for a similar set of bellows to expand as the reflector ones contract and vice versa.

X-cut quartz crystals are usually used as this cut gives the biggest amplitude of mechanical oscillation for a given electrical signal. The crystals have to be operated at their resonant frequencies and this varies inversely with their thickness. Thus one of the crystals used by Blythe was a 3·4 cm thick cylinder of diameter 2·5 cm, and had a resonant frequency of about 84 kc/s: it was held by three plastic pins which fitted into a groove

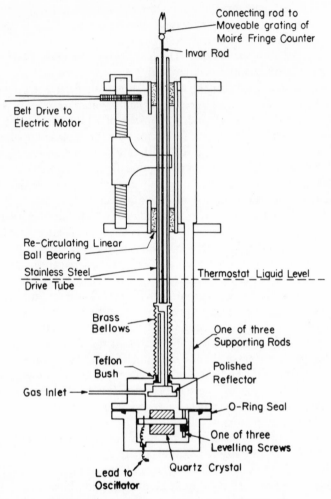

Figure 1.11   Acoustic interferometer of Blythe[49]

cut around the middle of the cylinder. The groove lies in the crystal's nodal plane and so the mounting does not interfere with the oscillation; this is more important with higher frequency crystals which are much thinner and for frequencies greater than a few megacycles per second they are too fragile to handle. Spurious frequencies can often be eliminated by bevelling the crystal's edges. Electrical connections are usually made to the gold-coated faces by soldering thin wires to tongues of gold on the bevelling. If there is mercury vapour in the gas-handling system there should be a gold trap between it and the interferometer in order to protect the gold plating on the crystal. The live lead to the crystal has to leave the interferometer via an insulated seal.

An alternative arrangement to that shown in Figure 1.11 has been designed by Stewart[52] for use with crystals whose resonant frequency exceeds a megacycle per second. It consists of a tube with a close fitting piston, the end of which acts as a reflector; the top is closed by the crystal which is only supported around its rim and is held in place by a small spring. Following Zartman[53] a fixed reflector can be placed at the back of the crystal and separated by a quarter wavelength so that sonic radiation is prevented from leaving the back face of the crystal.

The crystal is either driven by an oscillator, which uses the crystal itself to control the oscillator frequency, or by a second oscillator which has its frequency independently controlled but even so has to run at the resonant frequency of the interferometer crystal. When the separation between the crystal and reflector is an integral number of half wavelengths, standing waves are set up in the gas column which then resonates and the crystal impedance drops very sharply; this is observed by monitoring the crystal current. Thus one has to determine the distance between reflector positions at which the crystal current is a minimum. If the sonic wavelength is of the order of 1 mm or so then a micrometer screw is accurate enough; for higher frequencies and shorter wavelengths a lever arrangement with a velocity ratio of ten to one has been used[45,54]. Blythe, Cottrell and Day[50] have used a Moiré fringe counter to obtain very accurate measurements of the reflector travel.

Several precautions have to be taken to obtain accurate readings with an interferometer. Firstly it is important to have the opposing faces of the reflector and crystal polished flat and aligned exactly parallel and the mounting of the crystal should be designed to allow this. The faces are judged to be parallel when one obtains the biggest symmetrical dip in the crystal impedance at a standing wave position[55]. In order that plane waves are to be radiated and diffraction effects avoided the diameter of the crystal face should be large compared with the sonic wavelength. It is normal practice to calibrate an interferometer with a gas such as argon for which the velocity of sound is accurately known and it is often found that the measured wavelength exceeds the theoretical by a small amount $\delta\lambda$. This has been examined by Blythe, Cottrell and Day[50] and they tentatively suggest that it is due to the setting up of transverse waves. The have reviewed the results of several other workers and concluded that $(\delta\lambda/\lambda)$ is proportional to λ^2, so that for a particular interferometer, once the discrepancy has been determined for a gas with a known velocity of sound, an appropriate correction can be applied for other gases; this has been done for example by Cottrell and Matheson[56].

The wavelengths should be measured to an accuracy of 0·1 % and the crystal frequency can be easily measured to 0·01 % with an electronic counter. The pressure need only be known to within 1 % and a mercury manometer usually suffices for the pressure ranges over which interferometers are operated. One atmosphere is the usual maximum pressure. As the pressure is lowered in order to alter the effective frequency, readings become more difficult because the gas impedance drops and the standing wave positions become less pronounced as the signal to noise ratio decreases. Also readings are less easy to make in a relaxation zone because of the resultant absorption. Generally it is difficult to make measurements at values of the effective frequency much greater than 10^7 c/s atm though Sette, Busala and Hubbard[41] and Valley and Legvold[43] have made dispersion measurements up to 10^8 c/s atm. Some absorption measurements have been made with the acoustic interferometer[57]; to do this one must have a separately controlled oscillator coupled to the interferometer crystal and not only detect the minima in the crystal current but also measure the actual drop at each position (Hubbard[58], Stewart and Stewart[55]). For further general discussions of the acoustic interferometer see Blitz[59] and McSkimin[60].

1.2.2.2 Instruments with Two Transducers for use at High Effective Frequencies

As has been seen in the preceding section the main disadvantages of the acoustic interferometer are its limited effective frequency range and the difficulty in making accurate absorption measurements. Both of these defects can be overcome by using an apparatus with two transducers, one acting as transmitter, the other as receiver; absorption measurements are then made by determining the rate at which the received signal amplitude decreases with transducer separation and the velocity is deduced from the observed rate of phase change of the received sound with transducer separation. Both measurements demand that the phase, frequency and amplitude of the transmitted signal remain constant while readings are being taken for a given measurement.

If continuous waves are used when absorption of the sound by the gas is low and the transducers are close together, standing waves will be set up which will interfere with the measurements. If pulses of sound are transmitted instead, then in order to avoid standing waves the transducer separation need only exceed $x/2$ sonic wavelengths, where x is the number of waves in the pulse. However, at very low pressures where the absorption is high due to the classical absorption, the received signal is very weak and has to be amplified considerably. Pulses have to be amplified with a large bandwidth and this introduces noise, the amount being proportional to the bandwidth which varies approximately inversely with the number of waves in the pulse. For receiving very weak signals, when standing waves are damped out, one wants to use continuous waves, i.e. an infinitely long pulse, so that a very narrow bandwidth amplifier can be used and the noise level kept to a minimum. Thus the limitations of the simple interferometer can be overcome, the inevitable price being greater complexity of the apparatus.

Most double transducer instruments use condenser transducers which consist of a thin plastic membrane stretched lightly across a metal plate. On the outer side the membrane is coated with a film of metal such as aluminium which is earthed. A d.c. potential of about 100 V is applied between the coating and the backing plate. If an a.c. voltage is also applied it causes the membrane to oscillate, thus setting up a sound wave. Conversely, if a sound wave impinges upon the membrane it generates an a.c. signal. The transducers are often named after their inventor Sell[61] and were developed by Kuhl, Schodder and Schröder[62]. Recently Wright[63] has made a theoretical and practical study of their operating characteristics. Condenser transducers have an almost flat response curve, i.e. a large bandwidth; they can therefore be used in pairs without matching and for pulses. Condenser transducers are usually operated at frequencies around 100 kc/s, though they have been used at frequencies up to 800 kc/s[64].

The first condenser transducer apparatus was built by Meyer and Sessler[65] and it has been further developed at Göttingen by Hinsch[31] and Haebel[66]. The one shown in Figure 1.12, built by Stretton[28], is typical and similar to the one described by Holmes and Tempest[67]. The two transducers are contained in a vacuum-tight acoustic chamber. The transmitter transducer is fixed and mounted on a tripod while the receiver is movable and attached to the end of an Invar tube. This tube leads out of the chamber to a manually operated micrometer which measures

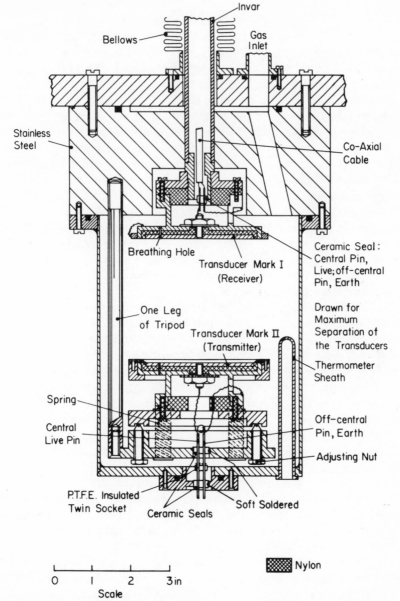

Figure 1.12 Drawing of acoustic chamber of double transducer apparatus of Stretton[28]

changes in the transducers' separation. The position of the transmitter can be altered in order to align it parallel to the receiver; for this the base of the transmitter is held firmly by two strong springs against a small ball-bearing and the ends of two adjusting screws. These three points form a right-angled triangle so that the two screws alter the alignment orthogonally. In order to minimize electrical 'pick-up' between the leads both transducers are carefully insulated from the rest of the apparatus by nylon bushes and the transmitter lead enters at one end of the apparatus whilst the receiver one leaves at the other via the inside of the Invar tube. Figure 1.13 shows the acoustic chamber with the mounted transducers. After assembly, the acoustic chamber is surrounded by a thermostated aluminium block.

The plastic membrane of the condenser transducers is made of I.C.I. Melinex about 6 μm thick and is coated with aluminium. In the Mark I transducer it is stretched lightly across the backing plate and held in position by a clamping ring. While this was found to be satisfactory for the receiver, with the transmitter aluminium tended to be lost from the circumference where the foil is bent; this was cured by adopting the clamping arrangement shown for the Mark II. It is important that, for good performance at all pressures, both backing plates are slightly roughened and not polished smooth. The transmitter is driven by about 100 V a.c. peak to peak and at atmospheric pressure the received signal is about 50 dB down on the transmitted one. The pulses contain at least ten waves with frequencies between 50 and 300 kc/s derived from a crystal-controlled oscillator. Even with transducer faces 7 cm in diameter there was evidence of a slight deviation from plane-wave behaviour at the lowest frequency. The received signals are amplified and displayed on the oscilloscope; provided that the amplification is linear, measurement of their amplitude as a function of the transducer separation gives the absorption. The velocity of the sound is measured via the wavelength. The latter is determined by comparing the phase of the waves in the received pulse with that of the continuous wave signal from which the transmitter pulses were produced. This is done using Lissajous figures again displayed on the oscilloscope; changing the transducer separation by half a wavelength causes the Lissajous figure to go from one collapsed position to the next. In order to avoid interference from multiple reflections of the transmitted pulse the oscilloscope is 'Z' pulsed, i.e. a specially delayed pulse is sent to the oscilloscope, in order to switch on the beam, only when the sonic pulse is being received; at all other times the beam is switched off.

Figure 1.13 View of acoustic chamber showing mounted transducers (from Stretton[28])

At low pressures the impedance of the gas has fallen to a small value and consequently the efficiency of the two transducers has decreased even though they have diaphragms of very thin plastic. In order to be able to amplify the received signal continuous waves have to be used and one has to face the problem of crosstalk. Crosstalk is any extraneous signal which is picked up by the receiver, or its attendant equipment, which did not pass through the test gas in the required manner. It can take two forms, mechanical or electrical. The former is due to sonic signals travelling around the apparatus and can be eliminated by careful design such as mounting the transducers on plastic bushes. Electrical crosstalk is more serious and is simply due to electrical 'pick-up'. This is instantaneous whereas the signal which has travelled through the gas is delayed so that with pulsing the two can be separated. Electrical pick-up can be minimized by careful earthing, all the parts of the apparatus should be earthed to a common point thus avoiding earth loops. Careful screening is also necessary; in particular the aluminium layer should be thick enough to prevent penetration by the radiation[68]—with such a thickness one cannot see through it.

An ingenious method for avoiding electrical crosstalk when using continuous waves is due to Haebel[66]. The d.c. polarization is replaced by an a.c. signal of low frequency, say 1 kc/s; then if the main signal has a frequency of 100 kc/s, *sonic* side bands are produced with frequencies of 99 and 101 kc/s; this is because the attractive force between the condenser transducer's plates depends upon the square of the total charge. Provided then that there is no electrical mixing in the transmitter, the crosstalk has only a frequency of 100 kc/s so that the sonic side bands are free of crosstalk. Using a highly selective receiver one of the side bands can be detected alone and amplified. A plot of its intensity against transducer separation gives the absorption as usual. For velocity measurements one can measure the rate of phase change with separation using a phase sensitive detector. The crosstalk is eliminated by taking advantage of the fact that it has a constant phase, so the reference signal is arranged to be out of phase with it[28].

Greenspan[69] and Greenspan and Thompson[70] built two instruments with pairs of quartz crystals operating at 1 Mc/s and 11 Mc/s respectively. These were designed to study classical absorption and translational dispersion in inert gases at very high effective frequencies and so continuous waves were used. Since quartz crystals only operate efficiently at their resonant frequencies the pairs of crystals have to be carefully matched, and the receiver crystal was tuned to the transmitter frequency by adjusting the thickness of its plated electrodes. In the 11 Mc/s apparatus in order to avoid electrical crosstalk to the receiver crystal from the transmitter one, the latter is mounted at one end of a steel coupling rod, 10 cm long, so that the sound waves are generated by the free end of the coupling rod which faces the receiver. The transmitter is moved by a synchronous motor driving a bench micrometer screw. To measure the sound velocity the phase difference between the transmitted and received signals is measured with a phase meter as a function of the transducer separation. The absorption constant was deduced directly from a logarithmic record of the receiver signal strength varying with transducer separation. Holmes and Tempest[67] built a double transducer apparatus to operate with pulses†. Quartz crystals were also used but their operating bandwidth is too narrow for pulsing: to overcome this they were damped by fixing them to pieces of polyethylene.

The double condenser-transducer instrument can be used for making both absorption and dispersion measurements over an effective frequency range of $10^{4.5}$ up to 10^9 c/s atm which is about the maximum needed for studying vibrational relaxation; in fact it was first used by Meyer and Sessler[65] to make measurements in the inert gases at values up to 10^{11} c/s atm. The main disadvantage is that the present condenser transducers are not suitable for making measurements at temperatures above about 100°C because of the plastic membrane.

The apparatus of Greenspan and Thompson with its two 11 Mc/s crystals could be used over a wider temperature range since quartz can be used as an effective piezoelectric transducer up to around 300°C. The operating frequency is much higher than with the condenser-transducer instruments, so that the wavelength is correspondingly shorter, and more difficult to measure accurately, and the lower effective frequency limit is about 10^7 c/s atm, unless the apparatus is pressurized. However, values of 10^9 c/s atm can be obtained with a gas pressure of 0·01 atm; the condenser-transducer apparatus has to use a pressure of about 10^{-4} atm for this. Then maintaining the gas purity is a big problem and frequent changes of the gas sample in the acoustic chamber are necessary to prevent contamination by desorption of impurities from the walls of the apparatus. The low pressures can be measured by a McLeod gauge for permanent

---

† Note added in proof: an adaptation of this apparatus for use at low temperatures has recently been described by Holmes and Stott[128].

gases[28] or by a capacitance manometer[71] or Alphatron[31,66] for condensible vapours.

1.2.2.3 Techniques for Low Effective Frequencies

One technique used for the effective frequency range 10^3 to 10^6 c/s atm is the 'tube' method first devised by Angona[72]. The apparatus consists of a glass tube, about 2 cm internal diameter, with a fixed receiver microphone at one end and a movable transmitter speaker at the other. The absorption is determined by measuring the received sound amplitude as a function of the distance between the microphone and speaker, whilst the velocity can be obtained by measuring the rate of phase change of the received signal with distance. The speaker is of the ribbon type; this consists of a piece of corrugated aluminium foil about 0·5 cm wide and 15 μm thick and it is placed in the field of a permanent magnet which is outside the tube; when an alternating current passes through the foil it vibrates and generates sound waves. The current is fed to the ribbon via an air-core transformer which has its primary coil outside and its secondary coil inside the tube. The transmitter is moved by means of a magnet system. Both ends of the tube are packed with an absorbing material such as glass wool in order to prevent the reflection of sound and the setting up of standing waves. In Angona's apparatus the microphone, of the condenser type, was attached to the tube wall and various precautions were taken to prevent crosstalk. Angona made absorption measurements in CO_2, CS_2 and C_2H_4O over the effective frequency range 10^4 to 10^6 c/s atm.

The tube method was developed by Shields and Lagemann[73] who replaced the condenser microphone with a second ribbon transducer. Not only do the absorption results have to be corrected for classical absorption but also for viscous and thermal losses at the tube walls, and they found that use of the Kirchhoff–Helmholtz equation for this was justified though the problem has been carefully reconsidered by Shields, Lee and Wiley[74]. Shields[75,76] has used the method for making measurements with the particularly corrosive halogen gases, even fluorine[76], and the work included measurements at elevated temperatures. This method is particularly suited for making measurements with corrosive substances because the only parts of the apparatus which have to withstand attack are the ribbons, secondary coils and tube; all the rest of the equipment is outside the tube. Shields and Lee have also made measurements with oxygen[77] and oxygen mixtures[78] at effective frequencies down to 10^3 c/s atm and the latest work has been carried out by Shields[124] on SO_2.

The other techniques for the lower effective frequency ranges are all based upon either the reverberation technique of Kneser[79] or upon the resonance bandwidth method as used by the Knötzels[80]. In the former method the absorption of the gas is determined from the time the sound field in the gas takes to decay; it was developed by Edmonds and Lamb[81] who intermittently excited a cylinder containing the test gas at one of its characteristic vibrations or eigen-frequencies and then, during the periods when the transmitter was switched off, the exponential rate of decay of the sound amplitude was measured. Their cylinder was a precision-bore Pyrex tube about 10 cm internal diameter and 60 cm long and in such tubes it is possible to excite longitudinal, radial and coupled fundamental modes together with their harmonics; the theory of this is discussed by Edmonds and Lamb[81] and by Fritsche[82]. The eigen-frequencies depend upon the test gas and cylinder dimensions and it is important to determine which mode is excited.

In order that the transmitter and receiver transducers do not appreciably affect the decay time they must be coupled only loosely to the resonator and they should have wide bandwidths. Edmonds and Lamb[81] used a ribbon loudspeaker placed at the mouth of a flared horn (see Figure 1.14).

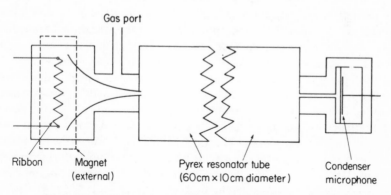

Gas port

Ribbon   Magnet   Pyrex resonator tube   Condenser
                (external)   (60 cm × 10 cm diameter)   microphone

Figure 1.14   Diagram of the resonator of Edmonds and Lamb[81]

The narrow end of the horn just protruded through a hole, diameter 1·5 mm, in the end plate of the resonator. At the other end the resonator is coupled via another small hole to a condenser microphone. The whole apparatus was immersed in a constant-temperature bath. Measurements were made with frequencies around 10 kc/s over the effective frequency range from 1 to 45 kc/s atm, the measured exponential-decay times ranging from 0·02 to 1·5 s. Corrections were made[83] for absorption due to wall losses, including the end plates, and for classical absorption. To calculate the results the sound velocity in the gas must be known; this can be determined knowing the length of the tube and the frequency of the longitudinal mode with the lowest frequency. Vibrational relaxation times for several gases were then obtained[83] using Equation (1.45).

The reverberation method has been further developed by Holmes, Smith and Tempest[84] in order to measure the vibrational relaxation time in oxygen. The peak of the absorption curve occurs around 10 c/s atm; to attain this low value of the effective frequency they worked at pressures up to 7 atm and used low frequencies down to around 100 c/s. The electronics for this has been described by Holmes and Tempest[67]. Only one transducer, of the moving coil type, was used which alternately acted as transmitter and receiver and was switched over every four seconds. It was coupled via a small hole to a resonator tube about half a metre in diameter and four metres long. The tube was operated at its resonant eigenfrequencies and was immersed in a water tank whose temperature was controlled to within $\pm 0·03°C$. Absorption measurements were made and it was found that the wall losses exceeded those calculated by the Kirchhoff–Helmholtz formula and so empirical allowances were made for these. About half the absorption curve for the vibrational relaxation of oxygen was covered and dispersion measurements were also made using the measured eigen-frequencies to give the sound velocity.

With the resonance bandwidth method, a resonator tube with receiver and transmitter transducers is used and operated over a narrow range around one of its resonant frequencies. A plot is made of received signal strength against frequency; this has a peak at the resonant frequency (f_r) and falls away on both sides. The width of the curve between the two points where the received signal is $\sqrt{\frac{1}{2}}$ of the peak, i.e. the width at the half-power points 3 dB below the peak, is the bandwidth δ measured in c/s. The theory of this is well explained by Parker[85] who shows that if $f_r \gg \delta$, then:

$$\pi\delta = W . \alpha_{total}$$

where W is the sound velocity and α_{total} is the total absorption including the classical and wall absorptions.

The cylindrical resonator built by Roesler[86] for the resonance bandwidth method is shown in Figure 1.15. One requires it to be long to give a

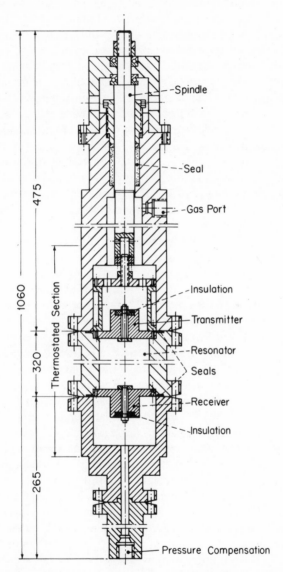

Figure 1.15 The resonator of Roesler[86] (dimensions in mm)

low fundamental axial frequency and to have a small diameter so as to have a high fundamental radial frequency. However it must be a practical size without too large a surface area which would make the wall absorption too big. The dimensions of the resonator shown in Figure 1.15 represent a compromise and it is just under 30 cm long and 7 cm in diameter with the walls honed smooth. The fundamental radial harmonic frequency is then about ten times greater than that of the axial fundamental. Two condenser transducers are used which form the ends of the tube and the transmitter can be withdrawn to enable the resonator to be evacuated and filled with gas; when the resonator has been filled to the required pressure the transmitter is forced down hard onto the end of the resonator tube. The receiver is fixed and the pressures on each side of it are equalized via a small by-pass tube. The resonator can operate with pressure up to 100 atm allowing relaxation measurements to be made at effective frequencies down to about 5 c/s atm[44,86]. Wall corrections were applied using the method of Fritsche[82] and good agreement was found between this theory and experiment for non-relaxing gases[82,86].

Other apparatus using the half-bandwidth principle have also been described. Parker and Swope[85,87,88] used a precision bore Pyrex tube, 75 cm long and 5 cm internal diameter, with an electromagnetically operated piston as transmitter at one end and a receiving condenser transducer at the other. This apparatus can be used at temperatures up to 200°C[88]. The resonator of Henderson and Donnelly[89] is of rectangular cross-section ($3\frac{1}{2} \times 1\frac{1}{2}$ cm) and about 117 cm long. It has a ribbon transducer at each end and the magnets for these are housed in the walls of the resonator which are thick enough to allow it to contain gases with pressures up to 1000 atm when the apparatus has a minimum effective frequency of about 5 c/s atm. Modifications to this apparatus have been described by Schnaus[90] and by Henderson, Clark and Lintz[91]. Parker and Swope and Henderson and Donnelly report that the absorption at the walls is greater than that calculated by the Kirchhoff–Helmholtz formula; this excess was allowed for when the absorption curves were calculated but the theory of Fritsche was not applied. Lukasik and Young[92] made measurements with the resonance bandwidth method on nitrogen in the temperature range 770 to 1190°K but unfortunately the experimental details were not reported. A variation of the resonance bandwidth method involving measurement of the impedance of the transmitter has been described by Smith, Harlow and Kitching[93,94].

1.2.2.4 *Ultrasonic Waves as Optical Diffraction Gratings*

Interest in this method has recently been revived by Martinez, Strauch and Decius[95,96] after a lapse of some twenty years. The apparatus consists of an acoustic chamber with a piezoelectric crystal and an optical system which allows a parallel beam of monochromatic light from a slit to cross the sound field perpendicularly just above the crystal and then be recorded on a photographic plate. It is important that stray light be eliminated and to do this a Lycot Coronagraph can be used[96]. Travelling waves are used with a frequency of around 1 Mc/s and the apparatus has to use pressures greater than about half an atmosphere in order to be able to photograph the diffraction pattern satisfactorily. The sound waves produce a grating with a spacing of one sonic wavelength moving with the speed of sound; since this is negligible compared with the speed of light the grating is effectively stationary.

The method is comparative, the pattern for the test gas being photographed on the same plate as the pattern for a gas such as argon or nitrogen whose sonic wavelength is known; it has been used to study[95] dispersion in nitrous oxide (N_2O) over the effective frequency range 10^5 to $10^{6.6}$ c/s atm and to measure[96] the non-ideality correction in ammonia (equation 1.18). Absorption interferes with the method since it destroys the grating, though Peterson[97] has measured the fading out of the diffraction pattern with distance away from the oscillating crystal and thus deduced the sonic absorption. The method has been reviewed by Martinez, Strauch and Decius[96].

1.3 AERODYNAMIC METHODS

Apart from the shock tube which is described and discussed by Dr P. Borrell in Chapter 3, there is only one other aerodynamic method which has been used for measuring vibrational relaxation times, namely the impact tube.

1.3.1 The Impact Tube

If say a monatomic gas is expanded through a nozzle from a reservoir where the static pressure is p', the enthalpy of the gas is converted to translational motion and its temperature falls. The moving gas is now brought to a halt isentropically at the nozzle of an impact tube, i.e. a tube of small diameter facing upstream and connected to a manometer; the expansion process is here reversed and the impact pressure equals the original p'. The experiment is now repeated with a gas having internal modes and thermal equilibrium is attained after the expansion; if the vibrational relaxation time is long compared with the period of compression at the impact tube, this process is now no longer isentropic and the stagnation temperature is higher than that in the reservoir and so the new impact pressure p'' is lower than p'. If the relaxation and compression times are comparable then the impact pressure is intermediate between p' and p''.

The compression time at the impact tube depends upon the diameter of the tube (d) and the gas velocity (v) and is approximately given by (d/v) s. Thus for a velocity of 150 m/s and a tube diameter of 0·15 mm, both easily realized in practice, the compression time is about 1 μs, which is comparable with many vibrational relaxation times. The method was devised by Kantrowitz and Huber[98] and developed by Griffith[99]. Tuesday and Boudart[100] used it for an important investigation of the relaxation times of oxygen mixed with water vapour. This method requires only simple apparatus but needs large quantities of gas and careful calibration.

1.4 OPTICAL METHODS

Optical methods are taken to be those in which light is used to excite the molecular vibrations; there are then various techniques which can be used to follow the resulting vibrational relaxation. With flash spectroscopy, successive absorption spectra are recorded in order to follow the population changes of the various vibrational levels. In the fluorescence technique, the collisional lifetime of an excited vibrational level is compared with its radiative lifetime, which has to be known and acts as an internal 'clock'. Then there is the spectrophone, based upon the optic-acoustic effect, and very recently there have been brief reports of novel optical methods involving the use of lasers.

With optical methods one can excite and study the relaxation of vibrations which cannot be investigated acoustically or aerodynamically because they have large quanta and hence their contribution to the vibrational specific heat is small. Also, if monochromatic light of the right frequency is used, one can excite a particular vibration provided that, and here is a limitation for optical methods, it is infrared active.

1.4.1   Flash Spectroscopy

This technique, devised by Lipscomb, Norrish and Thrush[101], uses a high energy, short duration flash to excite the test gas. The flash is produced by discharging a condenser bank through a krypton-filled discharge tube which lies alongside a tube filled with the test gas. The chemical reactions or energy-exchange processes initiated by the flash are followed by taking absorption spectra at known intervals after the flash. For this there is a spectroscopic flash tube which is triggered by the main flash after a preset delay time. The second flash travels the length of the test gas before passing into a spectrometer. In the latest flash spectroscopy apparatus[102,103] a microwave pulse is used as the short duration energy source. The reaction vessel is contained in a tuned cavity and the microwave field accelerates electrons which collide with the gas molecules and excite the vibrational or electronic degrees of freedom.

An example of the successive absorption spectra obtained is shown in Figure 1.16: this was photographed by Callear[104] in order to study the decay rate of vibrationally excited NO. A mixture of NO with a partial pressure of 5 mm Hg and N_2 with one of 600 mm was exposed to a 1600 Joule flash: this excites the NO from its electronic ground state ($X^2\Pi$) to higher ones, principally $A^2\Sigma^+$. These excited molecules then either radiate or are quenched to give vibrationally excited molecules in the electronic ground state and in this way the vibrational temperature of the NO is increased by several hundred degrees whereas the translational temperature is only raised by about 5°K. It appears that a Boltzmann distribution amongst the vibrational levels is rapidly achieved so that the majority of excited molecules occupy the first vibrational level, $X^2\Pi(v = 1)$; this is then depopulated by collisional deactivation, mainly by N_2 molecules, to the ground level ($v = 0$). The decay is followed by taking successive absorption spectra of the γ bands, in particular the (0, 1) transition from $X^2\Pi(v = 1)$ to $A^2\Sigma^+(v' = 0)$. A microphotometer is used to determine the intensity of the recorded lines for this transition and this gives a direct measure of the population of the $X^2\Pi(v = 1)$ state. As can be seen from Figure 1.16 this falls off with time and from a plot of intensity against time the relaxation time can be deduced.

By repeating the experiment with differing concentrations of NO the relative efficiencies of NO and N_2 in bringing about the vibrational de-excitation can be deduced. Deactivation of the NO by the N_2 is brought about by transferring the vibrational energy from the mode of one molecule

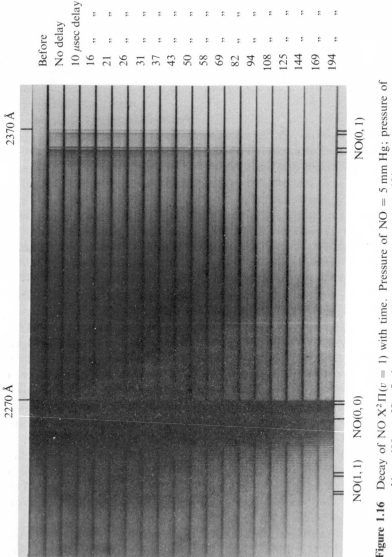

Figure 1.16   Decay of NO $X^2\Pi(v = 1)$ with time.   Pressure of NO = 5 mm Hg; pressure of N_2 = 600 mm Hg; flash energy = 1600 Joules (from Callear[104])

to that of the other, the balance being made up with translational energy; this is a complex process (see Chapter 2). Flash photolysis has proved to be a very useful method for studying such processes between NO and other simple molecules[103] and has even been used to study the vibrational energy exchange between different electronic states of NO^{10}.

1.4.2 Fluorescence Technique

A vibrationally excited carbon monoxide molecule, CO^*, can lose its vibrational energy by one of two parallel processes: either by the radiative process,

$$CO^* \leftrightarrows CO + hv \qquad (v = 2143 \, cm^{-1}) \qquad \text{(i)}$$

or in a process when it collides with some other molecule M,

$$CO^* + M \leftrightarrows CO + M + \text{kinetic energy} \qquad \text{(ii)}$$

At room temperature it so happens that if M = CO the collisional lifetime is well over a second and is much *greater* than the radiative one which is 33 ms[7,105]. With the vast majority of gases the reverse is true. This is one of the rare occasions where radiation is more important than collisions for the exchange of vibrational energy and use can be made of this to determine the efficiencies of various molecules in de-exciting the CO. If pure CO is irradiated with infrared light then there is an intense and easily measurable fluorescence signal as the CO re-emits the energy. If some other gas M is added which is efficient in bringing about reaction (ii) then the fluorescence will be increasingly quenched and when enough has been added to half-quench it, the rates of (i) and (ii) will be equal. The rate for (i) is known and so the collision efficiency of M can be deduced knowing its concentration.

This technique has been developed by Millikan[105,106] and the apparatus for it is shown in Figure 1.17. The secret is to use only very pure CO; the impurity level has to be reduced to below a few parts in a million. The CO is passed through two traps packed with alumina pellets and copper wool and immersed in a 'dry ice' bath and liquid oxygen, respectively[105]. The CO then issues out through a porous plug and is contained by surrounding it with an annular layer of argon flowing at the same velocity; in this way wall collisions are avoided. The CO is excited by an infrared source; in Figure 1.17 it consists of flame-heated zirconia but an alternative is to use the CO emission from a fuel rich CH_4/O_2 flame. Fluorescence is observed

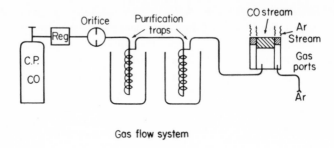

Gas flow system

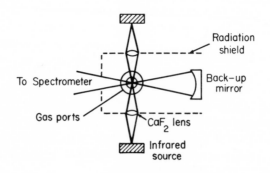

Top view of optical setup

Figure 1.17   The fluorescence apparatus of Millikan[106]

at right angles to the exciting light. Millikan showed that the CO gas,
which was usually diluted with argon to make it optically thin, could be
'heated' to a vibrational temperature around 1000°K whilst the rotational
and translational temperatures remained at 300°K. He examined various
deactivating partners for the CO including other molecules which partici-
pated in complex processes (see Chapter 2). NO is the only other molecule
to which the fluorescence principle has been applied and this work has
been reviewed by Callear[10].

1.4.3   The Spectrophone

The optic-acoustic effect, i.e. the absorption of chopped radiation by a gas
and its emergence as pressure bursts—a sound wave—after vibrational-
translational energy transfer, was first observed by Tyndall, Bell and
Röntgen[107,108] and can be used in two ways to measure vibrational

relaxation times. The first is the *amplitude* spectrophone; the principle of this method is similar to that of the fluorescence technique (see previous section). The light-excited molecules can either lose their vibrational energy by fluorescence radiation, reaction (i) of previous section, or else by collisional deactivation, reaction (ii). If the radiation is periodically interrupted by a mechanical chopper, the latter route leads to the production of a sound wave whose amplitude is proportional to the amount of energy which has been lost by the excited molecules by vibrational-translational energy transfer.

The first amplitude spectrophone was reported by Woodmansee and Decius[109]. They used a moving coil microphone to detect the sound and made a careful study of the resonant frequencies of the cell to ensure that these did not interfere. They measured the amplitude of the received sound as a function of the radial chopping frequency ω; this varies as $(1 + \omega^2 \tau^2)^{-1/2}$, where τ is the relaxation time. They made measurements with carbon monoxide and found that their results fitted best to a curve for which $\tau = 2$ ms. Unfortunately their sample contained sufficient H_2 and H_2O to have considerably affected the result and the experiment was repeated by Ferguson and Read[110] using the apparatus shown in Figure 1.18. Radiation from a Nernst filament is focussed through calcium aluminate glass into the cylindrical absorption cell which can be closed by a needle valve. The cell is acoustically linked via a passage to a condenser-transducer microphone. The apparatus was carefully baked out before each experiment and the CO was purified using Millikan's technique of cold traps[105] (Figure 1.17). The signal obtained with pure CO at a single chopping frequency was compared with that from CO containing about 15% H_2, which is a very efficient collision partner for deactivating CO. This procedure gave a relaxation time of 0·8 s for pure CO. However, as shown subsequently by Doyennette and Henry[111], account should be taken of the reabsorption of the fluorescent radiation and the corrected relaxation time is then 6 s assuming that there were no impurities affecting the result. Ferguson and Read[112] have also used the amplitude spectrophone to study relaxation of the hydrogen halides.

The second instrument based upon the optic-acoustic effect is the *phase* spectrophone; it was first suggested by Gorelik[113] as a method of measuring vibrational relaxation times and its development has been described by Delany[114]. This method measures the time lag between the absorption of the chopped radiation and its emergence as a sound wave; this lag appears as a phase difference between the modulated radiation and the

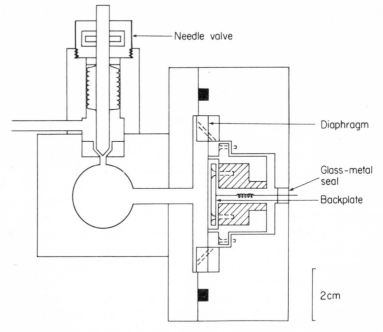

Figure 1.18   The amplitude spectrophone of Ferguson and Read[110]

emitted sound wave. This is the simple principle but in fact the experiment is difficult to perform because other effects such as absorption by the walls and thermal conductivity introduce various phase changes. These have been examined by Delany and the full theory is given by Delany[115] and by Cottrell and McCoubrey[9], Section 4.5.

The phase spectrophone built by Delany is shown in Figure 1.19. A cylindrical aluminium chamber is closed at both ends with rocksalt windows and two similar condenser microphones are fitted facing each other across the cylinder. Radiation from a suitable source such as a Nernst filament was modulated and focussed into the cell. A photo transistor placed outside the exit window of the absorption chamber was used to observe the modulated waveform and provide a reference signal for the phase measurements which were made with a phasemeter. The chamber walls were highly polished to minimize absorption of light and the chamber resonant frequency when filled with carbon monoxide was around 3 kc/s, well above the highest chopping frequency used, so that resonance effects

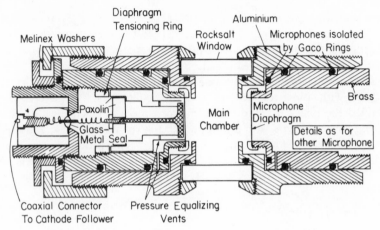

Figure 1.19   The phase spectrophone of Delany[115]

were avoided. There is a small port, not shown in Figure 1.19, for the introduction or removal of gas.

The optic-acoustic signal received by each microphone is measured both as to amplitude and phase relative to the modulated light; this gives the complex *ratio* of the microphone sensitivities. Then the light is switched off and one microphone is now used as a transmitter which sends a known signal to the receiver microphone opposite; from this second measurement one obtains the complex *product* of the two microphone sensitivities. These two measurements constitute the reciprocity technique which allows one to deduce the phase lag due to the microphone only so that the phase lag due to the optic-acoustic effect can be obtained by subtraction. With this method the microphone calibration can be determined for each set of operating conditions. If τ is the relaxation time at one atmosphere, p the pressure, ω the radial chopping frequency and ϕ the phase lag, then[115],

$$\tan \phi = \omega\tau/p$$

The construction and operation of another phase spectrophone has recently been reported by Cottrell, Macfarlane, Read and Young[116]. This incorporates a single condenser microphone and results reported earlier[117] are invalid because of spurious frequency responses. These were found to be due to the absorption cell acting as a Helmholtz resonator and the gas ports had to be modified to eliminate this. With a true response the phase lag varies linearly with the reciprocal of the test gas pressure and the lag due to the microphone and associated electronics is found by extrapolating the results to $(1/p) = 0$. Also a correction has to be applied to some gases for a thermal conduction phase shift. Lavercombe[118] has reported results for CO_2 obtained using a phase spectrophone which disagree with those of Cottrell and his colleagues[116]. Lavercombe obtained the phase lag due to the microphone and electronics by first making measurements with ethane, which relaxes very quickly, and then using this result for his experiments with CO_2. This procedure has been criticized by Macfarlane and Read[119] because no allowance was made for the different sound velocities in C_2H_6 and CO_2.

1.4.4   Laser Experiments

Yardley and Moore[120] have reported briefly an experiment which is a hybrid of the fluorescence and spectrophone techniques. They used a gas laser to excite the v_3 mode of methane (see Figure 2.16) and then at right angles to the laser they monitored the fluorescence of the v_4 mode, the exciting radiation being suitably chopped at frequency f. The radiative lifetime for v_4 is much longer than its collisional lifetime so that its fluorescence hardly affects the mode's population and is a measure of its modulated population. If the methane's pressure is p, then at values of (f/p) around $10^{5.5}$ c/s atm, the energy is rapidly transferred at collisions from v_3 to v_4 so that one can consider that it is effectively fed direct to v_4. When the chopper cuts off the radiation the vibrationally excited molecules loose their energy to translation at collisions and so the fluorescence level drops accordingly. Clearly the magnitude of the drop will depend on the vibrational–translational relaxation time and, from the ratio of the fluorescence level with and without the chopper intervening, i.e. the ratio of the out-of-phase and in-phase components $(\tan \theta)$ if the chopping action is sinusoidal, one can deduce the relaxation time. Yardley and Moore found that in this (f/p) range $\tan \theta$ varied linearly with (f/p) and they deduced a relaxation time in good agreement with those obtained ultrasonically.

After determining the relaxation time for v_4 Yardley and Moore increased the value of (f/p) so that during a chopper cycle collisional deactivation of v_4 is unimportant and the energy can only be lost by fluorescence. Comparison then of the phase lag between the in-going radiation to v_3 and the out-coming fluorescence of v_4 gives the relaxation time for the vibrational energy transfer between v_3 and v_4. Because this is a multilevel system the phase lag can exceed $90°$ as was also found by Lavercombe[118] with CO_2. The second half of Yardley and Moore's experiment is similar to the phase spectrophone except that there is no microphone*. This is a big advantage and, as pointed out by Read[121], it will be possible to apply the technique to many gases when tunable lasers are developed.

Hocker, Kovacs, Rhodes, Flynn and Javan[122] have carried out laser studies with carbon dioxide. They used pulses from a Q-switched, $10.6\ \mu m$ laser to produce sudden population changes of the two upper modes of CO_2; after the pulse the rate of decay of the fluorescence was used to obtain vibrational relaxation times. In addition experiments were carried out with a discharge across the sample tube which excited the molecules to higher vibrational levels†.

Another application of lasers has been to use them to stimulate Raman scattering. De Martini and Ducuing[123] have adapted this so as to measure the vibrational relaxation time of hydrogen at room temperature in a way analogous to flash spectroscopy (Section 1.4.1). A Q-switched, focussed laser with a pulse duration of 20 ns was used as the exciting flash which raises about 1% of the hydrogen molecules to the first vibrational level. Then a second (non-Q-switched) laser, whose output is a train of spikes lasting in all about 0.5 ms, was used to record a series of Raman spectra, the intensity of the anti-Stokes lines being a measure of the relaxing population in the first vibrational state. Experiments were carried out with the hydrogen under pressures ranging from about 20–60 atm and a relaxation time of 1.06 ± 0.1 ms was obtained. This method allows one to study the relaxation of Raman active modes and will be useful for other diatomic molecules; however for polyatomic molecules there will be the difficulty that several modes will be Raman excited simultaneously.

REFERENCES

As far as possible the Literature has been surveyed up to the end of 1966.

1. G. W. Pierce, *Proc. Nat. Acad. Sci. U.S.*, **60**, 271 (1925).
2. K. F. Herzfeld and F. O. Rice, *Phys. Rev.*, **31**, 691 (1928).
3. L. Landau and E. Teller, *Z. Phys. Sowjetunion*, **10**, 34 (1936).

* Note added in proof: an experiment, similar to that of Yardley and Moore[120], has just been reported by Houghton[125] for CO_2.

† Note added in proof: more detailed descriptions of laser experiments have been published by Moore and his colleagues[126].

4. J. Meixner, *Acustica*, **2**, 101 (1952); *Kolloid Z.*, **134**, 3 (1953).
5. R. Herman and K. E. Shuler, *J. Chem. Phys.*, **29**, 366 (1958).
6. S. J. Lukasik, *J. Acoust. Soc. Am.*, **28**, 455 (1956).
7. H.-J. Bauer, *Acustica*, **17**, 90 (1966).
8. B. Lambert and C. G. S. Phillips, *Phil. Trans. Roy. Soc. London, Ser. A*, **242**, 415 (1950).
9. T. L. Cottrell and J. C. McCoubrey, *Molecular Energy Transfer in Gases*, Butterworths, London, 1961.
10. A. B. Callear, *Appl. Opt.*, Suppl. No. 2, 145, (1965).
11. H. O. Kneser, Pt. IV of 'Relaxation Thermique dans les Gaz', in *Dispersion and Absorption of Sound by Molecular Processes* (Ed. D. Sette), Academic Press, London, 1963, pp. 64–68.
12. K. F. Herzfeld, *J. Acoust. Soc. Am.*, **39**, 813 (1966).
13. K. F. Herzfeld and T. A. Litovitz, *Absorption and Dispersion of Ultrasonic Waves*, Academic Press, London, 1959.
14. H.-J. Bauer, 'Theory of Relaxation Phenomena in Gases', in *Physical Acoustics* (Ed. W. P. Mason), Vol. 2, Pt. A, Academic Press, London, 1965, Chap. 2, pp. 47–131.
15. C. A. Coulson, *Waves*, 6th ed., Oliver and Boyd, Edinburgh, 1952, Chap. 6.
16. see J. R. Partington, *An Advanced Treatise on Physical Chemistry*, Vol. I, Longmans, London, 1949, p. 122.
17. A. van Itterbeek and W. H. Keesom, *Communication from Physics Lab. of Univ. of Leiden*, No. 209c (1930).
18. J. K. Roberts and A. R. Miller, *Heat and Thermodynamics*, 5th ed., Blackie, London, 1960.
19. T. L. Cottrell, I. M. Macfarlane and A. W. Read, *Trans. Faraday Soc.*, **61**, 1632 (1965).
20. R. O. Davies and J. Lamb, *Quart. Rev. (London)*, **11**, 134 (1957).
21. J. Lamb, 'Thermal Relaxation in Liquids', in *Physical Acoustics* (Ed. W. P. Mason), Vol. II, Pt. A, Academic Press, London, 1965, Chap. 4, pp. 203–280.
22. J. D. Lambert, 'Relaxation in Gases', in *Atomic and Molecular Processes* (Ed. D. R. Bates), Academic Press, London, 1962, Chap. 20, pp. 783–806.
23. F. I. Tanczos, *J. Chem. Phys.*, **25**, 439 (1956).
24. J. L. Stretton, *Trans. Faraday Soc.*, **61**, 1053 (1965).
25. G. G. Stokes, *Trans. Cambridge Phil. Soc.*, **8**, 287 (1845).
26. G. Kirchhoff, *Pogg. Ann. Phys.*, **134**, 177 (1868).
27. J. J. Markham, R. T. Beyer and R. B. Lindsay, *Rev. Mod. Phys.*, **23**, 353 (1951).
28. J. L. Stretton, *D. Phil. Thesis*, Oxford University, 1964.
29. D. G. Jones, private communication.
30. M. Greenspan, *J. Acoust. Soc. Am.*, **26**, 70 (1954).
31. H. Hinsch, *Acustica*, **11**, 230 (1961).
32. D. Burnett, *Proc. London Maths. Soc.*, **39**, 385; **40**, 382 (1935).
33. R. Holmes, G. R. Jones and N. Pusat, *Trans. Faraday Soc.*, **60**, 1220 (1964).
34. R. Holmes, G. R. Jones and N. Pusat, *J. Chem. Phys.*, **41**, 2512 (1964).
35. M. Greenspan, 'Transmission of Sound Waves in Gases at Very Low Pressures', in *Physical Acoustics* (Ed. W. P. Mason), Vol. II, Pt. A, Academic Press, London, 1965, Chap. 1, pp. 1–45.

36. C. Truesdell, *J. Rat. Mech. Anal.*, **2**, 643 (1953).
37. H.-J. Bauer and K. F. Sahm, *J. Chem. Phys.*, **42**, 3400 (1965).
38. K. S. Cole and R. H. Cole, *J. Chem. Phys.*, **9**, 341 (1941).
39. A. C. J. Johnson, *Proc. Phys. Soc. (London)*, **73**, 273 (1959).
40. J. D. Lambert and R. Salter, *Proc. Roy. Soc. (London)*, Ser. A, **243**, 78 (1957).
41. D. Sette, A. Busala and J. C. Hubbard, *J. Chem. Phys.*, **23**, 787 (1955).
42. J. D. Lambert and R. Salter, *Proc. Roy. Soc. (London)*, Ser. A, **253**, 277 (1959).
43. L. M. Valley and S. Legvold, *J. Chem. Phys.*, **33**, 627 (1960).
44. H.-J. Bauer and H. Roesler, 'Relaxation of the Vibrational Degrees of Freedom in Binary Mixtures of Diatomic Gases', in *Molecular Relaxation Processes* (Chem. Soc. Special Publication No. 20), Chem. Soc. and Academic Press, London, 1966, pp. 245–252.
45. J. D. Lambert, A. J. Edwards, D. Pemberton and J. L. Stretton, *Discussions Faraday Soc.*, **33**, 61 (1962).
46. H.-J. Bauer and E. Liska, *Z. Physik*, **181**, 356 (1964).
47. W. P. Mason, *Piezoelectric Crystals and their Application to Ultrasonics*, Van Nostrand, New York, 1950.
48. D. A. Berlincourt, D. R. Curran and H. Jaffe, 'Piezoelectric and Piezomagnetic Materials and their Functions in Transducers', in *Physical Acoustics* (Ed. W. P. Mason), Vol. 1, Pt. A, Academic Press, London, 1964, Chap. 3, pp. 169–270.
49. A. R. Blythe, *Ph.D Thesis*, University of Edinburgh, 1962.
50. A. R. Blythe, T. L. Cottrell and M. A. Day, *Acustica*, **16**, 118 (1966).
51. R. A. Walker, T. D. Rossing and S. Legvold, *N.A.C.A. Tech. Note, No. 3210*, Washington, 1954, (see Fig. 3.3 of Ref. No. 9).
52. J. L. Stewart, *Rev. Sci. Instr.*, **17**, 59 (1946).
53. I. F. Zartman, *J. Acoust. Soc. Am.*, **21**, 171 (1949).
54. R. Salter, *D.Phil. Thesis*, Oxford University, 1959.
55. J. L. Stewart and E. S. Stewart, *J. Acoust. Soc. Am.*, **24**, 22 and 194 (1952).
56. T. L. Cottrell and A. J. Matheson, *Trans. Faraday Soc.*, **59**, 824 (1963).
57. V. V. Voitonis and V. F. Yakolev, *Soviet Phys.—Acoust.*, **12**, 259 (1967).
58. J. C. Hubbard, *Phys. Rev.*, **38**, 1011 (1931); **41**, 523 (1932); **46**, 525 (1934).
59. J. Blitz, *Fundamentals of Ultrasonics*, Butterworths, London, 1963, Chaps. 2, 3 and 4.
60. H. J. McSkimin, 'Ultrasonic Methods for Measuring the Mechanical Properties of Liquids and Solids', in *Physical Acoustics* (Ed. W. P. Mason), Vol. 1, Pt. A, Academic Press, London, 1964, Chap. 4, pp. 271–334.
61. H. Sell, *Z. Tech. Physik*, **18**, 3 (1937).
62. W. Kuhl, G. R. Schodder and F.-K. Schröder, *Acustica*, **4**, 520 (1954).
63. W. M. Wright, *Tech. Memo. Nos. 47 and 48, Contract Nonr-1886 (24)*, Harvard Univ., 1962.
64. H. Hässler, *Thesis*, Stuttgart, 1963.
65. E. Meyer and G. Sessler, *Z. Physik*, **149**, 15 (1957).
66. E. U. Haebel, *Acustica*, **15**, 426 (1965).
67. R. Holmes and W. Tempest, *Lab. Pract.*, **10**, 774 (1961).
68. R. Holmes, private communication.
69. M. Greenspan, *J. Acoust. Soc. Am.*, **22**, 568 (1950).

70. M. Greenspan and M. C. Thompson Jr., *J. Acoust. Soc. Am.*, **25**, 92 (1953).
71. D. G. Parks-Smith, *D.Phil. Thesis*, Oxford University, 1964.
72. F. A. Angona, *J. Acoust. Soc. Am.*, **25**, 1111 and 1116 (1953).
73. F. D. Shields and R. T. Lagemann, *J. Acoust Soc. Am.*, **29**, 470 (1957).
74. F. D. Shields, K. P. Lee and W. J. Wiley, *J. Acoust. Soc. Am.*, **37**, 724 (1965).
75. F. D. Shields, *J. Acoust. Soc. Am.*, **32**, 180 (1960).
76. F. D. Shields, *J. Acoust. Soc. Am.*, **34**, 271 (1962).
77. F. D. Shields and K. P. Lee, *J. Acoust. Soc. Am.*, **35**, 251 (1963).
78. F. D. Shields and K. P. Lee, *J. Chem. Phys.*, **40**, 737 (1964).
79. H. O. Kneser, *J. Acoust. Soc. Am.*, **5**, 122 (1933).
80. H. Knötzel and L. Knötzel, *Ann. Physik*, **2**, 393 (1948).
81. P. D. Edmonds and J. Lamb, *Proc. Phys. Soc. (London)*, **71**, 17 (1958).
82. L. Fritsche, *Acustica*, **10**, 189, 199 (1960).
83. P. D. Edmonds and J. Lamb, *Proc. Phys. Soc. (London)*, **72**, 940 (1958).
84. R. Holmes, F. A. Smith and W. Tempest, *Proc. Phys. Soc. (London)*, **81**, 311 (1963).
85. J. G. Parker, *J. Chem. Phys.*, **34**, 1763 (1961).
86. H. Roesler, *Acustica*, **17**, 73 (1966).
87. J. G. Parker and R. H. Swope, *J. Chem. Phys.*, **43**, 4427 (1965).
88. J. G. Parker and R. H. Swope, *J. Acoust. Soc. Am.*, **37**, 718 (1965).
89. M. C. Henderson and G. J. Donnelly, *J. Acoust. Soc. Am.*, **34**, 349, 779 (1962).
90. U. E. Schnaus, *J. Acoust. Soc. Am.*, **37**, 1 (1965).
91. M. C. Henderson, A. V. Clark and P. R. Lintz, *J. Acoust. Soc. Am.*, **37**, 457 (1965).
92. S. J. Lukasik and J. E. Young, *J. Chem. Phys.*, **27**, 1149 (1957).
93. D. H. Smith and R. G. Harlow, *Brit. J. Appl. Phys.*, **14**, 102 (1963).
94. R. G. Harlow and R. Kitching, *J. Acoust. Soc. Am.*, **36**, 1100 (1964).
95. J. V. Martinez, J. G. Strauch, Jr. and J. C. Decius, *J. Chem. Phys.*, **40**, 186 (1964).
96. J. G. Strauch, Jr. and J. C. Decius, *J. Chem. Phys.*, **44**, 3319 (1966).
97. O. Petersen, *Z. Physik*, **41**, 29 (1940).
98. A. Kantrowitz, *J. Chem. Phys.*, **14**, 150 (1946); P. W. Huber and A. Kantrowitz, *J. Chem. Phys.*, **15**, 275 (1947).
99. W. Griffith, *J. Appl. Phys.*, **21**, 1319 (1950).
100. C. S. Tuesday and M. Boudart, *Tech. Note 7, Contract AF33 (038)-23976*, Princeton University, 1955.
101. F. J. Lipscomb, R. G. W. Norrish and B. A. Thrush, *Proc. Roy. Soc. (London)*, *Ser. A*, **233**, 455 (1956).
102. A. B. Callear, J. A. Green and G. J. Williams, *Trans. Faraday Soc.*, **61**, 1831 (1965).
103. A. B. Callear and G. J. Williams, *Trans. Faraday Soc.*, **62**, 2030 (1966).
104. A. B. Callear, private communication.
105. R. C. Millikan, *J. Chem. Phys.*, **38**, 2855 (1963); **40**, 2594 (1964).
106. R. C. Millikan, 'Relaxation Processes for Vibrational Energy in Gases', in *Molecular Relaxation Processes* (Chem. Soc. Special Publication No. 20), Chem. Soc. and Academic Press, London, 1966, pp. 219–234.
107. see J. Tyndall, *Proc. Roy. Soc. (London)*, **31**, 307, 408 (1881).

108. W. C. Röntgen, *Phil. Mag.*, **11**, 308 (1881).
109. W. E. Woodmansee and J. C. Decius, *J. Chem. Phys.*, **36**, 1831 (1962).
110. M. G. Ferguson and A. W. Read, *Trans. Faraday Soc.*, **61**, 1559 (1965).
111. L. Doyennette and L. Henry, *J. Phys. Radium*, **27**, 485 (1966).
112. M. C. Ferguson and A. W. Read, *Trans. Faraday Soc.*, **63**, 61 (1967).
113. G. Gorelik, *Dok. Akad. Nauk. SSSR*, **54**, 779 (1946).
114. M. E. Delany, *Sci. Progr.* (*London*), **47**, 459 (1959).
115. M. E. Delany, *Ph.D. Thesis*, University of London, 1958.
116. T. L. Cottrell, I. M. Macfarlane, A. W. Read and A. H. Young, *Trans. Faraday Soc.*, **62**, 2655 (1966).
117. T. L. Cottrell, T. F. Hunter and A. W. Read, *Proc. Chem. Soc.*, 1963, 272.
118. B. J. Lavercombe, *Nature*, **211**, 63; *Proc. Intern. Congr. Acoustics, 5th, Liège, 1965*, Paper C.26.
119. I. M. Macfarlane and A. W. Read, private communication.
120. J. T. Yardley and C. B. Moore, *J. Chem. Phys.*, **45**, 1066 (1966).
121. A. W. Read, private communication.
122. L. O. Hocker, M. A. Kovacs, C. K. Rhodes, G. W. Flynn and A. Javan, *Phys. Rev. Letters*, **17**, 233 (1966).
123. F. DeMartini and J. Ducuing, *Phys. Rev. Letters*, **17**, 117 (1966).
124. F. D. Shields, *J. Chem. Phys.*, **46**, 1063 (1967).
125. J. T. Houghton, *Proc. Phys. Soc.* (*London*), **91**, 439 (1967).
126. C. B. Moore, R. E. Wood, B-L Hu and J. T. Yardley, *J. Chem. Phys.*, **46**, 4222 (1967); J. T. Yardley and C. B. Moore, *J. Chem. Phys.*, **46**, 4491 (1967); J. T. Yardley and C. B. Moore, *J. Chem. Phys.*, **48**, 14 (1968).
127. R. Holmes and M. A. Stott, *J. Sound Vib.*, **5**, 449 (1967).
128. R. Holmes and M. A. Stott, *J. Sci. Instr.*, **44**, 136 (1967).
129. E. J. Morgan and R. A. Kern, *J. Acoust. Soc. Am.*, **43**, 859 (1968).

2

Vibrational Energy Exchange at Molecular Collisions

J. L. Stretton

2.1 INTRODUCTION

The transfer of energy between vibration and translation is essentially a collision process since it is only at molecular collisions that energy can be exchanged; then the molecular oscillators are perturbed and the energy converted. This is easy to visualize in terms of classical mechanics: the molecules are pictured as atoms held together by springs, and clearly collisions between the molecules will affect the oscillations of the constituent atoms. This probably gives the impression that energy transfer between vibration and translation is easy, but in fact it is a relatively inefficient process because, as will be seen, the duration of the collision is usually long compared with the period of the molecular vibration. It often appears that the energy transfer is rapid because of the high rate of collisions in a gas, about one every 10^{-10} second per molecule in a gas at N.T.P. Some molecular vibrations can convert their energy direct to electromagnetic radiation but this can generally be neglected as an energy exchange process in a gas because the radiative life times are usually long compared with collisional ones[1].

The vibrational energy is quantized so that only an integral number of quanta can be exchanged. The probability of transfer at a collision drops steeply as the amount of energy exchanged increases and so usually only one quantum is exchanged at a time. A vibrational quantum can be either converted entirely to translational energy (or vice versa)—this is a *simple process*—or else converted into a quantum of a different vibration—this is a *complex process*. The latter involves two vibrations and the difference between the two quanta is made up with translational energy which is effectively not quantized. The exchange of a quantum of the same vibration

58

between two different molecules achieves nothing, but there is the case where, perchance, two different vibrations have the same frequency, or one is an exact multiple of the other. An exchange between them is termed a *resonant complex process* and no energy is converted to or from translation; even so the transition probability is still less than unity as will be seen.

When two molecules approach each other their electron clouds first attract each other slightly due to the London dispersion forces plus the effects of any dipole interactions, etc. At closer range, the electron clouds tend to overlap and repulsion forces are set up in consequence of the Pauli Principle. These repulsion forces increase very rapidly at small separations and are the ones mainly responsible for perturbing the molecular vibrations and so causing exchange of energy. Usually the collision is not energetic enough for energy conversion and the molecules just 'bounce' off one another with translational energy conserved; this is an *elastic* collision. The opposite is an *inelastic* one where exchange takes place. It is found from the pressure dependence of the rate of vibrational energy exchange in gases (see Chapter 1) that only bimolecular collisions need be considered and this has been shown to be true even for liquids[2]. The rate of exchange of energy depends upon the product of the collision rate and the probability that at a collision exchange will take place; the probabilities are the same for collisions in gases or liquids, but in the latter the collision rate is very much greater†.

Since the probability per collision of energy exchange between vibration and translation is small, and often very small, many collisions will be required on average to achieve energy conversion. This number can vary from as few as ten for the low frequency modes of polyatomic molecules up to 10^{10} for diatomic molecules at room temperature; thus if the translational temperature is suddenly altered there will be a delay before the vibrational degrees of freedom reach equilibrium again. This delay can be characterized by a relaxation time as discussed in Chapter 1 where methods for measuring them are described. The first half of this chapter is concerned with the *a priori* calculation of vibrational relaxation times. In deriving theoretical values for the relaxation times the fundamental quantities are the transition probabilities for specific vibrational changes during a collision. These are fitted into relaxation equations which are solved to give relaxation times which can then be compared with experiment.

---

† Note added in proof: an alternative approach to considering successive, individual collisions has recently been proposed for *dense* media[269].

Such a comparison is carried out in the second half of this chapter and involves a review of the experimental results for pure gases and their binary mixtures; the latter are important because they can yield specific information about complex rather than simple processes. In all the gases considered it is assumed that there is neither ionization nor chemical reaction.

Two books[3,4] and several review articles[5-11] have recently been published on the subject of vibrational energy transfer. It is a branch of molecular or chemical physics and has tended to be treated from two aspects: firstly as the cause of relaxation phenomena and therefore affecting such processes as the propagation of sound, and secondly as something to be studied for the information which it can yield about subjects of interest to the physical chemist such as intermolecular potentials and molecular excitation prior to chemical reaction. The comprehensive treatise by Herzfeld and Litovitz[3] tends to take the first view and Cottrell and McCoubrey's book[4] adopts the second. In the last chapter of their book Clarke and McChesney[9] deal with the effects of vibrational energy transfer in relation to gas dynamics. The theoretical calculation of relaxation times has been reviewed by Takayanagi[8,11] and by Herzfeld[10] and surveys of the experimental results have been made by Kneser[5], Read[6] and Cottrell and McCoubrey[4]. Millikan[7] has discussed the subject with particular reference to diatomic molecules†.

2.2 CALCULATION OF THE TRANSITION PROBABILITIES

The transition probability (P) is the probability that at a collision a given process, either simple or complex, will take place; its inverse is the collision number (z), the average number of collisions required to bring about the process. P is defined as the ratio of the area of cross-section for the inelastic process to that for elastic scattering and from the latter one also obtains the molecular collision rate. The methods which have been used for the theoretical calculation of P fall into three groups: classical, semiclassical and quantum mechanical; with the classical treatments one assumes that both the translational and vibrational motions of the colliding molecules can be described by classical mechanics; in the semiclassical calculation the vibrations are treated quantum mechanically while the translational trajectories are still calculated classically, and finally there is the fully

† Note added in proof: a new journal has just appeared which covers molecular relaxation in general and has a review article concerned with vibrational-energy transfer[270].

quantum-mechanical approach. Even within these groups there are different methods which can be used and in describing them emphasis will be placed upon the assumptions and limitations of the treatments, rather than upon the mathematics involved.

The physical picture of energy transfer between translation and vibration caused by a molecular collision may be expressed in terms of Ehrenfest's adiabatic principle: if a changing external force acts on a periodic motion, the process will be *adiabatic* if the change of the force is small during a period of the motion. The criterion for a *non-adiabatic* process, i.e. one involving energy transfer, is thus a collision which is of short duration compared with the period of the oscillation. Thus if the collision is between an atom A and a diatomic molecule BC and A hits B end on, the collision forces the internuclear distance to contract and compress the 'spring' which is the molecular bond. When the spring expands again A must have rebounded clear of B if there is to be efficient energy transfer. Otherwise there will be a second collision in which energy will be transferred back to A with the result that there has been little or no nett transfer of kinetic energy from A to the vibrating mode of BC and the collision is therefore adiabatic. In quantum mechanical terms the collision is adiabatic if the relative change of the perturbing potential is small compared with $h/\Delta E^{10}$, where ΔE is the energy converted from translation to vibration.

If s is the distance travelled by A in the repulsive potential field of BC, u is the relative velocity at the collision and v is the oscillator frequency, then for the process to be non-adiabatic it is required that the product $(su^{-1}v)$ be of the same order as or less than unity. Take as an example the oxygen molecule: $v = 4.65 \times 10^{13}$ c/s ($\equiv 1554\,cm^{-1}$)† and from a consideration of molecular dimensions s is of the order of 1 Å ($\equiv 10^{-8}$ cm). Thus u has to be greater than about 5×10^5 cm/s which is an order of magnitude greater than the average thermal velocity at room temperature, and so very few collisions will be effective. From this simple treatment it can be seen that energy transfer is favoured by the colliding particles having small masses, a steep intermolecular repulsion potential and a low vibration frequency. Also the probability will increase with temperature

† It is more convenient to state vibrational frequencies in wave numbers (cm^{-1}). To convert frequencies in wave numbers to c/s, multiply by the speed of light (2.998×10^{10} cm/s); the inverse of frequency stated in wave numbers is the actual wavelength. Thus a vibration of 1000 cm^{-1} has a frequency of 3×10^{13} c/s and a wavelength of 10 μm ($\equiv 10^5$ Å). Such a vibration has an energy (hv) of 0.124 eV which is equivalent to 2.86 kcal/g mole (1 eV $\equiv$ 8,066 cm^{-1} $\equiv$ 23.06 kcal/g mole): its characteristic temperature (hv/k) is 1439°K.

because at higher temperatures there will be a larger number of molecules available with sufficient kinetic energy.

It must be stated here at the outset that an exact and rigorous calculation of P for any process, even a simple one, by any of the three methods is not possible at present: first because accurate, realistic intermolecular potentials are not yet available and secondly if they were it is very unlikely that an analytical solution would be possible and the equations would have to be solved numerically on a computer; even then the computing time needed could be prohibitive. Thus to ease the problem most of the treatments have been for the simple system of an atom colliding with a diatomic molecule in one dimension. The calculations for this system using the classical, semiclassical and quantum-mechanical approaches will be described first and then compared. After this the quantum-mechanical theory only will be developed for more complicated cases involving collisions in three dimensions and polyatomic molecules.

2.2.1 One-dimensional Collision between an Atom and a Diatomic Molecule

Figure 2.1 shows the geometry for an end-on collision between atom A and molecule BC; since the collision is in one dimension it has zero angular momentum and so rotation is not considered. BC is assumed to be a simple harmonic oscillator whose equilibrium internuclear distance is $2d_0$. The distance between A and B depends upon the distance between the centres of gravity of A and BC, i.e. r, and upon the state of oscillation of BC represented by χ, the displacement from the equilibrium position; thus $r = d + \chi + d_0$ and the potential energy of the system is V given by:

$$V = V(\chi, r) \tag{2.1}$$

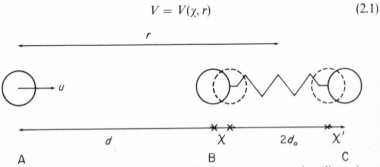

Figure 2.1   One-dimensional collision between an atom and a diatomic molecule

Unless V depends upon both the translational coordinate r and upon the vibrational one χ, there could be no transfer of energy between them. A discussion of the actual form of the intermolecular potential is deferred. A quantity common to all three treatments is the reduced mass of the colliding pair of particles, this equals the product of their masses divided by their sum, i.e. for the case represented in Figure 2.1:

$$\mu = \frac{m_A(m_B + m_C)}{m_A + m_B + m_C} \tag{2.2}$$

A similar expression holds for the reduced mass (M) of the diatomic oscillator BC.

In the absence of external forces the *total* energy of the two molecules is conserved throughout a collision and their centre of mass continues to move with a uniform velocity; it is only the *relative* velocity which is important and can be converted into vibrational energy and so BC is taken to be stationary. It is then easily shown[10] that the 'relative' kinetic energy equals $\frac{1}{2}\mu u^2$ where u is the relative velocity of approach. The full formulation of the collision problem is discussed by Takayanagi[8].

2.2.1.1 Classical Treatment

The two differential equations of motion which have to be solved are:

$$\mu \frac{d^2 r}{dt^2} = -\frac{\partial}{\partial r} V(\chi, r) \tag{2.3a}$$

and

$$M \frac{d^2 \chi}{dt^2} = -M\omega^2 \chi - \frac{\partial}{\partial \chi} V(\chi, r) \tag{2.3b}$$

In a series of papers Benson, Berend and Wu[12–15] solved them numerically with the aid of a computer. They obtained the trajectories associated with the collisions and were able to show that multiple collisions between A and B were associated with a low efficiency of energy exchange between vibration and translation in accordance with Ehrenfest's Adiabatic Principle. At first they assumed that the collision was only 'successful' if the energy transferred (ΔE) exceeded the magnitude of a vibrational quantum $(h\nu)$. This led to erroneous results[14]: for a self-consistent classical calculation account must be taken of all the energy transferred at a collision even if it is less than a quantum. Similar calculations have recently been carried out by Kelley and Wolfsberg[16] who also considered collisions between two diatomic molecules.

Parker[17], Rapp[18] and Takayanagi[11] have obtained analytical solutions to Equations (2.3a) and (2.3b) after making certain approximations. Firstly, the amplitude of the molecular vibration is small compared with the range of interaction of the intermolecular forces (s), e.g. for the oxygen molecule in its lowest vibrational state the amplitude is 4×10^{-2} Å whilst, as will be seen, s is about 1 Å; therefore to a good degree of approximation one can assume for the purposes of calculating the trajectories that the intermolecular distance between B and C is constant at $2d_0$ so that the right-hand side of Equation (2.3a) can be taken to be a function of r only. This is tantamount to saying that a change in the vibrational energy does not appreciably affect the relative motion of the two particles during the collision and this is true only when the kinetic energy of approach is much greater than the energy exchange, i.e. $\frac{1}{2}\mu u^2 \gg \Delta E$. Thus one has for the classical trajectory, $r(t)$,

$$-\frac{\partial}{\partial r} V(\chi, r) \simeq -\frac{\partial V}{\partial r}(d_0, r) = -\frac{\partial V}{\partial r} \qquad (2.4)$$

One now has to determine the interaction term $\partial V(\chi, r)/\partial \chi$ in equation (2.3b). In order to do this V is taken[11,16] to be a function of the distance between A and B, i.e.

$$V = V(d) = V(r - \chi - d_0) \qquad (2.5)$$

The classical trajectory $r(t)$ of the incident atom A is then used to obtain the transient driving force $F(t)$ on the oscillator.

It can be shown[19] that a simple harmonic oscillator of frequency v is only affected by a force of the same frequency; thus for the force $F(t)$ it is the Fourier component of frequency v which is important. For a force which varies slowly compared to the period of vibration, the required Fourier component increases with the rate of change of the force[10], which depends upon a combination of the variation of the potential with distance and the velocity of approach of atom A. One uses this to obtain the solution of Equation (2.3b) for the motion of the oscillator[19] at time t', in terms of angular frequency, ω,

$$\chi + \chi' = A \sin(\omega t' + \delta) + \frac{\Omega}{M\omega} \int_{-\infty}^{t'} F[r(t)] \sin[\omega(t' - t)] \, dt \qquad (2.6)$$

where $\Omega = m_C/(m_B + m_C) = \chi/(\chi + \chi')$. The first term represents the oscillation, if any, before the collision with amplitude A and phase δ; the second is the change brought about by the transient force. Takayanagi[11] has shown how the energy taken up by the oscillator is proportional to the difference between the squares of the vibrational amplitudes before and after collision averaged over the initial phase δ. If the oscillator has no energy before the collision, i.e. it is classically at rest, the result is:

$$(\Delta E)_{C1} = \tfrac{1}{2}M\omega^2 \left| \frac{\Omega}{M\omega} \int_{-\infty}^{\infty} F[r(t)]\exp[i\omega t]\,dt \right|^2 \qquad (2.7)$$

Thus in principle the problem can be solved for a given intermolecular potential. Usually only the repulsive forces are considered but Turner and Rapp[20] have shown how to take account of the attractive ones.

2.2.1.2 Semiclassical Treatment

The problem is to calculate the probability (P^{i-j}) that the molecular oscillator BC will change from vibrational state i to state j when subjected to a time-dependent perturbation provided by the molecular collision. The classical equation of motion (2.3a) is solved first for $r(t)$ as in the previous section (Equations 2.3a and 2.4) and this gives the time-dependent perturbing force $F(t)$. The time-dependent wave equation is:

$$(H + V)\Psi = i\hbar(\partial\Psi/\partial t) \qquad (2.8)†$$

where H is the Hamiltonian for the unperturbed oscillator for which:

$$H\psi_i = E_i\psi_i \quad \text{and} \quad E_i = (i + \tfrac{1}{2})h\nu \qquad (2.9)$$

Ψ is expanded in terms of the oscillator eigenfunctions,

$$\Psi = \sum_i a_i\psi_i \exp[itE_i/\hbar] \qquad (2.10)$$

The coefficients satisfy the coupled differential equations:

$$i\hbar\frac{da_i}{dt} = \sum_i a_iW_{ij}\exp[it\Delta E/\hbar] \qquad (2.11)$$

where $\Delta E = (E_j - E_i)$ and W_{ij} is the matrix element:

$$W_{ij} = \int_{-\infty}^{\infty} \psi^*_i F(t)\psi_j\,d\chi \qquad (2.12)$$

† $\hbar = h/2\pi$.

The first-order solution is obtained by substituting the zeroth-order one into the right-hand side of Equation (2.11). The details of the calculation have been given by Herzfeld[10], Takayanagi[11] and Rapp[21,22] who show that since the vibrations are assumed to be harmonic and the perturbation contains only the first-order term of a more general function of χ then transitions are only allowed if $|i-j|$ equals unity. It is also important that, in converting V from a function of distance (d) to one of time, one uses the arithmetic mean of the velocities of approach (u) and recession (v); this is called 'symmetrization', the importance of which was pointed out by Zener[23] who carried out the first semiclassical calculation.

The result is that the probability for the case where $i = 0$ and $j = 1$ is given by:

$$(P^{0-1})_{SC} = |a_1(t \to \infty)|^2 = \frac{\Omega^2}{2Mh\omega} \left| \int_{-\infty}^{\infty} F(t) \exp[i\omega t]\, dt \right|^2 \qquad (2.13)$$

If this is expressed as the average gain of vibrational energy per collision, i.e. $\hbar\omega(P^{0-1})_{SC}$, it is the same as the classical result.

A further approximate solution to Equation (2.11) has been proposed by Mies[24] and so called 'exact', but unsymmetrized, solutions have been obtained numerically by Kelley and Wolfsberg[25]. Rapp and Sharp[22,26,27] have investigated various approximations (see section 2.2.6). Cross[28] has discussed generally the semiclassical theory of inelastic scattering.

2.2.1.3   Quantum-mechanical Treatment

In contrast to the semiclassical calculation, which used time-dependent perturbation theory, the usual quantum-mechanical treatment considers the statistics of a uniform series of collisions on a time-independent basis. The scattering molecule BC is considered to be stationary and to have a beam of A atoms, represented by a plane wave, falling on it; thus this is a case of stationary perturbation theory. The Schrödinger equation for the system is

$$\left\{ -\frac{\hbar^2}{2\mu} \cdot \frac{d^2}{dr^2} + H + V \right\} \Psi = E\Psi(r, \chi) \qquad (2.14)$$

E and V are the total and potential energies of the system respectively, H is defined by Equation (2.9). Ψ may again be expanded in terms of the oscillator eigenfunctions, i.e.

$$\Psi = \sum_{n=0}^{n=\infty} f_n(r)\psi_n(\chi) \qquad (2.15)$$

At large separations the wave function is a superposition of the free particle functions. Two cases have to be considered: elastic scattering where the initial state $(f_i \cdot \psi_i)$ is unchanged and inelastic scattering to the jth vibrational state with the function $(f_j \cdot \psi_j)$. For the former f_i has the asymptotic form at large r,

$$f_i \sim e^{-ik_i r} + A_i e^{+ik_i r} \tag{2.16}$$

where k_i is the wave number defined by Equation (2.18). The first term represents the incoming wave with unit amplitude and the second the elastically scattered wave with amplitude A_i $(A_i < 1)$. There is no incident inelastically scattered wave so that f_j has the asymptotic form,

$$f_j \sim A_j \cdot e^{ik_j r} \tag{2.17}$$

The difference in translational energies between the inelastically scattered waves must be an exact multiple of hv, i.e.

$$-\Delta E = \frac{\mu}{2}(u^2 - v^2) = (j-i)hv = \frac{\hbar^2}{2\mu}(k_i{}^2 - k_j{}^2) \tag{2.18}$$

The probability for the transition P^{i-j} equals the ratio of the inelastic particle flux to the incident particle flux, thus;

$$P^{i-j} = (k_j/k_i)|A_j|^2 \tag{2.19}$$

The problem is to calculate A_j. Following the first treatment by Jackson and Mott[29] Equation (2.15) is substituted into Equation (2.14), multiplied by ψ_i, integrated over r and rearranged to give,

$$\left(\frac{d^2}{dr^2} + k_i{}^2 - V_{ii} \cdot \frac{2\mu}{\hbar^2} \cdot V(r)\right) f_i(r) = \frac{2\mu}{\hbar^2} \sum_{n \neq i} V_{in} \cdot V(r) f_j(r) \tag{2.20}$$

V_{ij} are the vibrational factors defined as:

$$V_{ij} = \int_{-\infty}^{\infty} \psi^*{}_i \cdot V(\chi) \cdot \psi_j \, d\chi \tag{2.21}$$

It will be shown later (Section 2.2.5) that the diagonal elements of the V matrix are approximately equal to unity and greatly exceed all the off-diagonal elements. There are then two common approximate methods for solving Equation (2.20), the Born approximation and the method of Distorted Waves. The former is suitable for very high energies of approach. The A atom is assumed to continue moving in a straight line with a *constant* velocity and the force acting on BC is calculated with this assumption. The Born approximation was used by Witteman[30] in calculations on the vibrational energy transfer processes for carbon dioxide. He has since found this method to be invalid and has recalculated[31] using the method of Distorted Waves. The two methods have been fully compared by Mott and Massey[32] and the Distorted Wave treatment is applicable to the present problem.

As a zeroth approximation, the off-diagonal elements of the V matrix are put equal to zero, i.e. the right-hand side of Equation (2.20) equals zero. This gives f_i. Then as a first approximation f_j is given by putting the value just obtained for f_i into the right-hand side of Equation (2.20) and neglecting other non-diagonal terms. Thus the transition probabilities are calculated using wave functions which are the solutions of the problem with elastic scattering only. (The waves are referred to as 'distorted' because the elastically scattered waves suffer a phase shift which causes the distortion.) It is assumed that the excitation of the final state j is due to a direct transition from the initial state i and that any transitions through intermediate quantum levels can be neglected. The detailed analysis for solving Equation (2.20) has been given elsewhere[3,4,29] and the quantum-mechanical calculation is continued in Sections 2.2.2 and 2.2.4.

2.2.1.4 *Comparison of the Classical, Semiclassical and Wave-mechanical Treatments*

Rapp[18,21] has carefully compared the three approaches and shown that for a given intermolecular potential $V(r, \chi)$ all three give the same result at the 'classical limit', i.e. when the kinetic energy of approach is infinite compared with the amount of energy converted from translation to vibration (ΔE). Cottrell and McCoubrey[4] have made a numerical comparison of the semiclassical and wave-mechanical treatments and found that with a repulsive exponential function for V there is little practical difference even for comparatively low velocity collisions, but, at the 'threshold level' where the kinetic energy of approach $\frac{1}{2}\mu u^2$ equals ΔE, the semiclassical method breaks down and incorrectly predicts a finite probability of energy transfer. Takayanagi[11] has made a similar numerical comparison using a Morse potential and found that the agreement between the two treatments is not so good. Parker[17] has compared the quantum-mechanical and classical treatments and shown that at the classical limit they give the same result, but below this the classical method overestimates the energy transfer and Treanor[33] has given an expression for the magnitude of this discrepancy. Other comparisons have been made by Bartlett and Moyal[34] and by Kerner[35] and errors in various treatments have been noted by Allen and Feuer[36].

All three treatments are essentially first-order perturbation methods and are invalid if the transition probabilities are large; thus in using them one assumes that the probabilities are small or, to put it another way, that the collisions are nearly adiabatic. Further, all three methods assume that the vibrational amplitude is small compared with the range of the intermolecular forces so that for all three treatments the collision trajectory is described in terms of the elastic motion. However, whereas the quantum-mechanical treatment uses the trajectory appropriate to the vibrational state of the oscillator, i.e. k_i and k_j for the approach and recession, account of this is only taken in the other two treatments if the relative velocity is symmetrized. The assumption that the collision trajectories can be calculated neglecting changes in the vibrational coordinate (Equation 2.4) has been questioned recently by Kelley and Wolfsberg[16]. Next, for the classical treatment of the collision trajectory to be valid, the de Broglie wavelength ($\lambda = h/\mu u$) must be small compared with s; if this were not so there would be no way of verifying that the colliding particles experienced a definite force which is necessary if the calculation is to be carried out classically[9]. For the hydrogen molecule with mean thermal energy at room temperature, $\lambda = 1\cdot1$ Å; for heavier molecules and the much higher velocities which are important for energy transfer, λ is considerably smaller than 1 Å so that the condition is fulfilled. This assumption that $s \gg \lambda$ is also made, as will be seen, when the quantum-mechanical treatment is evaluated. Finally, at the classical limit when ΔE is negligible compared with the kinetic energy of approach ($\frac{1}{2}\mu u^2$), all three methods give the same result; this is however somewhat artificial since it implies a very high energy of collision when all the first-order perturbation treatments will break down (see Section 2.2.6).

Having outlined and compared all three treatments, only the quantum-mechanical one will be developed further because it has been the most widely used and has the greatest applicability.

2.2.2 Three-dimensional Collision

Consider now a three-dimensional collision between the atom A and the diatomic molecule BC as shown in Figure 2.2. The relative motion of the

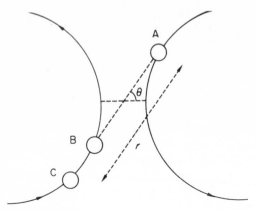

Figure 2.2 Diagram of a three-dimensional collision between an atom and a diatomic molecule

two centres of mass is defined by two coordinates r and θ. θ is the angle between r and the line joining the molecular centres at their smallest separation. The axis of B–C is assumed to be always pointing towards A, an orientation most favourable for the excitation of its vibration. Since in fact all orientations are possible a steric factor will be introduced later (Section 2.2.5). The intermolecular potential is of the same form as in Equation (2.1) and is taken to be spherically symmetric. The wave equation is of the form,

$$\left(-\frac{\hbar^2}{2\mu}\nabla^2 + H + V\right)\Psi = E\Psi \tag{2.22}$$

Ψ may be expanded in terms of the oscillator eigenfunctions and partial waves (see for example Section 38 of Mandl[37]) giving

$$\Psi = \sum_n \sum_\sigma {}_\sigma\Psi_n = \sum_n \sum_\sigma \frac{{}_\sigma f_n}{r} \cdot P_\sigma(\cos\theta) \cdot \psi_n(\chi) \tag{2.23}$$

where σ is the angular momentum quantum number for each partial wave; P_σ are the Legendre polynomials. Since V is independent of θ, angular momentum is conserved during the collision.

Substitution, etc., of Equation (2.23) into Equation (2.22) as with the one-dimensional case gives,

$$\left(\frac{d^2}{dr^2} + k_n{}^2 - \frac{\sigma(\sigma+1)}{r^2}\right){}_\sigma f_n(r) = \sum_n {}_\sigma f_n(r)V_{in} \cdot \frac{2\mu}{\hbar^2} \cdot V(r) \tag{2.24a}$$

then application of the distorted wave approximation for a transition from state i to state j gives as a zeroth approximation:

$$\left(\frac{d^2}{dr^2} + k_i{}^2 - \frac{\sigma(\sigma+1)}{r^2} - \frac{2\mu}{\hbar^2} \cdot V(r) \cdot V_{ii}\right)_\sigma f_i = 0 \qquad (2.24b)$$

and for a first approximation:

$$\left(\frac{d^2}{dr^2} + k_j{}^2 - \frac{\sigma(\sigma+1)}{r^2} - \frac{2\mu}{\hbar^2} \cdot V(r) \cdot V_{jj}\right)_\sigma f_i = \frac{2\mu}{\hbar^2} \cdot V(r) \cdot V_{ij} \cdot {}_\sigma f_i \quad (2.24c)$$

These equations are similar to the one-dimensional case except for the addition of the term $\sigma(\sigma+1)/r^2$ due to the centrifugal energy. Now this quantity changes much slower than $V(r)$ with decreasing r and it has been shown by de Wette and Slawsky[38] that such terms do not appreciably affect the transition probabilities directly, but merely alter the effective velocity of approach. The major contribution to the transition probability comes from the region $r \simeq r_c$, where r_c is the minimum separation or classical turning point. In this region the term may be replaced by $\sigma(\sigma+1)/r_c{}^2$, and so the problem is reduced to the one-dimensional one where the modified wave number is given by,

$$\bar{k}^2 = k^2 - \sigma(\sigma+1)/r_c{}^2 \qquad (2.25)$$

This modified wave-number method is due to Takayanagi[39,40]. In later papers[41,42] he has compared his calculations on oxygen, using this approximation, with the calculations of Salkoff and Bauer[43,44] who used numerical integration to solve the full distorted wave problem. He found that agreement was very good. Use of the modified wave number method avoids having to solve Equations (2.24) for a very large number of σ values for each value of k. The method was also proposed independently by Schwartz and Herzfeld[45]. It breaks down for very large values of σ when $\bar{k}^2$ can become negative; however a large σ means that the 'collision' involves only a very weak interaction between the two trajectories and the contribution to the transition probability from such a partial wave will be unimportant.

Solutions of Equations (2.24) are now found in terms of an effective velocity of approach $\bar{u}$ defined by:

$$\bar{u}^2 = u^2 - \frac{\hbar^2}{\mu} \cdot \frac{\sigma(\sigma+1)}{r_c{}^2} \qquad (2.26)$$

$\bar{u}$ is smaller than u since the centrifugal forces constitute a barrier which has to be surmounted and whose height increases with σ. The lower σ the more 'head-on' the collision, as is discussed fully elsewhere[4,45] and the full analysis for the solution of Equations (2.24) is given by Herzfeld and Litovitz[3]. Cottrell and McCoubrey[4] describe the semiclassical three-dimensional treatment and compare it with the quantum-mechanical one; classical calculations have only been carried out for two-dimensional space[46,47].

2.2.3   The Intermolecular Potential

So far all the equations relating to the calculation of the transition probability have been stated in terms of a general potential $V(r, \chi)$. Now the crux of the whole calculation has been reached, namely the substitution of an appropriate function for V. The problem is twofold. Firstly, present knowledge of intermolecular potentials is very imprecise and what is known has usually been deduced from measurements of the transport properties of gases; it is only relevant for comparatively small collision energies, below those which are responsible for vibrational energy exchange, and it also leads to functions which are independent of the collision angle. Secondly, if a realistic potential existed then the analytical solution of the problem would almost certainly be impossible and recourse would have to be made to numerical methods.

Calculations of transition probabilities are often criticized on the basis of an unsatisfactory potential; this is generally true but should not prevent one from proceeding with such calculations and comparing them with reliable experimental results, since they provide a very sensitive test for a given potential. Most of the calculations to date have assumed that V has an exponential form†, this allows an analytical solution of the problem; it is obtained by fitting an exponential curve to an experimentally-determined potential to determine the exponential repulsion constant. There is no real physical justification for using the exponential form but then neither is there much justification for using some of the other assumed potentials; as Longuet-Higgins[48] has aptly put it, determining their constants involved fitting the experimental results to a Procrustean bed!

The most commonly used intermolecular potential for non-polar molecules is the Lennard-Jones '12–6':‡

$$V(r) = 4\varepsilon \left[\left(\frac{r_0}{r} \right)^{12} - \left(\frac{r_0}{r} \right)^{6} \right] \tag{2.27}$$

† A list of the functions so far used for V is given by Takayanagi[8], section 1.4.1, p. 177.
‡ Note that there is an alternative form for this potential, see ref. 4, p. 138.

The first term represents the repulsion forces due to the interaction of the electron clouds of the colliding molecules whilst the second term covers the attractive London dispersion forces. For polar molecules this has been modified by Monchick and Mason[49] to take account of the dipole attraction:†

$$V(r) = 4\varepsilon\left[\left(\frac{r_0}{r}\right)^{12} - \left(\frac{r_0}{r}\right)^{6} - \delta^*\left(\frac{r_0}{r}\right)^{3}\right] \tag{2.28}$$

where δ^* is a function of the dipole moments of the colliding molecules. This '12–6–3' potential is shown diagrammatically in Figure 2.3; r_A is the

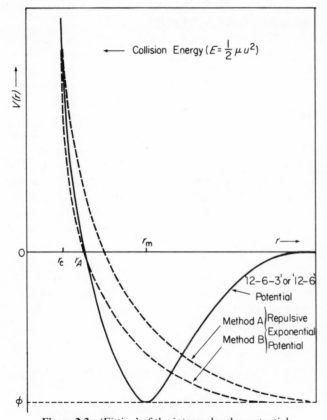

Figure 2.3 'Fitting' of the intermolecular potential

† This supersedes the similar Krieger potential which is incorrect[49].

value of r when $V = 0$, ϕ is the minimum value when $r = r_m$. With the '12–6' potential, $\delta^* = 0$, $r_A = r_0$ and at r_m, $\phi = -\varepsilon$. The parameters for these potentials are determined by fitting the experimental viscosity or second virial coefficient data as described by Hirschfelder, Curtiss and Bird[50] or by Monchick and Mason[49]. However the values of the parameters derived from the viscosity often differ considerably from those chosen to give the best agreement with the virial data. The latter give more weight to the attractive part of the potential, whilst the transport properties are characteristic of the repulsion part up to thermal energies, and so force constants derived from these are to be preferred.

It is now required to fit either Equation (2.27) or (2.28) to the repulsive exponential potential, the radial part of $V(r, \chi)$:

$$V(r) = U \cdot e^{-\alpha r} + \phi \tag{2.29}$$

Two conditions are needed. For the first, the two potentials are equated at the classical turning point; for the second, the exponential curve is made tangential to the curve at the classical turning point, method A, or else the two curves are equated at a second point r_A, method B. The necessary formulae for the fitting in order to determine α have been given by various authors[3,10,51]. Method B has been used more widely for the arbitrary reason that it gives a somewhat larger value of α which has usually led to better agreement with experiment. α is a function of the velocity of approach because the fitting depends on the value of r_c.

The above procedure only takes account of the repulsive forces at a collision—what of the attractive ones? It has been seen in the last section that terms varying slowly with distance do not alter the transition probability directly, but only by shifting the effective collision velocity. De Wette and Slawsky[38] have shown that, if the potential curve has a minimum value, this will increase the effective energy of the collision, and the number of molecular collisions with a given effective energy will be increased by the factor $\exp(-\phi/kT)$; thus calculated probabilities are increased by a similar factor. Note that with this method the attractive forces are only taken into account implicitly.

Explicit allowance can be made by using a double exponential potential of the Morse type:

$$V(r) = \phi \exp[-\alpha'(r - r_m)] - 2\phi \exp[-\tfrac{1}{2}\alpha'(r - r_m)] \tag{2.30}$$

As before ϕ is the well depth at $r = r_m$ and if $r_m \gg r$ then this potential has a similar form to Equation (2.29). The Morse potential has two exponential terms and allows an analytical solution to be obtained for the transition probability without curve fitting. Unfortunately the three force constants for this potential have only been obtained from the transport or virial data for a very few gases.

With a given intermolecular potential one can now define the elastic cross-section as πr_A^2 and use this for the calculation of the bimolecular collision rate Z,

$$Z = 4Nr_A^2(\pi kT/2\mu)^{1/2} \tag{2.31}$$

where N is the number of molecules per cm^3. The required transition probability is the ratio of the inelastic cross-section q_{ij} to the elastic one.

2.2.4 Completion of the Quantum-mechanical Calculation of the Inelastic Cross-section: the SSH–Tanczos Formula

To obtain the inelastic cross-section one has to solve Equation (2.24); this is done in terms of functions such as $_\sigma F_i(r)$ which are normalized solutions of the homogeneous equation,

$$\left(\frac{d^2}{dr^2} + \bar{k}_i^2 - V_{ii} \cdot \frac{2\mu}{\hbar^2} \cdot V(r)\right)_\sigma F_i(r) = 0 \tag{2.32}$$

This is a form of Bessel's Equation. Following Jackson and Mott[4,29], one obtains for the amplitude $(_\sigma A_j)$ of the inelastic partial wave scattered to the jth state with angular momentum quantum number σ,

$$_\sigma A_j = \frac{4U\mu}{\bar{k}_i\hbar^2} \cdot V_{ij} \int_{-\infty}^{\infty} {_\sigma F^*_i} \, e^{-\alpha r} \cdot {_\sigma F_j} \, dr \tag{2.33}$$

where substitution has been made for the intermolecular potential $V(r)$ using Equation (2.29). One could also have used the Morse potential Equation (2.30), for it is the integral in Equation (2.33) which can only be evaluated in closed form for an exponential potential; Jackson and Mott[29] have evaluated it for the simple repulsive exponential, equation (2.30), and Devonshire[52] has performed a similar calculation for the double exponential Morse potential. Only the former result will now be used and Jackson and Mott's solution can be simplified by assuming that the de Broglie wavelength (λ) for the collision is small so that $\alpha\lambda$ is small compared to $4\pi^2$ for both the approaching and the receding trajectories[3,10].

To give the probability of energy exchange, the inelastic cross-section q_{ij} is used. This is the number of particles scattered into the jth state per second from an initial beam of unit intensity. For an initial effective velocity of approach $\bar{u}$ and integrating over a sphere,

$$q_{ij}(\bar{u}) = \frac{\pi k_j}{k_i^2} \sum_{\sigma} (2\sigma + 1) |_\sigma A_j|^2 \qquad (2.34)$$

The sum is over all partial waves, each of which has a different value of $_\sigma A_j$. The sum over σ contains a large number of terms and can be replaced by an integral[3,53] in terms of $\bar{u}$ (Equation 2.26).

This inelastic cross-section is characteristic of a given velocity of approach. What is required is the probability at a given temperature T. If the distribution of the molecular velocities is Maxwellian the fraction of the molecular flux with velocities in the range $u \to u + du$ is given by,

$$\frac{dN}{N} = 2\left(\frac{\mu}{2kT}\right)^2 u^3 \cdot \exp[-\mu u^2/2kT] \, du \qquad (2.35)$$

Evaluation of the integral over the partial waves, substitution of Jackson and Mott's simplified solution for the integral in Equation (2.33) and integrating over the Maxwellian distribution gives for the inelastic cross-section

$$q_{ij}(T) = 2\pi b \int_0^\infty \frac{\bar{u} r_c^2}{4\pi^2} \exp[-b\bar{u}^2] \, V_{ij}^2 (L'^2 - L^2)^2 \frac{\sinh L \sinh L'}{(\cosh L' - \cosh L)^2} \, d\bar{u}$$

$$(2.36)$$

where

$$b = \mu/2kT, \qquad L = \frac{4\pi^2 \mu}{\alpha h} \cdot \bar{u} \quad \text{and} \quad L' = \frac{4\pi^2 \mu}{\alpha h} \cdot \bar{v}$$

The lower limit of the integral is at the threshold energy where the collision energy equals ΔE, the energy transferred between translation and vibration. Also if $\sigma > 0$ there is a centrifugal barrier to be surmounted which rises in height as the collision becomes less 'head-on' and σ increases; thus $\bar{u}$ and $\bar{v}$ are now the effective velocities of approach and recession.

So far the collision between an atom and a diatomic molecule has been considered with the molecule in the most favourable orientation for a transition. This is now to be extended to a collision between two polyatomic molecules in which the vibrational mode a in one molecule changes its quantum state from i to j while simultaneously a second vibrational mode b in the other molecule changes its quantum state from

k to l. The probability for this process will be denoted by $P_{k-l(b)}^{i-j(a)}$. Applying the principle of the conservation of energy for the complex process,

$$-\Delta E = \frac{\mu}{2}(u^2 - v^2) = hv_a(j-i) + hv_b(l-k) \tag{2.37}$$

Since a second mode is involved another vibrational factor $V^{kl}(b)$ will have to be introduced and because the two molecules might not collide with the most favourable orientation two steric factors $P_0(a)$ and $P_0(b)$ are introduced. Now from Equation (2.36) substitution of the value for the elastic cross-section, taking implicit account of the attractive forces with the term $\exp[-\phi/kT]$, gives for the transition probability:

$$P_{k-l(b)}^{i-j(a)} = P_0(a) \cdot P_0(b) \cdot [V^{ij}(a)]^2 [V^{kl}(b)]^2$$

$$\times (4\mu/kT) \exp[-\phi/kT] \left(\frac{8\pi^3 \mu \Delta E}{h^2}\right)^2 \int_0^\infty f(\bar{u}) \, d\bar{u} \tag{2.38}$$

where

$$f(\bar{u}) = \frac{\bar{u}}{\alpha^4} \cdot \left(\frac{r_c}{r_A}\right)^2 \exp\left[-\frac{\mu \bar{u}^2}{2kT}\right] \left\{\frac{\exp[L-L']}{(1-\exp[L-L'])^2}\right\}$$

The terms in L and L' in Equation (2.36) have been simplified by assuming that the de Broglie wavelength is short for the collision so that the hyperbolic functions can be replaced by exponential terms. As pointed out by Jones[54] the probability, which is always positive, is independent of the signs of ΔE and $(L-L')$.

Calculations have been performed using Equation (2.38) and carrying out the integration numerically[55,56], but it can be done analytically for two extreme cases. Firstly, for a resonant transfer, i.e. $\Delta E = 0$ there being no nett transfer of energy from translation to vibration, then:

$$P_{k-l(b)}^{i-j(a)} = P_0(a) \cdot P_0(b) [V^{ij}(a) \cdot V^{kl}(b)]^2 \frac{64\pi^2 \mu kT}{\alpha^2 h^2} \exp[-\phi/kT] \tag{2.39}$$

where α corresponds to the thermal average approach velocity. In Equation (2.39) there is a translational transition factor of less than unity even though no exchange of translational energy is involved, because, for the vibrations to be induced to change their quantum states, they must be perturbed by a time-dependent force[3].

If the nett transfer of energy is large, the transition probability for a given velocity is small for values of $\bar{u}$ corresponding to the mean thermal energy and rises rapidly with increase of $\bar{u}$; since the number of collisions with such a value of $\bar{u}$ is then falling rapidly, the integrand goes through a sharp maximum at $\bar{u}^*$ say. Under these conditions the integral can be found to a good approximation by the method of steepest descent in terms of ξ which is given by[3,57]:

$$\xi = \frac{E^*}{kT} = \frac{\mu\bar{u}^{*2}}{2kT} = \left(\frac{\Delta E^2 \mu \pi^2}{2\alpha^{*2} \hbar^2 kT}\right)^{1/3} \tag{2.40}$$

α^* and E^* are the values of α and E corresponding to $\bar{u}^*$, and iteration has to be used to obtain α^*. The approximate value of the integral is then:

$$P_{k-l(b)}^{i-j(a)} = P_0(a) \cdot P_0(b) \left(\frac{r^*_c}{r_A}\right)^2 [V^{ij}(a) \cdot V^{kl}(b)]^2 8 \left(\frac{\pi}{3}\right)^{1/2}$$

$$\times \left[\frac{8\pi^3 \mu \Delta E}{\alpha^{*2} \hbar^2}\right]^2 \xi^{1/2} \exp\left[-\left(3\xi - \frac{\Delta E}{2kT} + \frac{\phi}{kT}\right)\right] \tag{2.41}$$

This formula is due to Tanczos[58] and has since been used widely: it was derived as an extension to the three-dimensional quantum-mechanical formula of Schwartz, Slawsky and Herzfeld[45,57] (SSH), which is fully discussed in references 3 and 11. For values of ΔE in the range 0–200 cm^{-1} the probability has been found by interpolation using Equations (2.39) and (2.41) or by direct integration of Equation (2.38). For the reverse process, $P_{l-k(b)}^{j-i(a)}$, Equation (2.41) is the same except that the second term in the exponential would have the opposite sign so that the two probabilities would differ by a factor of $\exp[\Delta E/kT]$ in accordance with the principle of detailed balancing. In the semiclassical treatment this term only occurs when the velocity is symmetrized, whereas in the quantum-mechanical calculation it arises from having k_i^2 in Equation (2.24b) but k_j^2 in Equation (2.24c).

In Equation (2.41) the exponential terms dominate, and the important one is the 3ξ; for the first-order perturbation approach to be justified, ξ must exceed unity. Also it should be noted that the form of ξ is as predicted from consideration of Ehrenfest's adiabatic principle. If allowance is made for the changes with temperature of the second and third exponential terms and for the temperature dependence of $\xi^{1/2}$ (which contains $T^{-1/6}$), the theory predicts that $\log P$ will vary as $T^{-1/3}$: from such a plot against $T^{-1/3}$ the slope leads to the value of α^* (see Dickens and Ripamonti[51] for examples of this). The smaller the probability, the less steep is the temperature dependence for small values of ϕ:

the effect of ϕ is to increase the probability but, unlike the first two terms of the exponential in Equation (2.41), its effect decreases with temperature and for large values of ϕ, P can vary inversely with T.

Landau and Teller[59] first obtained the result that log P varies as $T^{-1/3}$ from a purely classical treatment in which they studied the integral:

$$\int_{-\infty}^{\infty} F[r(t)] \exp(i\omega t)\, dt$$

which is similar to Equation (2.7). This then gives for P, using a purely repulsive exponential potential,

$$P(u) \sim \exp(-2\pi\omega/\alpha u)$$

which when averaged over the thermal distribution of velocities gives $P \sim T^{-1/3}$.

In Equation (2.41) the vibrational factors $V^{ij}(a)$ and $V^{kl}(b)$ are independent of each other and for a simple process $k = l$, $V^{kl}(b) = 1$ and the steric factor $P_0(b)$ is unity. If $|i - j| = 1$ this is a single-quantum jump of the a mode, while for a multi-quantum jump $|i - j| > 1$.

2.2.5 Vibrational and Steric Factors

The vibrational factors were defined in Equation (2.21) and there is one for each vibration which changes its quantum state. For a collision between two molecules undergoing a complex process the full repulsive exponential potential is:

$$V = U \exp \alpha[-r + \chi(a) + \chi(b)] + \phi \tag{2.42}$$

this being an extension of Equation (2.1). Since Equation (2.42) factorizes into separate terms in $\chi(a)$ and $\chi(b)$, the vibrational factors for each of the two modes are independent. Each is calculated by expressing the displacement of the vibrational coordinate χ in terms of the normal coordinate (Q) of the vibration,

$$\chi = \bar{A}Q \tag{2.43}$$

where $\bar{A}$ is the average movement of the atoms for unit change of Q. Then with the exponential term of Equation (2.21) expanded, the vibrational factors are calculated using:

$$V^{i-j} = \sum_n \left[\frac{\alpha^n (\overline{A^n})}{n!} \int_{-\infty}^{\infty} \psi^*_i \cdot Q^n \cdot \psi_j\, dQ \right] \tag{2.44}$$

The solutions of the integrals in Equation (2.44) are well known, e.g. see appendix III of Wilson, Decius and Cross[60]; for V^{i-i} the integrals are zero except when n is zero or even; thus:

$$V^{i-i} = 1 + \frac{\alpha^2(\overline{A^2})}{2\gamma}(i+\tfrac{1}{2}) + \frac{3\alpha^4(\overline{A^4})}{4!4\gamma^2}(2i^2+2i+1) + ... \qquad (2.45)$$

where $\gamma = 4\pi^2\nu/h$. The terms in α in Equation (2.45) are small compared to unity, e.g. for the lowest mode of CH_4 ($\nu = 1306 \, cm^{-1}$) for which $\overline{A}$ is large since only light H atoms are involved, $\alpha^2(\overline{A^2})/2\gamma \simeq 5 \times 10^{-2}$; thus $V^{i-i} \simeq 1$. Similarly one obtains,

$$[V^{(i+1)-i}]^2 = [V^{i-(i+1)}]^2 = \alpha^2(\overline{A^2})(i+1)/2\gamma$$

and

$$[V^{(i+2)-i}]^2 = [V^{i-(i+2)}]^2 = \alpha^4(\overline{A^4})(i+1)(i+2)/16\gamma^2 \qquad (2.46)$$

The expression for the double jumps differs slightly from that given by Tanczos[58]. The assumption made in the distorted wave approximation that the off-diagonal matrix elements are small compared to the diagonal ones is justified. Also for a harmonic oscillator, where the levels are equally spaced, the only difference between the expressions for P^{2-1} and P^{1-0} is in the vibrational factors; thus $P^{2-1} = 2P^{1-0}$ from Equation (2.46), an example of the Landau–Teller[59] relations which are used in deriving the relaxation equations (see Section 2.3.1). It should also be noted that, compared with single jumps, double quantum jumps involve higher powers of α and $\overline{A}$ (compare with Section 2.2.1.2), and the values of the vibrational factors fall sharply.

$\overline{A}$ is the average Cartesian displacement of the surface atoms of the molecule for unit change of the normal coordinate. The individual displacements are obtained from a normal coordinate analysis of the molecular vibrations which is fully described in the book by Wilson, Decius and Cross[60]. Methods for obtaining the displacements from such an analysis are given by Tanczos[58] and Stretton[55]. The averaging is carried out according to the 'breathing sphere' model, i.e. the full displacements of the surface atoms are taken and the appropriate powers averaged. This assumes that all the displacements are normal to the surface, but this is not necessarily true and the steric factors are introduced to compensate for this.

For diatomic molecules determination of $\overline{A^2}$ is trivial: in the hetero-nuclear molecule BC the value of A^2 for atom B is given by

$$A_B^2 = \frac{m_C}{m_B(m_B + m_C)} \qquad (2.47)$$

The corresponding equation is used to obtain the value for atom C and the two are averaged to give $\overline{A^2}$. With a diatomic molecule the steric factor is usually taken as $\frac{1}{3}$, the average of $\cos^2 \theta$ taken over a sphere[61]. Herzfeld[10] has recently propounded a more detailed theory† for the steric factor of linear molecules, and introduced an angular dependence into the interaction between the colliding molecules which couples rotational transitions to the vibrational ones. He obtains for P_0,

$$P_0 = 1 - 1/\alpha d_0 \qquad (2.48)$$

where $2d_0$ is the bond length of the diatomic molecule, e.g. for O_2: $2d_0 = 1.21$ Å and $\alpha = 4.9$ Å^{-1}, $\therefore$ $P_0 = 0.66$. For the bending modes of linear molecules, Herzfeld[61] suggests a factor of $\frac{2}{3}$ (the average of $\sin^2 \theta$ over a sphere); however, as pointed out by Witteman[31] such modes are always doubly degenerate so that the steric factor should be $\frac{1}{3}$ for each mode and account should be taken of the degeneracy in the relaxation equations.

With non-linear polyatomic molecules only a rather arbitrary factor can be used—usually $\frac{2}{3}$ following Tanczos[58]. This rather high value is justified by the fact that usually nearly all the surface atoms are oscillating appreciably. Values for $\overline{A^2}$ for vibrations of some polyatomic molecules have been tabulated by Tanczos[58] and Stretton[55,62]. Generally the values are much larger for vibrations which are predominantly concerned with the movement of hydrogen atoms, e.g. with CH_3F: ν_3, the mode with the lowest frequency (1049 cm^{-1}), is essentially a stretch of the carbon–halogen bond and involves only a slight movement of the hydrogen atoms; for this $\overline{A^2} = 0.022$ a.m.u.$^{-1}$, whereas it is 0.197 a.m.u.$^{-1}$ for ν_6 (1200 cm^{-1}), which involves a rocking motion of the hydrogen atoms of the CH_3 group.

2.2.6 Discussion of the Methods of Calculation

The SSH–Tanczos method, which has been outlined in the preceding sections, and the simple classical and semi-classical perturbation treatments are the ones which have been most widely used for theoretical

† This supersedes the erroneous treatment of the same subject in section 64 of reference 3.

calculation of the transition probabilities. However recently several papers have been published describing more advanced calculations and some of the assumptions made in the foregoing must now be examined in the light of these new studies.

It has been assumed all through that the molecular vibrations are harmonic; the effects of anharmonicity are to decrease the spacing between the vibrational levels and increase the transition probabilities, affecting in particular multi-quantum jumps. Bauer and Cummings[63] considered changes between the vibrational states of electronically excited nitrogen molecules whose anharmonic molecular vibrations were represented by a Morse oscillator. A three-dimensional modified wave number calculation was carried out and it was found that, with the degree of anharmonicity assumed, the probabilities of multiple-quantum jumps were not greatly enhanced and were still several orders of magnitude smaller than those for single jumps.

Mies[24,64] has considered the effect of anharmonicity on the vibrational matrix elements. It was assumed for the distorted wave approximation that the diagonal elements V_{ii} all equalled unity. However consider for example O_2, if it is assumed to be harmonic $V_{00} = 1.0042$ and $V_{11} = 1.0127$, while if account is taken of the anharmonicity then $V_{00} = 1.0164$ and $V_{11} = 1.0503$. These values seem close enough to the assumed: $V_{00} = V_{11} = 1$, but Mies, after solving the Jackson and Mott integral for diagonal matrix elements differing from unity, found that the correction produced a pre-exponential factor varying between 10^{-1} and 10^{-2} for diatomic molecules such as oxygen; even smaller values were obtained for the hydrogen halides. Not all this correction is due to anharmonicity as even with harmonic oscillators the diagonal elements differ slightly from unity. In view of the potential assumed, Takayanagi[8] considers that Mies' results are only qualitative, but even so the point raised must be borne in mind.

In a further paper Mies[65] has critically examined the quantum-mechanical treatment for the excitation of H_2 in collision with He atoms. This was chosen because it is possible with this comparatively simple system to *calculate* a full, angle-dependent intermolecular potential using quantum mechanics, as has been done by Krauss and Mies[66]. The results were used as data for a distorted wave and modified wave number calculation ignoring rotation–vibration coupling. For this system Mies again concluded that the calculated probability is strongly dependent on the ratio of the diagonal matrix elements and further that the assumption of

additive exponential potentials acting between atom centres did not fit
the results obtained from the exact potential. Finally he found that the
greatest contribution to the collision cross-section came from collisions
having a collision angle of 40° whereas it had been hitherto assumed that
head-on collisions (zero collision angle) were the most effective, though
Jameson[67] and Kelly and Wolfsberg[47] have suggested that the most
effective collision angle depends upon the relative masses of the colliding
atom and molecule. However the steric factor obtained by Mies was a $\frac{1}{4}$
which is close to the usually assumed $\frac{1}{3}$: these compare with a value of 0·2
calculated using Herzfeld's formula[10] (Equation 2.48) taking $\alpha = 3\cdot4$ Å^{-1}
and $d_0 = 0\cdot37$ Å. Although the system is as yet of only theoretical interest
it raises doubts about the approximations usually made particularly with
regard to that part of the calculation which is dependent upon the vibrational coordinate.

In two papers Rapp and Sharp[22,26] have considered high energy
collisions in order to test the validity of using first-order perturbation
theory†. All the calculations are one-dimensional, semiclassical and
involve numerical solutions of the time-dependent Schrödinger equation
(Equation 2.8) for an intermolecular potential similar to the exponential
repulsive one. The solution given in Section 2.2.1.2 is termed by them the
first-order perturbation approximation (FOPA) and only terms corresponding to the initial and final states (i and j) are used. This is applicable to
small transition probabilities; it can be extended to the two state approximation (TSA) by allowing transitions back and forth between the states i
and j during a collision, and then further to an Nth order perturbation
approximation (NOPA) which takes account of the states between i and j
for an N quantum jump, i.e. $|i-j| = N$. Finally there is the Nth state
approximation (NSA) when account is taken of N states which include i
and j and intervening states together with other ones outside this range.
During a collision 'virtual' transitions are allowed between all these states
so that account is taken of a large number of W matrix elements of Equation (2.8), e.g., for a $0 \rightarrow 1$ transition, a virtual one during the collision would
be $0 \rightarrow 2 \rightarrow 1$ or $0 \rightarrow 1 \rightarrow 2 \rightarrow 1$, both of which involve state 2. The NSA
calculation is 'exact' if the result is unchanged after increasing N further.

The calculations were carried out for collisions between two nitrogen
molecules as a function of the approach velocity u. For the transition
P^{0-1} the results are shown in Figure 2.4, taken from Sharp and Rapp[26]. At
higher velocities of approach the FOPA result rises towards and then

† Note added in proof: a summary paper has recently appeared[27].

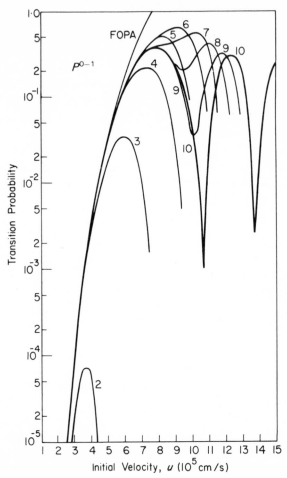

Figure 2.4 The results of the calculations of Rapp and Sharp[26] for the single quantum jump transition (P^{0-1})

exceeds unity. The various N state approximations are shown for values of N up to 10 and for small u all tend to the FOPA value. All the N state results oscillate at high u and as more states are included they tend to coincide; these results are in good agreement with the classical calculations of Treanor[33]. For a probability < 0.1 the FOPA result is a good approximation, but at higher values the upper states play a major rôle and

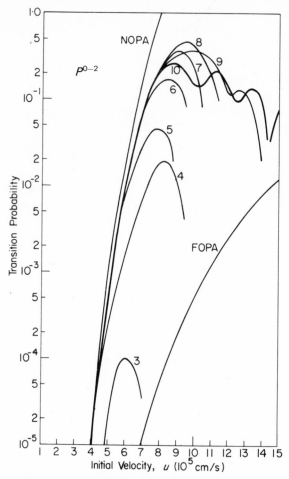

Figure 2.5 The results of the calculations of Rapp and Sharp[26] for the double quantum jump transition (P^{0-2})

it was concluded that for the P^{0-1} transition the FOPA result is valid up to $5000°$K for N_2.

Figure 2.5 (also taken from Sharp and Rapp[26]) shows the calculated probabilities for a P^{0-2} transition. The FOPA value is very low and all the N state calculations converge for small u to the second-order perturbation approximation (NOPA with $N = 2$) where account is taken of the zero and two states and the intervening one state. For all the calculations the velocity was not symmetrized, which could alter the results considerably[4]. Rapp and Englander-Golden[68] have applied the FOPA treatment to the transfer of energy between two diatomic molecules; a formula for resonant transfer ($\Delta E = 0$) is given which is the semiclassical analogue of Equation (2.39).

In a series of papers[69-73] Marriott has carried out a full quantum-mechanical calculation for collisions in three dimensions by solving the analogue of Equation (2.24a) numerically. Thus, although the problem is solved in terms of the unperturbed wave functions, the distorted wave and modified wave approximations are avoided and account can be taken of many vibrational states, so that it is an N state approximation as with Sharp and Rapp[26]. An exponential intermolecular potential was fitted to the Lennard-Jones '12-6' using Method A and, in the first paper[69], the calculations were carried out for carbon monoxide. It was found, similar to Sharp and Rapp[26], that virtual transitions affect the probability. Thus for CO–CO collisions having an energy of 1·5 eV†, which is in the region of the most favourable energy for the transition P^{0-1} at temperatures in the range 1000–3000°K, calculations were carried out with two coupled levels ($N = 0, 1$), three coupled levels ($N = 0, 1, 2$) and 4 coupled ones ($N = 0, 1, 2, 3$); the values obtained for the inelastic cross-sections were $0\cdot110 \times 10^{-2}$, $0\cdot167 \times 10^{-1}$ and $0\cdot223 \times 10^{-1}$ respectively, all in units of πr_A^2. The inelastic cross-sections were converted to probabilities, averaged over the Maxwellian distribution of the molecular velocities to give $P^{1-0}(T)$, and then the relaxation time was calculated (see Section 2.3). The results obtained were in agreement with experiment and with the calculations of Stretton[62] who used the SSH method (Equation 2.38). A fuller comparison of the calculated and experimental results for CO will be made in Section 2.4.

In a very recent paper Secrest and Johnson[74] have carried out a so-called 'exact' quantum-mechanical calculation for an atom colliding with a diatomic molecule by solving the Schrödinger equation numerically using their method of amplitude density functions. The molecular oscillation was assumed to be harmonic and for an exponential interaction potential the results were compared with those obtained by the distorted wave approximation. It was found that the latter could give higher values, even for low transition probabilities when the two methods would be expected to converge.

† Equivalent to $u = 4\cdot5 \times 10^5$ cm/s.

Summing up, the problem of calculating the inelastic cross-section revolves around solving a series of coupled differential equations; these are either Equation (2.11) (one-dimension, semiclassical) or Equation (2.20) (one dimension, fully quantum-mechanical) or Equation (2.24a) (three dimensions, fully quantum-mechanical). They are coupled by the vibrational matrix elements W_{ij} for Equation (2.11) or V_{ij} for Equations (2.20) and (2.20a). Posing the problem in three instead of one dimension only introduces the term in σ which is not a function of the matrix elements. Notice also that the equations are first order in t for the semiclassical case but second order in r for the fully quantum-mechanical one. Consider now the possible methods of solving either Equation (2.24) or (2.24a), there being an analogous procedure for Equation (2.11)[26]. Using the distorted wave approximation for the transition from state i to state j, V_{ii} and V_{jj} are set equal to unity (though in fact even for harmonic oscillators this is not quite true) and V_{ij} ($= V_{ji}$) is used to provide the perturbation, all other matrix elements being neglected. This is a first-order approximation which Mies[64,65] then tried to improve by taking account of the fact that V_{ii} and V_{jj} differ from unity. Thus he appears to have carried out the quantum-mechanical analogue of what Sharp and Rapp[26] call a two-state approximation. Mies' results for the inelastic cross-sections were much smaller than the distorted wave ones, as was found semi-classically by Sharp and Rapp (see Figure 2.4). Marriott[69] carried out the quantum-mechanical N state approximation for values of N up to 4, i.e. he took account of some sixteen matrix elements and hence the virtual transitions between four states. He found that increasing N increased the calculated probability, similar to Sharp and Rapp[26].

From the results of Sharp and Rapp[26] it appears that for small values of the transition probability, i.e. $P(u) < 0.1$, the first-order perturbation treatment is valid and is in agreement with the N state approximation. However, according to the calculations of Secrest and Johnson[74], this is not necessarily true for the quantum-mechanical case, nor is it true for the classical case according to the results of Kelley and Wolfsburg[16]. Secrest and Johnson found that it broke down if the mass ratio parameter for the collision (see Figure 2.1):

$$\tilde{m} = \frac{m_A m_C}{m_B(m_A + m_B + m_C)}$$

is large; e.g. $\tilde{m} = 0.588$ for an argon atom and nitrogen molecule and $\tilde{m} = 12$ for a neon atom colliding with the hydrogen atom of HCl.

Kelley and Wolfsburg[16] also found that there was a correlation between $\tilde{m}$ and the difference between their numerical solution and the first-order perturbation solutions of the classical problem; they point out that the same is true for the results of Mies[64], which are equivalent to the two-state calculations of Sharp and Rapp. Large values of $\tilde{m}$ correspond to a large displacement of the internal coordinate χ (see Figure 2.1) which would tend to invalidate the first-order perturbation approach (see Section 2.2.1); Sharp and Rapp only considered collisions between two nitrogen molecules for which $\tilde{m} = 0\cdot5$. More work is required to sort out these discrepancies for the solution of the one-dimensional collision problem with a low transition probability. Care should be taken to see that similar forms of the intermolecular potential, etc., are used and that if possible the velocities are symmetrized.

For values of the transition probability greater than $0\cdot1$ the N state approximation has to be used (Marriott[69] or Sharp and Rapp[26]) and $P(u)$ becomes an oscillating function of u and virtual transitions are important; then there is said to be strong coupling between the vibrational states rather than the weak coupling for which the first-order perturbation approaches are valid. Correction of the distorted wave method to take account of the diagonal elements differing from unity produces a two-state calculation, which for the approach velocities of interest for diatomic molecules (u between 1 and 5×10^5 cm/s) gives inelastic cross-sections which are too small. Rapp and Sharp[22] explain the inadequacy of the two-state system as being due to resonances between the states. The transition from weak to strong couplings has been studied classically by Alterman and Wilson[75] for Br_2 molecules colliding with Xe atoms; they found that with strong coupling multi-jump transitions become prevalent.

So far in both the semiclassical and fully quantum-mechanical calculations the problem has been solved in terms of the unperturbed oscillator wave functions. This has been appropriate because the duration of fruitful collisions is not long compared with the period of vibration, and in view of the large vibrational force constants the distortion of the vibrational wave functions during a collision is slight[76]. If it is appreciable then the problem must be solved in terms of perturbed wave functions, the so-called method of Perturbed Stationary States† which has been applied to other collision problems[32] but not yet to vibrational energy transfer. However, it would be applicable, for instance, in the case of collisions between H_2 and H^+

† In reference 10, p. 286 and 287 there is confusion between 'Perturbed Stationary States' and 'Stationary Perturbation Theory' which is used in Section 2.2.1.3.

for which Korobkin and Slawsky[77] found there was considerable vibrational distortion due to the incipient formation of H_3^+. Perturbed Stationary States might also have to be used for anharmonic vibrations or for highly polar polyatomic molecules where there is strong mutual attraction; in such cases multiquantum jumps become probable such as are observed by Rabinovitch and his colleagues[78].

The probability for vibrational–translational energy transfer can be affected by other degrees of freedom such as rotation which is specially considered in Section 2.5. Another possibility is the interaction at collisions with electronic energy levels. If two electronic states are close together, Nikitin[79] has suggested that simultaneous electronic and vibrational collisions in opposite directions might enhance the inelastic cross-section. Such electronic states occur in NO and this theory could explain the experimental results obtained by Wray[80] (see Section 2.4.1.1.4).

In all the calculations it has not been possible to take full account of the intermolecular potential. In a series of papers Widom[81] and Shin[82–84] have demonstrated the dependence of the calculated probability on the precise form of $V(r)$ as given by various potentials such as the Morse or Lennard-Jones†. They find that exponential correction terms dependent upon the well depth ϕ should be inserted and that the simple term in (ϕ/kT) in Equation (2.41) is not enough. No pre-exponential terms are given and corrections of this type should await the determination of more detailed intermolecular potentials. Mahan[85] has considered the effect of the attractive forces upon resonant complex processes and suggested that dipole interaction could considerably enhance the probability of such energy transfer between infrared active modes.

Further consideration will be given to these and other points in Sections 2.4 and 2.5 when the results of theoretical calculations are compared with experimental values.

2.3 CALCULATION OF RELAXATION TIMES FROM TRANSITION PROBABILITIES

It is not yet possible to observe transition probabilities directly, and most experiments result in a measured relaxation time (see Chapter 1). For diatomic molecules there is a simple relationship between the relaxation

---

† Note added in proof: Hartmann and Slawsky[271] have recently reported calculations using the Lennard-Jones potential directly, and compared the results with the SSH method which was generally justified.

time and the transition probability, P^{1-0}, for the sole vibrational mode. However for cases where there are several modes, such as polyatomic molecules or gas mixtures, complex as well as simple transitions have to be considered and the situation can become complicated, especially where acoustical relaxation times have to be calculated for comparison with the experimental results.

Three approaches have been used to obtain relaxation times from transition rates. The SSH–Tanczos method[3,57,58] sets up rate equations for the temperature changes of the modes using the 'acoustical approximations'. The second method, based on irreversible thermodynamics (see Meixner[86], Davies and Lamb[87,88]), was largely developed by Bauer[89,90] and is part of a general theory of relaxation phenomena in gases. Finally, Shuler and his colleagues[9,91,92] have used statistical methods to investigate relaxation processes†. In the last sub-section are discussed two models which have been widely used for the reverse process, namely deducing transition probabilities from experimental results.

2.3.1   The SSH–Tanczos Method of Rate Equations

2.3.1.1   *Diatomic Molecules*

Consider a diatomic molecule with only one mode; in anticipation of the extension of the treatment, this is termed the ath mode. As always, it is assumed that energy transfer takes place only at binary collisions—each molecule making Z such collisions per second (see Equation 2.31). If there are N molecules/cm^3 and the oscillator's frequency is v, then define:

$$a = hv/kT, \qquad K_a = Nhv, \qquad \theta_a = 1 - e^{-a} \qquad (2.49)$$

If $\overline{T}$ is the equilibrium temperature of the gas and T' is the instantaneous temperature after a small temperature change, then the corresponding vibrational energies are as follows:

$$\varepsilon_a(\overline{T}) = \bar{\varepsilon}_a = K_a e^{-a}/\theta_a \qquad \text{(Einstein's formula)} \qquad (2.50a)$$

and

$$\varepsilon_a(T') = \varepsilon_a = K_a \sum_\lambda \lambda n_\lambda / N \qquad (2.50b)$$

where n_λ is the number of molecules/cm^3 in the λth state.

The assumption for the present is that $\Delta\lambda$ is restricted to values of ± 1.

† Note added in proof: another review of this work has just appeared[272].

This is to be expected from the discussion in the previous section where it was shown that multi-quantum jumps are much less likely than single ones and there is also experimental evidence for there being only step-wise deactivation, e.g. see Callear and Smith[93]. The rate of energy change of the mode is then the algebraic sum of the activation and deactivation rates:

$$\frac{d\varepsilon_a}{dt} = ZK_a\left[\sum_\lambda (P^{\lambda-(\lambda+1)}n_\lambda/N) - \sum_\lambda (P^{\lambda-(\lambda-1)}n_\lambda/N)\right] \quad (2.51)$$

The different probabilities (P) all refer to the same $\Delta E(= h\nu_a)$ and so they have the same translational factor (see Equation 2.41) and differ only in their vibrational matrix elements (Section 2.2.5). Inspection of these (Equation 2.46) shows that,

$$P^{\lambda-(\lambda-1)} = \lambda P^{1-0}$$

and

$$P^{\lambda-(\lambda+1)} = (\lambda+1)P^{0-1} \quad (2.52)$$

Equation (2.52) is a statement of the Landau–Teller relations for a simple harmonic oscillator; these have been demonstrated experimentally by, *inter alia*, Roth[94] and Callear and Smith[93]. P^{1-0} and P^{0-1} are connected by the principle of detailed balancing (see Equation 2.46) so that:

$$e^{-a}P^{1-0} = P^{0-1} \quad (2.53)$$

Then from Equations (2.51), (2.52) and (2.53) one obtains:

$$\frac{d\varepsilon_a}{dt} = ZK_aP^{1-0}\left[\sum_\lambda (e^{-a}(\lambda+1)n_\lambda/N) - \sum_\lambda (\lambda n_\lambda/N)\right] \quad (2.54)$$

But

$$\sum_\lambda (n_\lambda/N) = 1 \quad \text{and from Equation (2.50)} \quad \sum_\lambda (\lambda n_\lambda/N) = \varepsilon_a/K_a$$

Thus using Equations (2.50), Equation (2.54) becomes:

$$\frac{d\varepsilon_a}{dt} = ZP^{1-0}\theta_a(\bar{\varepsilon}_a - \varepsilon_a) \quad (2.55)$$

This is an energy relaxation equation and has now to be transformed into a temperature relaxation equation using the SSH acoustical approximations[58] (care should be exercised in applying them for other experimental techniques). Note that Equation (2.55) is independent of N.

In a sound wave the translational temperature T_{tr} varies sinusoidally about a mean value $\bar{T}$. If T' is the instantaneous temperature of the vibrational mode one can write for very small deviations about $\bar{T}$, such as are produced in a sound wave,

$$T_{tr} = \bar{T} + \Delta T_{tr}$$

$$T' = \bar{T} + \Delta T_a \qquad (2.56)$$

$$\Delta \varepsilon_a = C_a \Delta T_a$$

where C_a is the static molar specific heat of the mode. Using these relations, equation (2.55) becomes:

$$\frac{dT'}{dt} = ZP^{1-0}\theta_a(\Delta T_{tr} - \Delta T_a) = ZP^{1-0}\theta_a(T_{tr} - T'_a) \qquad (2.57)$$

Comparison of Equation (2.57) with Equation (1.23) shows that the required relation between the transition probability and the isothermal relaxation time† for a molecule with *one* vibration is:

$$\frac{1}{\tau_{pT}} = ZP^{1-0}(1 - \exp[-h\nu/kT]) \qquad (2.58)$$

Implicit in this derivation is the assumption that at all times the distribution of the energy amongst the various oscillator levels is 'Boltzmann', i.e. it corresponds to a definite vibrational temperature.

2.3.1.2 *Polyatomic Molecules and Gas Mixtures*

Consider now a hypothetical polyatomic molecule with two modes a and b having frequencies ν_a and ν_b and degeneracies g_a and g_b respectively. If $\nu_a \approx \nu_b$ complex transitions of the type $P_{0-1(b)}^{1-0(a)}$ are likely to be favourable and will serve to couple the two modes. The flow of energy in and out of mode a by this process only is:

$$\frac{d\varepsilon_a}{dt} = ZK_a\left[\sum_\lambda \sum_\mu g_a g_b P_{\mu-(\mu-1)(b)}^{\lambda-(\lambda+1)(a)} \frac{n_\lambda}{N} \frac{n_\mu}{N} - \sum_\lambda \sum_\mu g_a g_b P_{\mu-(\mu+1)(b)}^{\lambda-(\lambda-1)(a)} \frac{n_\lambda}{N} \frac{n_\mu}{N}\right] \quad (2.59)$$

There are Landau–Teller relations for these transitions which can be deduced from Equation (2.46). Application of these together with the principle of detailed balancing and the acoustic approximations gives,

† The different relaxation times are discussed in section 1.2.1.3.

after much manipulation,

$$\frac{dT'_a}{dt} = g_b Z P_{0-1(b)}^{1-0(a)}(\theta_a/\theta_b)[(\Delta T_{tr} - \Delta T_a) - (v_b/v_a)(\Delta T_{tr} - \Delta T_b)] \quad (2.60)$$

In deriving Equation (2.60) it is assumed that $(e^{-a}\theta_a)/(n_\lambda/N)$ and $\varepsilon_a/\bar{\varepsilon}_a$ can both be equated to unity; neither is strictly correct but they are justified within the same range as the acoustic approximations. Note that, according to Equation (2.60), the energy flow into and out of mode a via this complex process depends upon the degeneracy of mode b (but *not* upon g_a) and upon the instantaneous temperatures of *both* modes. Equation (2.60) can be written for the general complex transition $P_{0-y(b)}^{x-0(a)}$ as:

$$\frac{dT'_a}{dt} = k_{0-y(b)}^{x-0(a)}(\Delta T_{tr} - \Delta T_a) - k'^{x-0(a)}_{0-y(b)}(\Delta T_{tr} - \Delta T_b), \quad (2.61)$$

where the transition rates (k's and k''s) are given by the general expressions[55]:

$$k_{0-y(b)}^{x-0(a)} = g_b Z P_{0-y(b)}^{x-0(a)}(x\theta_a/\theta_b^y)\exp[(x-1)(-hv_a/kT)]$$
$$k'^{x-0(a)}_{0-y(b)} = (v_b/v_a)k_{0-y(b)}^{x-0(a)}(\theta_b^{(y-1)}/\theta_a^{(x-1)}) \quad (2.62)$$

The formulae given by Tanczos[58] contain an error as pointed out by Schofield[56]; those given here agree with Herzfeld[95].

For a given mode the transition rates are then combined,

$$k_{aa} = k_{(a)}^{1-0} + \sum_{b \neq a} [k_{0-1(b)}^{1-0(a)} + k_{0-2(b)}^{1-0(a)} + k_{0-1(b)}^{2-0(a)} + ...] \quad (2.63a)$$

and

$$k_{ab} = k_{0-1(b)}^{'1-0(a)} + k_{0-2(b)}^{'1-0(a)} + k_{0-1(b)}^{'2-0(a)} + ... \quad b \neq a \quad (2.63b)$$

where $k_{(a)}^{1-0}$ $(= ZP^{1-0}(a)\theta_a)$ is the transition rate for a simple process (see Equation 2.57). k_{aa} and k_{ab} are the coefficients of the temperature relaxation equation of the ath mode. There is a similar equation for the bth and if the hypothetical polyatomic molecule, so far considered as having only two modes, is now taken to have n modes, the sum in Equation (2.63a) is over $(n-1)$ terms and there are $(n-1)$ equations like Equation (2.63b). Thus including the terms like k_{aa}, the temperature relaxation equation for each mode has n terms and, as there is one such equation for each mode, the array of k's forms a square matrix **k**. This is unsymmetrical and has all its off-diagonal elements negative[58].

Setting up the **k** matrix for a polyatomic molecule is usually much simpler than would appear from the above. Firstly, there are often several modes with such high frequencies that they can be neglected because they do not contribute appreciably to the specific heat. Secondly, between any two modes there is usually only one type of transition which has an appreciable probability. Consider some examples. Acetylene, C_2H_2, has five normal modes; of these the three with frequencies around 2000 cm^{-1} or above can be neglected, leaving v_4 (612 cm^{-1}) and v_5 (729 cm^{-1}); thus there is only a 2×2 **k** matrix. Three transitions suffice: simple deactivation of each mode and the complex one $P_{0-1(v_5)}^{1-0(v_4)}$. Figure 2.6 shows transition schemes for CD_3F and SF_6[62]; the latter involves transitions of the type P_{0-2}^{1-0}.

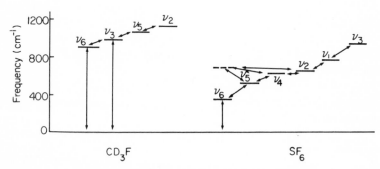

Figure 2.6   Transition schemes for CD_3F and SF_6

In order to obtain the relaxation times from the **k** matrix, Tanczos[58] proceeded as follows. A second $n \times n$ square matrix **H** is defined with each element of the ith column equalling the specific heat of the ith mode divided by the *total* specific heat at constant volume. A third matrix $\mathbf{b} = \mathbf{H} - \mathbf{I}$ (**I** is the unit matrix) is obtained and then the latent roots of the eigenvalue equation:

$$|\mathbf{b}^{-1} . \mathbf{k} - y\mathbf{I}| = 0 \qquad (2.64)$$

equal the negative, reciprocal relaxation times (i.e. $y = -1/\tau_{vs}$). There are n latent roots giving n relaxation times corresponding to the number of relaxing modes. The relaxation times are adiabatic ones and for comparison with acoustic experiments the dispersion and absorption curves (see

Chapter 1) are calculated using:

$$\left(\frac{W}{W_0}\right)^2 = 1 + \sum_{i=1}^{i=n} \frac{D_i(\omega\tau_{vsi})^2}{1+(\omega\tau_{vsi})^2} \tag{2.65}$$

and

$$\alpha\lambda = \pi\left(\frac{W_0}{W}\right)^2 \sum_{i=1}^{i=n} \frac{D_i(\omega\tau_{vsi})}{1+(\omega\tau_{vsi})^2} \tag{2.66}$$

respectively. (Equation (2.65) is the analogue of Equation (1.40b)). The D_i's are the relaxation constants, obtained by comparing coefficients[58]. They are related directly to the relaxation strengths (see Section 1.2.1.3) and are a measure of the 'importance' of each relaxation time, e.g. if a particular D_i is very small, it and its relaxation time τ_{vsi} have no influence on the dispersion or absorption curve according to Equations (2.65) or (2.66). With most polyatomic gases or mixtures only one or perhaps two relaxation times have a significant rôle in determining the dispersion or absorption curve (for example see Table 2.3). Although there are n modes and n relaxation times, one can only relate a particular relaxation time to a particular mode by inspection of the transition rates.

Tanczos' method of calculating relaxation times can easily be extended for mixtures of relaxing gases[62,96]. Consider a binary mixture of diatomic molecules A and B, with modes a and b respectively; simple deactivation of say mode a can occur either at collision with other A molecules or with B molecules. The overall rate will be the sum of the two processes, each of which has its own probability; thus:

$$k_{(a)}^{1-0} = \theta_a(m_A \cdot {}^{AA}Z \cdot {}^{AA}P_{(a)}^{1-0} + m_B \cdot {}^{AB}Z \cdot {}^{AB}P_{(a)}^{1-0}) \tag{2.67}$$

where m_A and m_B are the constituents' mole fractions. The pre-superscripts denote the collision pair; ${}^{AA}Z$ is the number of collisions experienced per second by an A molecule in pure A and ${}^{AB}Z$ is the corresponding rate for an A molecule in otherwise pure B. For calculating either ${}^{AB}P_{(a)}^{1-0}$ or ${}^{AB}Z(= {}^{BA}Z)$ heteromolecular force constants have to be used; if none are available direct from experimental data, they have to be calculated from the corresponding homomolecular ones using the usual combining rules[50]. Complex exchange of vibrational energy between the two molecules can only take place at heteromolecular collisions, therefore:

$$k_{0-1(b)}^{1-0(a)} = m_B \cdot {}^{AB}Z \cdot {}^{AB}P_{0-1(b)}^{1-0(a)}(\theta_a/\theta_b) \quad \text{and} \quad k_{0-1(b)}^{\prime 1-0(a)} = (v_b/v_a)k_{0-1(b)}^{1-0(a)} \tag{2.68}$$

Equations similar to (2.67) and (2.68) can be written for mode b. The **H** matrix is computed as described above except that the vibrational specific heat of a mode is multiplied by the appropriate mole fraction and the total specific heat is that for the whole mixture. The eigenvalue problem (Equation 2.64) can thus be solved for the relaxation times and constants at any desired concentration. The relaxation times are concentration-dependent because the relative frequencies of the various types of collision change with composition.

Equation (2.64) can be simply solved for the case when $n = 2$, since it then reduces to a quadratic equation in y; for large values of n it is best to solve the equation numerically using a computer. To be physically meaningful the roots must all be real and negative. Tanczos[58] quotes Meixner[97] as giving a proof for this but Meixner only discusses the problem generally in terms of irreversible thermodynamics (see next sub-section) and gives no explicit proof based upon the properties of the **b** and **k** matrices. Jones[98] has investigated these and shown that a general proof can only be obtained if the **k** matrix contains only terms concerned with simple or complex processes involving single quantum jumps; with multi-quantum jumps such a proof is not possible and Bauer[99] has suggested that in these cases it is not strictly permissible to define a 'vibrational temperature' for a relaxing mode. The SSH–Tanczos method of rate equations has been used most widely and in the majority of cases only single quantum transitions are important. Further work is needed to establish the method's range of validity by careful comparison with the more rigorous but complicated techniques which are described in the next two sub-sections.

2.3.2   The Bauer Method using Irreversible Thermodynamics†

The basis of this treatment is that if irreversible reactions are occurring in a system then the entropy production is given, according to the Second Law of Thermodynamics, by the inequality:

$$dS > dQ/T$$

Introduction of the concept of the irreversible entropy $d_i S$ allows the inequality to become

$$dS = dQ/T + d_i S$$

† The notation of Bauer[90] is used in this sub-section.

and the unit of irreversibility is defined as $\sigma = d_i S/dt$, which is always positive. During a reversible process, a system passes through a continuous series of states of equilibrium: during an irreversible one there is displacement from equilibrium and the degree of irreversibility determines the rate at which the system seeks to return to equilibrium. σ is a function of the 'forces' (X_i), which depend upon the non-equilibrium state of the system and the fluxes (J_i) which represent the velocity of each irreversible process. The two are connected by the so-called 'phenomenological coefficients' (L_{ik}),

$$J_i = \sum_k L_{ik} X_k$$

then the rate of entropy production takes on the quadratic form,

$$\sigma = \sum_{i,k} L_{ik} X_i X_k$$

Bauer[90] shows that the forces are represented by the reaction parameters (ξ), which are measures of the degree to which each reaction has proceeded, whilst the fluxes are represented by the affinities (A), defined for a given reaction:

$$A_\alpha = -\left(\frac{\partial u}{\partial \xi}\right)_{vs}$$

where u is the internal energy. The affinities are closely connected to the chemical potentials. The parameters V (volume), S and ξ are a complete set of variables for the system, or alternatively one can use the thermodynamically conjugate set: p (pressure), T and $-A$. The reaction parameters and affinities are connected by the matrix of coefficients $\mathbf{L}$,

$$\xi = \mathbf{L} A \tag{2.70}$$

On this basis Bauer[90] builds up a general theory for relaxation in a gas due to the passage of a sound wave, including the case when a chemical reaction is proceeding, e.g. the dissociation of N_2O_4 which has been much studied[5]. One is here only concerned with vibrational relaxation, and each energy exchange process between two vibrational levels is considered as a 'reaction'. This is a much broader definition compared with the usual chemical concept and individual processes such as say P^{1-0}, P^{2-1}, P_{0-1}^{1-0} and P_{1-2}^{2-3} are all individual reactions. Thus account is taken of each vibrational *level* rather than of each vibrational *mode* as was done by SSH.

Bauer's method of calculating the relaxation times involves the setting up of three matrices. In order to illustrate this, consider a mixture of two gases A and B[90]. A has two energy levels A_0 and A_1 and B is an inert gas which only provides an alternative collision partner for A. This system has two components (A_0 and A_1) and there are just two 'reactions' for the two-state oscillator of A:

$$A + A_0 \rightleftharpoons A + A_1 \tag{i}$$

$$B + A_0 \rightleftharpoons B + A_1 \tag{ii}$$

The first matrix $\mathbf{v}$ contains just stoichiometric coefficients; these are ± 1 depending upon whether the particular component has been created or destroyed by a reaction going from right to left. It is simpler to construct the transpose of $\mathbf{v}$ ($\tilde{\mathbf{v}}$), in which reaction has a row and each component a column

$$\therefore \quad \tilde{\mathbf{v}} = \begin{bmatrix} -1 & 1 \\ -1 & 1 \end{bmatrix} \tag{2.71}$$

The second matrix $\mathbf{c}^{-1}$ is diagonal; each element is the inverse of the components concentration.

$$\therefore \quad \mathbf{c}^{-1} = \begin{bmatrix} 1/c_0 & 0 \\ 0 & 1/c_1 \end{bmatrix} \tag{2.72}$$

The third matrix $\mathbf{R}$ is also diagonal and its elements are the rates of the two reactions going right to left.

$$\therefore \quad \mathbf{R} = \begin{bmatrix} k_{AA}c_1c_A & 0 \\ 0 & k_{AB}c_1c_B \end{bmatrix} \tag{2.73}$$

As can be seen the rates are computed according to the Law of Mass Action; k_{AA} is the rate constant for reaction (i), k_{AB} is for (ii).

From these three matrices two further ones are defined:

$$\mathbf{G} = \tilde{\mathbf{v}}\mathbf{c}^{-1}\mathbf{v} \quad \text{and} \quad \mathbf{L} = \frac{V}{RT}\mathbf{R} \tag{2.74}$$

In terms of these two matrices Bauer states the eigenvalue equation for the relaxation times:

$$|\mathbf{GL}^{-1} - z\mathbf{I}| = 0 \tag{2.75}$$

The eigenvalues z are the inverse of the isothermal relaxation times (τ_{pT}, see Chapter 1) and there is one for each reaction in contrast to the SSH method where there is one for each vibration.

Bauer[90] has applied this method to consider the relaxation of all the levels of a harmonic oscillator and has found that, with only single quantum jumps governed by the Landau–Teller relations, all the levels relax together with the same relaxation time as is given by Equation (2.58). The method is also applicable to mixtures and has been used by Bauer and Roesler[90,100] for mixtures of diatomic gases and for O_2/N_2 mixtures it has been compared by Jones[98] with the SSH method, see Section 2.4.2.1. For a discussion of the application of irreversible thermodynamics to relaxation processes and full details of this method, recourse should be had to the excellent exposition by Bauer[90].

2.3.3 Shuler's Investigations of Relaxation

Equation (2.51) expressed the rate of change of energy of a vibrational mode in terms of the rates of activation and deactivation of the levels of that mode; this can be written more generally for each level as:

$$\frac{dn_\lambda}{dt} = \sum_{\mu=0}^{\mu=\infty} \gamma_{\mu\lambda} \cdot n_\mu - \sum_{\lambda=0}^{\lambda=\infty} \gamma_{\lambda\mu} \cdot n_\lambda \tag{2.76}$$

The first term on the right-hand side represents the transitions from the μth state to the λth state which increase the population of the latter state whereas the second sum is for terms which cause a decrease. The rate coefficients $\gamma_{\mu\lambda}$ simply equal $ZP^{\mu-\lambda}$ and Equation (2.76) can be rearranged to give:

$$\frac{dn_\lambda}{dt} = \sum_{\lambda=0}^{\lambda=\infty} \Gamma_{\mu\lambda} \cdot n_\lambda \tag{2.77}$$

where $\Gamma_{\mu\lambda} = \gamma_{\mu\lambda}$ and $\Gamma_{\lambda\lambda} = -\sum_\mu \gamma_{\lambda\mu}$, $\mu \neq \lambda$. An equation similar to Equation (2.77) could be written for each level so that the Γ's form a square matrix analogous to the $\mathbf{k}$ matrix (Equation 2.63) except that it refers to levels rather than to modes. Equation (2.77) is the so-called 'Master Equation' and its solution has been studied for many initial conditions[9,91]; only those directly relevant to vibrational relaxation are considered here.

Rubin and Shuler[101a] considered the solution of Equation (2.77) for diatomic molecules in a heat bath of inert gas. The concentration of the diatomic molecules is very low so that their vibrational relaxation does not affect the translational temperature and is therefore isothermal. They found that, after an abrupt change in the translational temperature, the oscillators relaxed through a continuous series of Boltzmann distributions to the new translational temperature. This result was obtained analytically assuming that the oscillators were harmonic, obeyed the Landau–Teller relations and only exchanged energy in single-quantum jumps. This result justifies the assumption, made in the SSH–Tanczos treatment, that a relaxing mode can be assigned a definite vibrational temperature, though again there is the restriction to single-quantum jumps. The relaxation time obtained by Rubin and Shuler is as given by Equation (2.58). The analysis has been extended by Rankin and Light[101b]; they showed that an arbitrary distribution of vibrational states (i.e. non-Boltzmann) relaxed at least twice as fast towards a time-dependent Boltzmann distribution as the relaxation of the time-dependent Boltzmann distribution proceeded towards final equilibrium. This was for a very dilute solution of the oscillators in a heat bath; at higher concentrations the relaxation to a Boltzmann distribution was much faster†. Shuler and Weiss[92] have further shown that, with a mixture of two harmonic oscillators in a heat bath exchanging energy *via* a process of the type P_{0-1}^{1-0}, both relax through a series of Boltzmann distributions and so each can be said to have its own vibrational temperature.

Marriott in his calculations on CO_2[70] considered relaxation of the three fundamental modes of this linear triatomic molecule behind a shock wave. Account was taken of only the low-lying levels of the modes since these are the only ones which are significantly populated. Equation (2.77) was solved numerically for these few states, the solutions being of the form:

$$n_\lambda(t) = \sum_\mu a_{\lambda\mu} \exp[-t/\tau_{vs}] \qquad (2.78)$$

Marriott found that there were considerable deviations from Boltzmann distributions for the vibrations and calculated values were compared with the shock tube results of Witteman[102]. The SSH acoustical approximations are invalid here because of the large deviations from equilibrium.

2.3.4 The Calculation of Transition Probabilities from Experimental Results: Series and Parallel Excitation.

In the previous sub-sections methods have been described for obtaining the relaxation times from the calculated transition probabilities. Then, for

† Note added in proof: similar studies have just been reported by Rich and Rehm[273].

say acoustical measurements, the theoretical absorption or dispersion curves can be computed for a direct comparison with experiment. This is now the best way of comparing the two and should be used for mixtures or polyatomic molecules. However in the past it has been the general practice to try to calculate the transition probability, or its inverse the collision number (z), from the experimental results. For diatomic molecules this is simple enough using Equation (2.58) but one still has to calculate the collision rate (Z). This involves assuming a value for the elastic cross-section which, for a fair comparison, must be the same as that used in the theoretical calculation. This difficulty is avoided if the theoretical calculation is carried through to the relaxation time, because, for example, Equation (2.41) contains a reference factor $(r_c/r_A)^2$. This relates the distance of closest approach r_c to the separation at zero energy r_A which is also used to calculate Z (Equation 2.41 and so cancels out. Thus one has avoided having to define exactly what is meant by a collision[103]. The corollary to this is that if a transition probability or collision number is quoted the corresponding value of Z must also be stated. By convention, relaxation times are stated for a pressure of one atmosphere.

The experimental dispersion or absorption results for the majority of polyatomic molecules can be fitted to curves calculated for only *one* relaxation time (see Chapter 1). Thus all the vibrational specific heat appears to relax together and it is assumed, somewhat arbitrarily, that the rate-determining step for this is the relaxation of the lowest lying mode, i.e. the one with the minimum frequency (v_{min}), with all the other modes exchanging energy with translation via v_{min}. It is easily shown (e.g. see Section 2.10 of Cottrell and McCoubrey[4]) that if all the other modes exchange energy much more rapidly with v_{min} than v_{min} exchanges with translation, then:

$$\bar{\beta} = \beta_{min}\left(\sum_{i=1}^{i=n} C_i\right) \Big/ C_{min} \qquad (2.79)$$

where $\bar{\beta}$ is the observed isothermal relaxation time†, C_i is the specific heat of the ith mode, C_{min} is that for the lowest and β_{min} is the 'true' relaxation time for the lowest mode, i.e. the one it would have in the absence of all the others, so that the transition probability for the lowest mode can then be obtained using Equation (2.58). Equation (2.79) simply states that the more energy the lowest mode has to exchange with translation the longer it will take *pro rata*.

† β_{min} is denoted by β by Cottrell and McCoubrey[4].

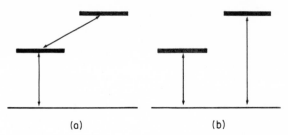

(a)   (b)

Figure 2.7   Two modes relaxing with (a) series and (b)
parallel mechanisms

The series mechanism is illustrated in Figure 2.7 for just two modes. It represents an extreme view because complex processes between the upper modes often involve partial exchange of energy with translation and may not be very much faster than the simple process for v_{min}. Thus the series mechanism is questionable and its assumption can affect conclusions drawn from say temperature dependence measurements[55].

In contrast to a series mechanism is a parallel one, see Figure 2.7; in a polyatomic molecule two modes could exchange energy independently with translation—another way the series model can be invalidated. A case where the parallel mechanism definitely does operate is a mixture of a diatomic gas A with an inert one B which was the example used in Section 2.3.2. Energy exchange can occur at either AA or AB collisions and it is found[4] that the overall relaxation time for the mixture ($\bar{\tau}_{pT}$) is given by:

$$\frac{1}{\bar{\tau}_{pT}} = \frac{m_A}{^{AA}\tau_{pT}} + \frac{m_B}{^{AB}\tau_{pT}} \qquad (2.80)$$

i.e. the relaxation times are summed like electrical resistances in parallel after weighting by the mole fractions m_A and m_B. $^{AA}\tau_{pT}$ is the relaxation time for the pure gas ($m_A = 1$) and $^{AB}\tau_{pT}$ that for one molecule of A in otherwise pure B ($m_B = 1$). $\bar{\tau}_{pT}$ has a linear dependence upon concentration so that experimental results obtained with different mixtures can be extrapolated to give $^{AB}\tau_{pT}$. In the next section, where experimental and theoretical results will be presented and compared, such mixtures with only one vibrationally relaxing component will be dealt with at the same time as the pure gas, since the difference between $^{AA}\tau_{pT}$ and $^{AB}\tau_{pT}$ depends only upon the different reduced mass of the colliding pair and the different intermolecular force constants—the process and vibration concerned are the same.

Series and parallel processes are fully discussed by Herzfeld and Litovitz[3], Chapter 2. For polyatomic molecules use of the SSH or Bauer method for calculating relaxation times obviates the need for using these two models. As can be seen from Figure 2.6, the transition schemes involve both concepts. If two modes have no complex exchange of energy between them, they must relax in parallel and the **k** matrix will have only diagonal elements and can be factored.

2.4 REVIEW OF EXPERIMENTAL RELAXATION TIMES AND COMPARISON WITH THEORETICAL ONES

The principal aim of this section is comparison. Since many reviews of experimental results have appeared in the last few years reference will often be made to them rather than repeat a list of experimental results. This is not intended to be an exhaustive review of the experimental results though results which have not been reviewed before will be mentioned. In general the relaxation times referred to are the isothermal ones (τ_{pT}), see Table 1.1 and Section 1.2.1, and stated for a pressure of one atmosphere. The experimental methods are described and discussed in Chapter 1, except for the shock tube which is the subject of a separate chapter by Dr P. Borrell (Chapter 5) and a recent review by Bauer[104].

In the first half of this section pure gases will be dealt with in order of increasing molecular complexity. Also included are binary mixtures with only one relaxing component, i.e. the second component is say an inert gas with no vibrational modes, so that these mixtures provide information concerning simple processes at heteromolecular collisions. In the second half, binary mixtures will be considered where both components have relaxing vibrational modes and the emphasis is on complex processes at heteromolecular collisions, i.e. intermolecular vibration–vibration transfer.

2.4.1 Pure Gases and Mixtures with only One Relaxing Component

2.4.1.1 *Diatomic Molecules*

These have only *one* vibrational mode and so are the easiest to study theoretically. Also they are often stable enough to be investigated experimentally over a wide temperature range. At the high temperatures the relaxation can be conveniently observed with a shock tube whilst acoustical methods may be used for measurements around room temperature. Special developments have been necessary for the latter because the vibrational specific heat is often very small at such temperatures and hence the relaxation is difficult to detect, whilst with the shock tube there have been troubles over interpretation of the shock wave profiles. The maintenance of a high standard of gas purity has been a common problem. However many of these difficulties have been overcome for several gases and there is now consistent agreement between the various results over wide temperature ranges.

2.4.1.1.1 *Hydrogen and deuterium.* These two molecules have extremely small specific heats because of their very high vibrational frequencies (4160 cm^{-1} for H_2 and 2990 cm^{-1} for D_2). Hence their relaxation can only be studied after optical activation or at high temperatures and these experiments have only recently been successfully carried out.

Kiefer and Lutz[106] used a shock tube with a refined Schlieren technique involving a laser beam to detect the small effect of the vibrational relaxation of H_2 and D_2. For pure deuterium the relaxation time is 11 μs at $1000°K$ falling to 0·6 μs at $3000°K$ and the corresponding values for H_2 are 8·5 μs and 0·5 μs. Moreno[107] has measured the relaxation times for D_2 only and his results agree with Kiefer and Lutz who also made measurements with para enriched D_2[108]; the results obtained were similar to those for the normal mixture. Using a laser technique (see Section 1.4.4) De-Martini and Ducuing[109] have measured the vibrational relaxation times of H_2 at $300°K$, they obtained a value of $1·06 \pm 0·1$ ms and found that this was close to an extrapolation of the shock tube results which fit an equation of the form $\log \tau_{pT} = A + BT^{-1/3}$.

The H_2 and D_2 molecules are too light for the quantum-mechanical SSH calculation to be applicable and the transition probability is overestimated[106], by several orders of magnitude at room temperature[109]. Parker[110] has carried out full calculations using his classical theory[17,111] and compared them with the shock tube results. Good agreement with experiment was obtained for mixtures of H_2 and D_2 with the inert gases but not with the pure gases themselves; then the theory considerably overestimates the transition probability, the discrepancy increasing with decreasing temperature†.

2.4.1.1.2 *Oxygen, carbon monoxide and nitrogen.* These three gases have received much attention, both theoretical and experimental, and they are best considered together. Reliable experimental results have now been obtained over a wide range of temperature and they provide

---

† Note added in proof: Kiefer[274] has just reported a classical calculation of the ratio of the relaxation times of H_2 and D_2.

some of the best opportunities for a critical comparison of theory and experiment.

Dickens and Ripamonti[51] carried out calculations with the SSH–Tanczos formula (Equation 2.41) using Method B fitting (see Section 2.2.3) and compared their results with the experimental relaxation times then available. Fuller SSH calculations have been carried out[62] using Equation (2.38) for both methods of fitting the intermolecular potential and with two sets of intermolecular force constants for oxygen; the data used and the results are set out in Table 2.1. The most recent calculations for the three molecules have been presented by Calvert and Amme[112]; they used the semiclassical approach for three dimensions with 'symmetrization' (see Section 2.2.1.2) and a Morse intermolecular potential (equation 2.30). Parker[111] calculated the relaxation times completely classically using his method[17] (see Section 2.2.1.1) and Benson and Berend[15] have reported results obtained by solving the classical equations of motion numerically.

There is now a wealth of experimental data available for the vibrational relaxation time of oxygen. At high temperatures the shock tube results of Generalov[113] and of White and Millikan[114] are in excellent agreement, see Figure 2.8, and have recently been confirmed by Lutz and Kiefer[115] who used an improved method for analysing the experimental data. They also applied this method to the pre-1961 work and found that it brought some of these earlier results into line with their own. At low temperatures the acoustic results of Shields and Lee[116] and of Parker and Swope[117] fit well to an extrapolation of the shock tube results. At the bottom end of their temperature range the results of Parker and Swope are a little longer than the values obtained by Henderson[118] and Roesler[119]. At room temperature the results of Holmes, Smith and Tempest[120], Parker [121], Roesler[119], Bauer and Roesler[100] on pure oxygen and the extrapolations to pure oxygen (using Equation 2.80) of oxygen mixtures examined by Holmes, Smith and Tempest[122] all agree that the vibrational relaxation time of oxygen is about 20 ms. Thus, as can be seen in Figure 2.8, the acoustic results curve away below the linearly extrapolated shock tube results. As pointed out in Section 2.2.4, one does not necessarily expect a straight line logarithmic plot against $T^{-1/3}$; all the theoretical calculations predict a slight curvature which becomes more pronounced at the low temperatures in the same direction as the experimental results. It is of course possible that, in spite of all the precautions taken, traces of an impurity such as water vapour are shortening the relaxation time; the

Table 2.1

SSH calculations on diatomic molecules (from Stretton[62])

	$T(°K)$	Z (ns^{-1})	Method A			Method B		
			α^* (Å^{-1})	$P^{1-0} \times 10^4$	$\tau_{pT}(\mu s)$	α^* (Å^{-1})	$P^{1-0} \times 10^4$	$\tau_{pT}(\mu s)$
N$_2$	476	5·78	4·635	5·43, −7	3·20, 6	4·999	3·12, −6	5·56, 5
	1000	3·99	4·697	1·99, −4	1·30, 4	4·961	5·15, −4	5·05, 3
	2000	2·82	4·778	2·36, −2	1·85, 2	4·939	3·68, −2	1·19, 2
	5000	1·78	4·914	3·70	3·10	4·929	3·88	2·96
CO	286	7·28	4·658	6·31, −8	2·18, 7	5·114	6·81, −7	2·02, 6
	500	5·51	4·678	4·86, −6	3·74, 5	5·070	2·59, −5	7·04, 4
	1000	3·89	4·728	8·20, −4	3·28, 3	5·029	2·18, −3	1·23, 3
	1500	3·18	4·769	1·23, −2	2·94, 2	5·012	2·40, −2	1·50, 2
	2200	2·63	4·814	1·25, −1	4·05, 1	5·000	1·94, −1	2·61, 1
	3000	2·25	4·855	6·85, −1	1·01, 1	4·993	9·16, −1	7·58
	5000	1·74	4·932	8·17	1·53	4·987	9·07	1·38
O$_2$	303	6·04	4·872	2·04, −5	8·13, 4	5·377	1·37, −4	1·21, 4
	600	4·29	4·890	1·66, −3	1·44, 3	5·314	5·79, −3	4·12, 2
Set I	1372	2·84	4·950	2·07, −1	2·12, 1	5·263	3·99, −1	1·10, 1
	1953	2·38	4·987	1·25	4·95	5·247	2·00	3·08
	3000	1·92	5·040	8·85	1·12	5·233	11·9	0·833
	5000	1·49	5·113	66·7	0·280	5·222	76·7	0·243
O$_2$	303	5·68	—	—	—	5·593	3·07, −4	5·75, 3
	600	4·04	—	—	—	5·518	1·04, −2	2·45, 2
Set II	1372	2·67	—	—	—	5·452	5·95, −1	7·84
	1953	2·24	—	—	—	5·431	2·81	2·33
	3000	1·80	—	—	—	5·411	15·7	0·671
	5000	1·40	—	—	—	5·395	96·0	0·207

Data

Vibration Frequencies: N$_2$: 2331 cm^{-1}, CO: 2143 cm^{-1}, O$_2$: 1554 cm^{-1}

Force constants[50]:

	r_0(Å)	ε/k(°K)
N$_2$	3·749	79·8
CO	3·706	88
O$_2$	3·541	

presence of only 0·01 mole % H_2O would account for the discrepancy (see Section 2.4.2.2)!

Parker and Swope[117] have suggested that the low temperature curvature is due to a 'low temperature state' of oxygen which has a shorter relaxation time. There seems to be no physical basis for this second 'state' and the alleged anomaly only really shows up on their plot of the relaxation time against $T^{-1/2}$ (see Figure 3 of reference 117). It relies for its effect upon the two results of Shields and Lee[116], whereas according to Fig. 2.8 one of these two appears to be too short and generally the measurements of Parker and Swope agree well with the extrapolated shock tube results. It is a pity that Parker and Swope[117], who have obtained the longest relaxation time for O_2 at around 100°C and hence had perhaps the purest oxygen sample, did not extend their range of measurements to room temperature.

However, in spite of this slight doubt at room temperature, the oxygen results still offer an excellent chance for comparison with theory. In Figure 2.8 the SSH calculated relaxation times from Table 2.1 are plotted. The Set II force constants were obtained[50] by fitting the Lennard-Jones '12–6' potential to viscosity data obtained over the temperature range 80–300°K, whilst the Set I values were similarly derived but from data for the range 300–1000°K and are therefore more appropriate (see Section 2.2.3). Use of the Set II constants, as was done by Holmes, Smith and Tempest[120], gives relaxation times which are too short for the whole temperature range, but serve to demonstrate the sensitive dependence of the calculations upon the intermolecular force constants. With the Set I values, method A gives a smaller α^* and hence longer relaxation times than method B. It should be noted that α^* increases with temperature using method A but that the reverse is true for method B; because of this and the smaller values of α^*, method A predicts a steeper temperature dependence for the relaxation time compared with method B. Both theoretical curves give good agreement with the experimental results, that for method A is slightly better and would be improved if the larger steric factor recently proposed by Herzfeld[10] (see Equation 2.48) were used. For O_2 this would give a value of 0·66 for P_0 instead of the value used, 0·33. At temperatures above about 5000°K the relaxation times are under-estimated because of the breakdown of the perturbation theory (see Section 2.2.6).

Parker[111] has shown that his classical theory[17] fits the experimental results well over most of the temperature range if his exponential repulsion parameter for the Morse potential $(\alpha)_P = 3·86\ \text{Å}^{-1}$. Benson and Berend[15] found good agreement over the whole temperature range between their

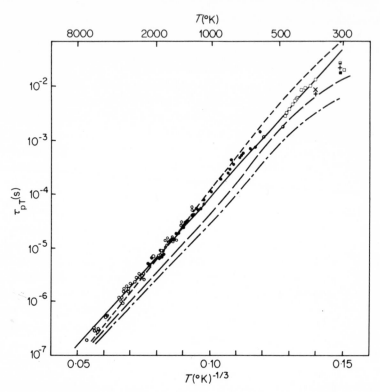

Figure 2.8 Experimental and theoretical results for oxygen

Key
Experimental results:

  Shock tube: ○—Generalov[113], ●—White and Millikan[114]
  Ultrasonic: ◑—Shields and Lee[116], □—Parker and Swope[117], ■—Holmes, Smith and Tempest[120], ▣—Holmes, Smith and Tempest[122], +—Roesler[119], Bauer and Roesler[100], ×—Henderson[118]

———— Straight line drawn by Millikan[7] through post 1961 data (agrees with later one of Lutz and Kiefer[115]), equation:

$$\log_{10} \tau_{pT} = 54 \cdot 7 T^{-1/3} - 9 \cdot 54$$

SSH Theoretical Results:
  (See Table 2.1)
  .... Set I Force Constants (Method A);
  — — — Set I Force Constants (Method B);
  —·—·— Set II Force Constants (Method B)

classical calculations and experiment with their exponential repulsion parameter $(\alpha)_{BB} = 3 \cdot 8 \, \text{Å}^{-1}$†. Calvert and Amme[112] adjusted their repulsion parameter $(\alpha)_{CA}$ to a value of $4 \cdot 7 \, \text{Å}^{-1}$ which gave good agreement with experiment over the whole temperature range. This value is close to that for α in the SSH calculation (Table 2.1) but about $1 \, \text{Å}^{-1}$ greater than the values for the classical calculations, though with the different approaches and the intermolecular potentials they are not necessarily comparable.

For carbon monoxide there is good agreement between the shock tube results of Gaydon and Hurle[123], Matthews[124] and Hooker and Millikan[125]. In Figure 2.9 their combined results are represented by a solid line which spans the temperature range 5000–1000°K. There is then a long gap before the lone result obtained by Ferguson and Read[126] using the amplitude spectrophone which, after correction by Doyennette and Henry[127] for self-absorption, gives a relaxation time of six seconds.

The classical calculations of Parker and of Benson and Berend fit the shock tube results very well with $(\alpha)_P = 4 \cdot 03 \, \text{Å}^{-1}$ and $(\alpha)_{BB} = 4 \cdot 0 \, \text{Å}^{-1}$, but the room temperature result, which was not available to them, does not fit the classical calculations. They both predict probabilities which are an order of magnitude too large at room temperature. Calvert and Amme found that there was good agreement between their semiclassical calculations and the shock tube results if they set $(\alpha)_{CA} = 4 \cdot 5 \, \text{Å}^{-1}$. Also shown in Figure 2.9 are the calculated values from Table 2.1. The correct temperature dependence is predicted and the method A and method B curves bracket the spectrophone result. Method A gives the slightly better agreement and this would be improved if the new Herzfeld steric factor was used. Marriott[69] carried out a quantum-mechanical N-state calculation (see Section 2.2.6) for CO over the temperature range 1000–5000°K; he used the same force constants as in Table 2.1 with a method A fitting. The results plotted in Figure 2.9 are a little low, but have the correct temperature dependence.

Nitrogen has the highest vibrational frequency of this trio and hence the longest relaxation times. Hurle[128] has critically reviewed, and in some cases corrected, the earlier shock tube results of several workers; he drew a 'best straight line' through some sixty results of his own and other workers. This is shown in Figure 2.10‡ with the later shock tube results of Tsuchiya[129] and Cary[130]§. The acoustic measurements of Lukasik and

† Benson and Berend[15] denote their parameter by 'a'; the connection is: $(\alpha) = 2a$.

‡ Results substantially different to all these have been obtained in non-equilibrium nozzle expansion flows[133].

§ Note added in proof: new results of Appleton have since appeared[275].

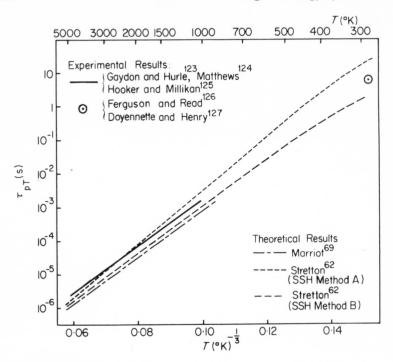

Figure 2.9 Experimental and theoretical results for carbon monoxide

Young[131] are in complete disagreement with the shock tube results; Henderson's[132] value is only a lower limit.

Parker found that Hurle's line is too steep to fit his classical calculation and fitted a curve to Henderson's result and the high temperature shock tube data; for this $(\alpha)_P = 4.07 \text{Å}^{-1}$. Benson and Berend were able to fit the shock tube data with a curve for which $(\alpha)_{BB} = 3.8 \text{ Å}^{-1}$. Blythe, Cottrell and Read[134] carried out an early semiclassical calculation using a Morse potential and, although an error has since been noted[36] in their analysis, the calculated results agreed with the then available shock tube data in magnitude but predicted a less steep temperature dependence. Calvert and Amme were able to fit their semiclassical curve to the experimental data with a value of 4.8 Å^{-1} for $(\alpha)_{CA}$, but the fit is less satisfactory than those for O_2 and CO. They note that a value of 3.1 Å^{-1} is obtained for $(\alpha)_{CA}$ from the Morse potential constants of Konowalow and Hirschfelder[135] which were derived from the crystal structure and virial data. As

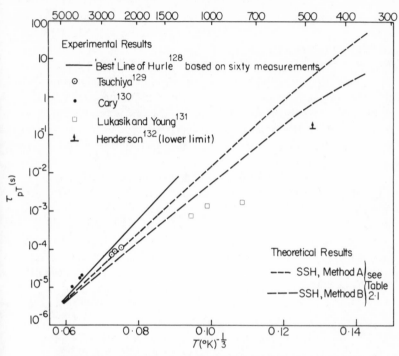

Figure 2.10 Experimental and theoretical results for nitrogen

can be seen from Figure 2.10, though the SSH curves coincide with Hurle's line they predict a less steep temperature dependence. Thus with N_2 there is a disagreement between theory and experiment; more experimental results are needed, particularly at the lower temperatures around 1000°K. At temperatures much below this the relaxation time is probably too long to measure and it could be as long as a minute at room temperature!

Apart from some difficulty with N_2 there is generally good agreement between theory and experiment for this trio of diatomic molecules. With one adjustable parameter the classical and semiclassical calculations can be fitted to the experimental results so as to give simultaneous agreement as to magnitude and temperature dependence. This parameter is a measure of the steepness of the intermolecular potential; with their classical calculations Benson and Berend and Parker obtain good fits with values around 4 Å^{-1}. The semiclassical and quantum-mechanical SSH calculations use values nearer 5 Å^{-1}. The SSH calculations obtain the parameter *a priori* from an intermolecular potential whose constants are derived from the measured transport properties. This gives fairly successful agreement though the fitting procedure is somewhat crude; values of α^* in the region of $5\,\text{Å}^{-1}$ are calculated and α^* itself has a small temperature dependence. Providing that the intermolecular potential to which the simple exponential one is fitted is itself correct, method A is correct in principle as the fitting method and gives better agreement with experiment than method B. The latter has been more widely used because it gave larger values of α^* which led to better agreement with the earlier experimental results.

A large amount of work has been carried out on binary mixtures of one of these three diatomic gases with a non-relaxing component such as an inert gas or H_2; often the results are reported along with those for the pure gas. There have been earlier collations and reviews[4,5,6,7] and generally the results fit the empirical correlation of Millikan and White[136] (Equation 2.81). Calculations have been reported by Benson and Berend[15] and by Calvert and Amme[112] for mixtures with oxygen; the calculated curves fitted well with the experimental ones. Recently results have been reported for oxygen mixtures by Shields and Lee[137], by Holmes, Smith and Tempest[122] and by White[138], for nitrogen mixtures by Cary[130] and White[139], and for carbon monoxide mixtures by Millikan[140,141] and White[142].

2.4.1.1.3   *The hydrogen halides.* At room temperature three of the halides HCl ($v = 2886\,\text{cm}^{-1}$), DCl ($v = 2091\,\text{cm}^{-1}$) and HBr ($v = 2560\,\text{cm}^{-1}$) have been investigated by Ferguson and Read[143] using the amplitude spectrophone. The radiative life times have to be known and only lower limits could be quoted for the relaxation times (11, $\sim$10 and 1·5 ms respectively), because no corrections were made for reabsorption[127]. Borrell[144] has made high temperature measurements of the relaxation times of HCl and HBr by observing the infrared emission behind a shock wave; for HCl he obtained a relaxation time of 8 μs at 2000°C which was almost independent of the temperature though there was wide experimental scatter. For HBr the average result was 16 μs at 1000°K with again almost no temperature dependence. Chow and Greene[145] observed the relaxation behind a shock wave of HI in collision with argon atoms; at 2000°K the relaxation time was found to be 5·4 μs.

In all cases the measured relaxation times of the hydrogen halides are several orders of magnitude shorter than predicted by the SSH theory (even allowing for the dipole moments[54,143]) or expected from the empirical correlation for diatomic molecules of Millikan and White[136] which has otherwise been found to be very successful (see Section 2.4.1.6). Further, with such high vibration frequencies one would expect steep temperature dependences. There is still doubt about the experimental results, as shown by the disagreement between the high and low temperature measurements, but even so there appears to be wide disagreement between experiment and conventional theory. This could be because, as suggested by Secrest and Johnson[74] (see Section 2.2.6), the theoretical treatments are not applicable to diatomic molecules whose constituent atoms have vastly different masses or because the influence of molecular rotation is important; the latter point will be discussed in Section 2.5.1. Conclusive experimental results are needed to test these possibilities.

2.4.1.1.4 *Nitric oxide.* NO has been fully investigated acoustically at room temperature by Bauer and Sahm[146] and at high temperatures by Wray[80] using the shock tube; earlier results have been reviewed by Read[6]. In the electronic ground state $(X^2\Pi)$ the vibrational frequency is 1876 cm^{-1}. Therefore, since the molecular weights are similar, one would expect it to have a relaxation time somewhere between the values for oxygen and carbon monoxide and at room temperature SSH calculations predict[98] a relaxation time of 92 ms. The '12–6–3' potential was used with force constants $r_0 = 3.51$ Å, $\varepsilon/k = 113.5°$K and $\delta^* = 0.0167$, though the small dipole moment has very little effect. However the experimental relaxation time is about 2 μs at room temperature for pure NO. Measurements on mixtures of NO with helium[147] and argon[148] show that inert gas/NO collisions are much less effective for vibrational energy transfer than NO/NO collisions, though with helium one would expect quite the reverse because of its very low atomic weight. Further evidence that the NO/NO collisions are anomalous is provided by the fact that good agreement is obtained between theory and experiment for the complex transfer of vibrational energy from NO to other simple molecules (see Section 2.4.2.3).

The NO molecule has an unpaired electron in the π orbital and when two molecules collide two electronic potential energy states $^3\Sigma$ and $^1\Sigma$ arise. Nikitin[79] has suggested that the spacing between the two states is similar to the vibrational spacing and resonance between them aids vibrational energy transfer; he has backed this with an order of magnitude calculation which agrees with experiment. However this mechanism has a small temperature dependence and at high temperatures the SSH theory gives a faster rate. Wray[80] has shown that, by adding the contributions of the two theories, agreement can be obtained with experiment over the entire temperature range.

2.4.1.1.5 *The halogens.* The halogens are diatomic molecules with low vibrational frequencies; hence they have appreciable vibrational energy at room temperature and relaxation times shorter than those for other diatomic molecules. Shields has measured acoustically using the tube method the relaxation times of all four halogens: F_2 $(v = 895 \, cm^{-1})$[149], Cl_2 $(v = 557 \, cm^{-1})$[150], Br_2 $(v = 321 \, cm^{-1})$[150] and I_2 $(v = 213 \, cm^{-1})$[150]; the results are shown plotted in Figure 2.11. The experiments are difficult to carry out because of the corrosive nature of the halogens, particularly fluorine. Shield's relaxation times for Cl_2 agree with the room temperature results of Sittig[151], who used an acoustic interferometer. At higher temperatures there are the shock tube results of Smiley and Winkler[152]; where these nearly overlap the results of Shields they are shorter, as are the earlier measurements[153,154] at room temperature. Note the trends shown by the 'best' lines drawn through the acoustically obtained experimental data. The higher the vibrational frequency the steeper the temperature dependence and the longer the relaxation times, though the latter trend is counteracted by the fact that the molecules with the lower frequencies have the larger molecular weights.

For chlorine, the SSH calculations of Dickens and Ripamonti[51] agree well with experiment around room temperature, as shown in Figure 2.11. Although the predicted temperature dependence is a little too steep in this range, it agrees with the trend of the shock tube measurements. α^*, calculated from the Lennard-Jones '12–6' potential, has an average value of $4.9 \, \text{Å}^{-1}$ over the temperature range. Shields[149] proposed modifications to the SSH theory in order to improve agreement with his experimental results for the four halogens; however these are based upon the erroneous steric factor of Herzfeld and Litovitz[3,10] and give poor agreement between experiment and theory for O_2 and N_2. Calvert and Amme[112] fitted the results of their semiclassical calculations to the experimental values and found that they obtained reasonably good fits for Cl_2, Br_2 and I_2 with the same value, $4.1 \, \text{Å}^{-1}$, for their repulsion parameter $(\alpha)_{CA}$. However, in order to fit his classical calculations to the experimental results, Parker[111] had to use a range of values for his parameter $(\alpha)_P$ varying from $4.06 \, \text{Å}^{-1}$ for F_2 to $2.65 \, \text{Å}^{-1}$ for I_2. A similar trend was noted by Benson and Berend[15] with their classical calculations; they found that for the best fits to the available experimental data, $(\alpha)_{BB}$ varied from $4.0 \, \text{Å}^{-1}$ for F_2 to about $2.8 \, \text{Å}^{-1}$ for I_2.

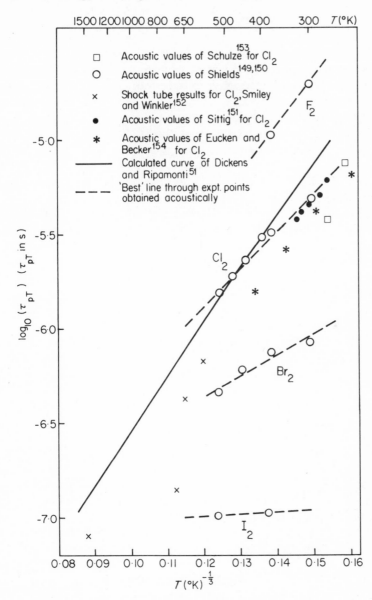

Figure 2.11 Experimental and calculated relaxation times for the halogens

2.4.1.2   Triatomic Molecules

These are the simplest polyatomic molecules; they all have three fundamental vibrations and, in addition to the simple exchange of energy with translation, complex processes involving vibration–vibration energy transfer between the modes have to be considered.

2.4.1.2.1   *Linear molecules*. Five molecules fall in this group: carbon dioxide (CO_2), carbon oxysulphide (COS), carbon disulphide (CS_2), nitrous oxide (N_2O) and hydrogen cyanide (HCN). All have three types of vibration: a doubly degenerate bending (v_2) which has the lowest frequency and so contributes the majority of the vibrational specific heat, a symmetrical valence bond vibration (v_1) and a corresponding antisymmetrical one (v_3).

With CO_2 the frequencies of the modes are: $v_2 = 672.2 \, cm^{-1}$, $v_1 = 1351 \, cm^{-1}$ and $v_3 = 2349 \, cm^{-1}$; note the close correspondence between the first harmonic of v_2 and the fundamental of v_1. The respective heat capacities of the modes are 1.8, 0.14 and 0.002 cal mole^{-1} (°K)$^{-1}$ at 300°K and 3.7, 1.45 and 0.8 cal mole^{-1} (°K)$^{-1}$ at 1000°K. A large number of acoustic determinations of the room temperature relaxation times have been made and these have been reviewed by Cottrell and McCoubrey[4] and by Read[6] who conclude that for the pure gas the value of τ_{pT} is 6–7 μs. A Cole–Cole plot (see Section 1.2.1.4) has been constructed[5] from the results of Kneser and Roesler[155] and of Bauer and Liska[156] and it shows that all the vibrational specific heat is relaxing together, at these temperatures only v_2 and v_1 are contributing significantly to the specific heat and they would be expected to relax together because of the favourable process $P_{0-2(v_2)}^{1-0(v_1)}$ which is near resonant.

The specific heat contribution of v_3 is negligible at room temperature, but it is infrared active and can easily be excited optically and then its relaxation followed. One technique for doing this is the spectrophone which is designed to measure the average time interval between the absorption of a light quantum by a mode (here the v_3 one) and its subsequent emergence as translational energy; Cottrell, MacFarlane, Read and Young[157] obtained a value of $7.0 \pm 0.5 \, \mu$s for this interval. Values of 12 μs and 11 μs were obtained earlier by Slobodskaia and Gasilevich[158] and by Delany[159] respectively; however the measurements were made on mixtures of CO_2 with nitrogen. Cottrell, Macfarlane and Read[160] have shown that linear extrapolation of these results to give a value for pure CO_2 (according to equation 2.80) is erroneous because of rapid exchange of energy between N_2 and the v_3 mode of CO_2. Slobodskaia[161] has repeated her experiment with pure CO_2 and obtained a value of 4·2 μs, a little shorter than that of Cottrell and his colleagues[157]. Houghton[162], using a fluorescence technique, obtained a value of 7 ± 1 μs for the relaxation time of the v_3 mode. In this experiment, the level of the fluorescence signal for the v_3 mode is a measure of its population and the decay with time gives the relaxation time; thus in principle this experiment is different from the spectrophone. Houghton used a Nernst filament to excite the v_3 mode; a variation is to use a laser pulse. This has been done by Hocker, Kovacs, Rhodes, Flynn and Javan[163] and by Moore, Wood, Hu and Yardley[164] who obtained subsequent relaxation times of 3·4 and 3·8 μs, respectively. It is not clear why these values are shorter than Houghton's but it may be due to the fact that the laser pulse produces a very violent redistribution amongst the energy levels. Thus the available evidence points to the fact that the v_3 mode relaxes in a few microseconds; because of its high frequency it is very unlikely to exchange energy direct with translation at a fast enough rate and so it will relax via a complex process involving the other modes. The close correspondence between the relaxation times of 6–7 μs determined acoustically and the results of Houghton and Cottrell's spectrophone suggests that this exchange takes place via v_1 and v_2 and that the simple deactivation of v_2 is the rate-determining step in the conversion to translational energy.

Simpson, Bridgman and Chandler[165] have recently investigated the relaxation of pure CO_2 in a shock tube over the temperature range 330–1600°K. As the energy flows into the vibrational modes in the shock-heated gas the translational temperature falls and this in turn affects the vibrational relaxation time. Taking account of this effect they found that the earlier shock tube results of Johannesen, Zienkiewicz, Blythe and Gerrard[166] could be brought into line with their own. At elevated temperatures, when all three modes of CO_2 have significant specific heat, the results of Simpson and his colleagues for the density changes across the shock front show that all three modes are relaxing together. They have carefully reviewed other high-temperature work on CO_2 and their values also agree with the acoustic results of Shields[167] who made measurements up to 580°K. Their values for the relaxation time of CO_2 vary almost linearly with $\exp[T^{-1/3}]$ from 7 μs at 300°K to 1·3 μs at 1000°K and 0·6 μs at 2000°K†.

† Note added in proof: further results of Weaner, Roach and Smith have since appeared[276].

Theoretical calculations on the relaxation of CO_2 have been reported by Herzfeld[95]†, Witteman[30,31], Schofield[56] and Marriott[70], all using three-dimensional quantum-mechanical treatments. Herzfeld predicts a relaxation time about three times shorter than that observed at room temperature, Schofield's result is five times too short, but Witteman's[31] value is twenty times too long. Marriott's calculations, though in good agreement with the shock tube results of Witteman[31,102], have a much steeper temperature dependence than the measurements reported by Simpson and coworkers[165], a tendency which is shared by all the other theoretical calculations.

The theoretical calculations agree that v_1 and v_2 should relax together. Much attention has been given to the relaxation of v_3; Herzfeld[95] and Witteman[30,31] have examined various routes in detail and have tried to take account of the fact that, with complex processes between degenerate modes involving multi-quantum jumps, there are several ways in which the quanta can be assembled—for example, for a double-quantum jump from a doubly degenerate mode each mode could contribute one quantum or both quanta could come from the same mode. To allow for these possibilities the various modes have to be weighted as discussed by Herzfeld and Witteman, but the big snag is to assign the steric factors, since these will differ for each permutation of the quanta. The original Tanczos[58] treatment for a complex collision assumed that the exchanging modes were in different molecules; evidence that such an *intermolecular* exchange is much more probable than an *intramolecular* one is provided by the work of Yardley and Moore[164,168] who found that inert gas/CO_2 collisions, when only intramolecular complex processes are possible, were about an order of magnitude less efficient for the deactivation of the v_3 mode compared with CO_2/CO_2 collisions. However, even allowing for the extra possibilities, Herzfeld and Witteman conclude that the v_3 mode should relax separately from the other two which is directly contrary to the shock tube results[165]. This disagreement could be due to the fact that the simple perturbation calculations seriously underestimate the probability of a multi-quantum jump as suggested by Sharp and Rapp (Figure 2.5), or that complex vibrational energy exchange is promoted by mixing of the vibrational states due to anharmonicity and Coriolis coupling[168].

† Note added in proof: Herzfeld[277] has recently carried out more refined calculations for CO_2 taking account of the Fermi resonance between v_1 and v_2 and using an improved steric factor. This gives much improved agreement between theory and experiment for the deactivation of v_2 but the wide disagreement concerning v_3 remains.

Thus agreement between theory and experiment is rather poor for carbon dioxide, both as regards relaxation of the lowest mode by a simple process and complex transfer between the three modes. Whilst agreed values for the relaxation time of the lowest mode as a function of temperature are emerging, there is still confusion concerning the measurement of the rates of energy exchange between the modes. Careful consideration needs to be given as to exactly what is being measured. Perhaps the laser technique of De Martini and Ducuing[109] could be applied to Raman excite the v_1 mode and then use the spectrophone effect and fluorescence of the v_3 mode to determine simultaneously the energy transfer rates of the v_2 and v_3 modes respectively. The accumulation of precise experimental results will aid and stimulate renewed theoretical efforts.

The vibrational frequencies of COS, CS_2 and N_2O each have a similar pattern to those of CO_2. All the acoustic results have been reviewed before[4,5,6] and the only new work has been carried out with the spectrophone and shock tube. With the former, Cottrell and his colleagues[157] found that the v_3 mode of COS behaved similarly to the one in CO_2, i.e. the spectrophone relaxation time equalled the acoustic one, while with N_2O there was an indication that the v_3 mode relaxed slightly more slowly than the lowest mode. Simpson, Bridgman and Chandler[169] have studied the relaxation of nitrous oxide behind a shock wave and found the density changes to be consistent with all three modes relaxing together at elevated temperatures; at room temperature they obtained a value of 0·9 μs for the relaxation time which agrees well with the acoustic results. Holmes, Parbrook and Tempest[170] made a careful acoustic study of the relaxation of N_2O; they found no sign of multiple relaxation and deduced that the two lowest modes ($v_2 = 589\,\text{cm}^{-1}$, $v_1 = 1285\,\text{cm}^{-1}$) relaxed together. The specific heat of the highest mode ($v_3 = 2224\,\text{cm}^{-1}$) is then too small to be detected. Schofield[56] has carried out SSH calculations for N_2O; his predicted relaxation times are too short by a factor of three and the calculated relaxation time for v_3 is much larger than that for v_1 and v_2 which relax together. This disagreement between theory and experiment is similar to that for CO_2.

A large amount of work has been done on mixtures of these triatomic gases with non-relaxing components such as the inert gases, H_2, D_2 and so forth. Lists of references are given by Kneser[3] and many of the results have been reviewed by Cottrell and McCoubrey[4] and by Read[6]. The most recent work is that of Bauer and Liska[171], Cottrell and Day[172], Lavercombe[173], Lewis and Shields[174] and Moore and his colleagues[164,168] on mixtures of CO_2 with the inert gases. The ratios of the efficiencies of the various gases as collision partners for CO_2 are at variance with simple theoretical considerations and as yet there is no explanation[172]. Addition of an inert gas such as helium is important for the action of a CO_2 gas laser because its small mass aids the deactivation of the lowest mode but hinders that of the v_3 mode, and so helps to preserve the inverted population of v_3 necessary for the laser action[164,168].

Hydrogen cyanide (HCN) has been investigated by Cottrell and his colleagues[157,175], but they could only deduce that its relaxation time was less than 0·1 μs.

2.4.1.2.2 *Non-linear triatomic molecules.* There are five molecules to be considered in this group: H_2O, D_2O, H_2S, D_2S and SO_2. All have three non-degenerate vibrations and because of their shape they have dipole moments which can considerably increase the attractive forces at molecular collisions. Nitrogen dioxide (NO_2) is also a member of this group but its vibrational relaxation cannot be studied because it only occurs in the gas phase in dynamic equilibrium with its dimer N_2O_4; however this equilibrium can be investigated acoustically[5].

The relaxation time of light water (H_2O) is very short and this coupled with its small vibrational specific heat makes the relaxation time difficult to determine because the vibrational relaxation is nearly obscured by the rotational and translational (or viscothermal) relaxation; making allowance for these effects is discussed in Chapter 1. Only two investigations have managed to define the relaxation zone. Roesler and Sahm[176] made absorption measurements with a double condenser-transducer apparatus and obtained a relaxation time of $6·0 \pm 2$ ns at 50°C. In a similar experiment Yamada and Fujii[177] obtained a value of $8·0 \pm 2·5$ ns at 137°C for pure H_2O. The relaxation time can be deduced from measurements made of the relaxation times of O_2/H_2O mixtures, see section 1.4.2.2. Stretton[62] obtained a value of 1·5 ns at 27°C from the results of Tuesday and Boudart[178] and Henderson and Herzfeld obtained a value of 1·0 ns at 25°C from the results of Henderson, Clark and Lintz[179]. Both of these deduced values are markedly shorter than the directly measured results, unless there is a very strong temperature dependence. Jones, Lambert and Stretton[180] have pointed out that with such a large attractive term in the intermolecular potential the relaxation time could lengthen with increasing temperature (see Equation 1.41). Some support for this is given by the above directly measured values[176,177] and those of Fujii, Lindsay and Urushihara[181] who made absorption measurements at several temperatures but were only able to cover part of the relaxation zone. Even so the relaxation time is unlikely to increase by a factor of at least four over the range 300–410°K.

The lowest fundamental vibration of H_2O has a frequency of 1596 cm^{-1}, the other two modes have frequencies above 3000 cm^{-1} and so their contributions to the specific heat can be neglected. Double isotopic substitution gives heavy water (D_2O) with a lowest frequency of 1178 cm^{-1}; Yamada and Fujii[177] have obtained a value of 7.0 ± 1.0 ns for its relaxation time at 137°C. The mixtures of D_2O with O_2 behave differently to those of H_2O and O_2 and it can only be inferred[62] from them that the relaxation time is of the same order as that for H_2O.

Dickens and Ripamonti[51] using Equation (2.41) have calculated a relaxation time of 1.45 ns for pure H_2O at 585°K, but with such a large probability use of simple perturbation theory is hardly justified and is liable to overestimate the ease of energy transfer. Marriott[70] using his N state method (see Section 2.2.6) has obtained a value of about 200 ns which is independent of temperature.

Geide[182] made absorption and dispersion measurements of H_2S at 25°C and measured the rotational relaxation time but did not detect the vibrational relaxation in the effective frequency range 10^6–10^9 Hz atm^{-1}. The lowest frequency of H_2S is 1188 cm^{-1} but it is heavier and has a smaller dipole moment than H_2O (0.92 as against 1.83 Debye units) and with '12–6–3' force constants[49] $r_0 = 3.49$ Å, $\varepsilon/k = 343$°K and $\delta^* = 0.21$; Jones[98] has calculated, using Equation (2.38), a relaxation time of 1.2 μs at 298°K. If this is correct and rotation does not play an important rôle (see Section 2.5.1) then Geide would not have covered the relaxation zone. H_2S has a much longer relaxation time than H_2O mainly because of the effect of the dipole of H_2O for which $\delta^* = 1.2$[49]. Jones[98] predicts that the relaxation time of D_2S, whose lowest mode has a frequency of 855 cm^{-1}, will be 0.089 μs at 298°K. No experimental work has been carried out on D_2S and measurements of the vibrational relaxation times of H_2S and D_2S are now needed.

Sulphur dioxide provides one of the few well-characterized examples of multiple relaxation. Figure 2.12 shows the dispersion results of Lambert and Salter[183] for SO_2 at 102°C plotted as effective specific heat against the logarithm of the effective frequency. The dotted curve is the theoretical one calculated assuming that all the specific heat relaxes together; the solid curve is for two relaxation times, the shorter one ($\tau_{pT_1} = 89$ ns) corresponding to the relaxation of the specific heat of the lowest mode

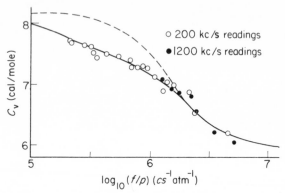

Figure 2.12   The dispersion results of Lambert and Salter[183]
for sulphur dioxide at 102°C

($v_1 = 519 \, \text{cm}^{-1}$; $C_V = 1\cdot45 \, \text{cal mole}^{-1} \, (°K)^{-1}$) and the longer one ($\tau_{pT_2} = 890 \, \text{ns}$) corresponding to the relaxation of the upper two modes ($v_2 = 1151 \, \text{cm}^{-1}$ and $v_3 = 1300 \, \text{cm}^{-1}$; $C_V = 0\cdot78 \, \text{cal mole}^{-1} \, (°K)^{-1}$). Further measurements have been made on SO_2 at several temperatures[184,185,186], both relaxation times first decrease with rising temperature and then increase[184]. With dispersion measurements, careful correction of the experimental readings is required for non-ideality of the gas (see Section 1.2) because SO_2 is a polar gas[187,188]. Double relaxation of the vibrational specific heat has also been observed in liquid SO_2[189,190], where of course the same collision processes operate but the collision rate is much greater.

Dickens and Linnett[191] carried out theoretical calculations using the SSH–Tanczos Method (Equation 2.41) and showed that the double relaxation is due to the slow exchange of energy between v_1 and v_2. Even the rate of the most favourable complex process for this exchange, $P^{2-0(v_1)}_{0-1(v_2)}$ is much less than the simple energy exchange between translation and v_1 using the process $P^{1-0}_{(v_1)}$, see Figure 2.13. The two upper modes rapidly exchange energy between themselves. Parks-Smith[192] carried out further SSH calculations using Equation (2.38), the potential constants of Monchick and Mason[49] and method B fitting. He found that the calculated relaxation times were 2–3 times longer than the experimental results at room temperature[185,192] ($\tau_{pT_1} = 0\cdot06 \, \mu s$, $\tau_{pT_2} = 1\cdot2 \, \mu s$) and the calculated ratio of the two relaxation times differed from the experimental one by less than a factor of two; this represents good agreement.

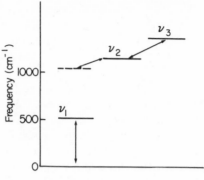

Figure 2.13 Transition scheme for sulphur dioxide

Shields[186] has recently reported ultrasonic absorption measurements on SO_2 and its mixtures with argon. He analyzed the results in order to try and distinguish experimentally between series and parallel excitation (see Section 2.3.4) of the upper two modes but found that this was not possible. If the ratio of the two relaxation times is less than five then Dickens[193] has shown that it might just be possible to distinguish between series and parallel excitations from the absorption curves but even then the difference is only slight. Shields also analyzed his results in order to try and determine the diagonal and codiagonal elements of Tanczos' **k** matrix of rate constants for SO_2 (see Section 2.3.1.2); he found that there was no unique set of values which fitted the experimental results though additional information was provided by invoking the results for the mixtures with argon. Shields' results confirm earlier ones, in particular the unusual temperature dependence of the relaxation times, and he also found that argon was an inefficient collision partner for all processes, particularly complex ones (compare with CO_2).

2.4.1.3   Tetra-atomic Molecules

These are also divided into two groups—linear and non-linear molecules; with the latter only the Group V hydrides are important though BF_3 is also a member but has been adequately reviewed previously[4].

2.4.1.3.1   *Linear molecules.*   Two substances fall into this group— acetylene and cyanogen. Stretton[55] carried out ultrasonic velocity measurements on two acetylenes, C_2H_2 and C_2D_2, at two temperatures; he obtained the following values for τ_{pT}: for C_2H_2 0·082 and 0·079, for

C_2D_2, 0·072 and 0·072 μs at 298 and 335°K respectively. Jones[98] has carried out dispersion and absorption measurements at 298°K and confirmed the relaxation time for C_2H_2 but obtained a shorter one, 0·061 μs, for C_2D_2. Earlier results for C_2H_2 have been reviewed by Cottrell and McCoubrey[4].

Stretton[55] carried out SSH calculations using Equation (2.38) and compared them with his experimental results. With force constants derived from viscosity measurements[50] giving values of α^* about 4·8 Å^{-1}, there was close agreement. The calculated transition probabilities for C_2D_2 showed that the simplest deactivation of the lowest mode v_4 (511 cm^{-1}) is slower than that for the next mode up ($v_5 = 539$ cm^{-1}) because, while the modes have nearly equal frequencies, the vibrational factor (see Section 2.2.5) is larger for v_5 and so C_2D_2 would appear to be an exception to the general postulate that, with a single relaxation time, vibrational relaxation occurs via the lowest mode. Dickens and Ripamonti[51] made an earlier SSH calculation for C_2H_2 using Equation (2.41).

The frequencies of cyanogen are shown in Figure 2.14. Dickens[194] carried out an SSH calculation and predicted multiple relaxation.

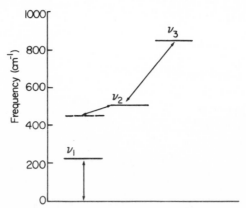

Figure 2.14 The   vibration   frequencies   of cyanogen and a possible transition scheme

Provisional absorption and dispersion results of Parks-Smith[192] over the effective frequency range 10^6–$10^{8\cdot3}$ Hz atm^{-1} have confirmed this and there is the possibility that there is a separate relaxation zone for each of the three modes. C_2N_2 fulfils the usual empirical requirement for multiple relaxation that the frequency of the lowest mode should be much less than half of the next one up.

2.4.1.3.2 *Non-linear molecules—The Group V hydrides.* Although the lowest frequency of ammonia (NH_3) is 948 cm^{-1}, until recently all the work indicated that the relaxation time of the vibrational energy was very short (<1 ns) and beyond the range of most apparatus. Careful correction for non-ideality of the gas is essential because of the very small vibrational specific heat; this and other aspects of the work on NH_3 have been well reviewed by Strauch and Decius and by Cottrell and his colleagues[175,195,196]. Jones[98] has just reported the results of absorption and dispersion measurements made with a condenser-transducer apparatus up to an effective frequency of $10^{8.3}$ Hz atm^{-1}; he found that there was relaxation but to a greater extent than would be expected for the vibrational energy alone. This must be rotational relaxation which is of such a magnitude as to swamp any vibration relaxation so that it would appear that the rotational energy has a relaxation time equal to or longer than the one for the vibrational energy. Cottrell and Matheson[196] have suggested that the very short vibrational relaxation time for NH_3 is somehow connected with the inversion of the molecule; both NH_3 and ND_3 show inversion but the rate is very much greater for NH_3. ND_3, with a lowest mode of 749 cm^{-1}, has a measured vibrational relaxational time of 13 ns[196].

Cottrell and Matheson[196] and Cottrell, Dobbie, McLain and Read[197] have measured the relaxation times of six Group V hydrides and their results are summarized in Table 2.2. They provide evidence that molecular rotation is concerned in vibrational energy exchange and discussion of this is deferred until Section 2.5.1.

2.4.1.4 *Penta-atomic Molecules*

The majority of the molecules in this group are carbon compounds with one carbon atom surrounded by four atoms which are either halogens or hydrogen; the simplest and most studied case is methane (CH_4). All the molecules have nine modes of vibration but some are degenerate with the more symmetrical molecules. Special interest is attached to the methylene halides (CH_2X_2) as they show well-defined multiple relaxation of their vibrational specific heat. Compounds of the other Group IV elements in their quadrivalent state are considered together with their carbon analogues.

Table 2.2

Relaxation times of the group V hydrides at 30°C

	Frequency of lowest mode (cm^{-1})	τ_{pT} (ns)	Ref.
NH_3	950 (mean)	< 10	98, 195
ND_3	749	13	196
PH_3	992	196	196
PD_3	728	343	196
AsH_3	905	310	197
AsD_3	660	220	197

2.4.1.4.1 *Methanes.* A considerable amount of effort has been devoted to studying the vibrational relaxation of CH_4 and its deuterated derivatives. Acoustic measurements on CH_4 made prior to 1965 have been reviewed by Read[6]. There is a single relaxation time for all the vibrational energy and he concluded that at room temperature τ_{pT} had a value of about 1·8 μs based upon the reverberation experiments of Edmonds and Lamb[198] and the dispersion measurements obtained with an interferometer by Cottrell and Martin[199] and by Cottrell and Matheson[200]. More recent values have been obtained using a more advanced interferometer by Cottrell and Day[201] (1·7 μs) and by Read[202] (1·7 μs), whilst lower values have been reported by Gravitt, Whetstone and Lagemann[203], who deduced a value of 0·94 μs from their ultrasonic absorption measurements and by Parke and Swope[204] who derived a value of 1·08 μs from their resonator measurements. Madigosky[205] made dispersion measurements in high density methane below room temperature and showed that binary collisions were still responsible for the energy transfer. At elevated temperatures there are the values of τ_{pT} from the early dispersion measurements of Eucken and Aybar[206] and the recent shock tube results of Richards and Sigafoos[207]. The experimental results are shown plotted against temperature in Figure 2.15† and it can be seen that, whereas the two sets of high temperature results agree between themselves, those of Eucken and Aybar extrapolate to a room temperature relaxation time about half the generally accepted value of 1·8 μs. It is strange that the results fall into two groups, those around 1·8 μs and those at 1·0 μs; in the absence of other evidence it is assumed that the longer values are correct

† Note added in proof: further ultrasonic absorption results over the temperature range 243–393°K have just been reported by Hess and Hinsch[278].

since impurities, the most likely source of error, tend to shorten the relaxation time. However, the experimental results for CH_4 cover a wide temperature range, greater than that for any other penta-atomic molecule.

Cottrell and Ream[208] carried out the first calculation of the temperature dependence of the relaxation time of methane; they used a semiclassical approach, extended to handle a polyatomic molecule, but even after symmetrization the predicted temperature dependence is too steep. Tanczos[58] calculated the relaxation time of CH_4 at room temperature using the SSH theory (Equation 2.41) and obtained a similar result to Stretton[55] who used Equation (2.38) and covered the temperature range up to 600°K. The latter calculations have been extended by Jones and Stretton[209] to cover the whole range of the experimental results and the results are plotted in Figure 2.15. There is good agreement at room temperature but the predicted temperature dependence is too steep.

Also shown in Figure 2.15 are the experimental results of Richards and Sigafoos[207] for the relaxation of methane in collision with argon atoms at high temperatures and the two room temperature values of Yardley and Moore[210] and of Parker and Swope[204]. Jones and Stretton[209] have carried out corresponding calculations assuming that the relaxation is controlled by the activation and deactivation of the lowest mode; they used the usual combining rules[50] to obtain the force constants $\varepsilon/k = 136°K$ and $r_0 = 3\cdot59$ Å for the heteromolecular CH_4/Ar collisions, compared with $\varepsilon/k = 148°K$ and $r_0 = 3\cdot76$ Å for the homomolecular CH_4/CH_4 collisions. The predicted relaxation times for CH_4/Ar are longer than those for pure CH_4, in agreement with experiment, but their temperature dependence is too steep and the actual values are much shorter than the experimental ones.

Around room temperature only the lowest two modes of CH_4 (see figure 2.16) contribute significantly to the specific heat. The acoustic measurements show that there is only one relaxation time; this implies that energy transfer between ν_2 and ν_4 is faster than energy exchange between ν_4 and translation. This was confirmed by the calculation of Tanczos[58]. The ν_3 mode is infrared active and its relaxation can be studied after it has been optically activated (see Section 1.4). This was first attempted by Cottrell, Hunter and Read[211] who found that the relaxation time of the ν_3 mode was much longer than the acoustically determined one for CH_4. Stretton[55] pointed out that this was at variance with the calculated relaxation time for ν_3. Subsequent experimental investigations by Cottrell, Macfarlane, Read and Young[157] using the spectrophone

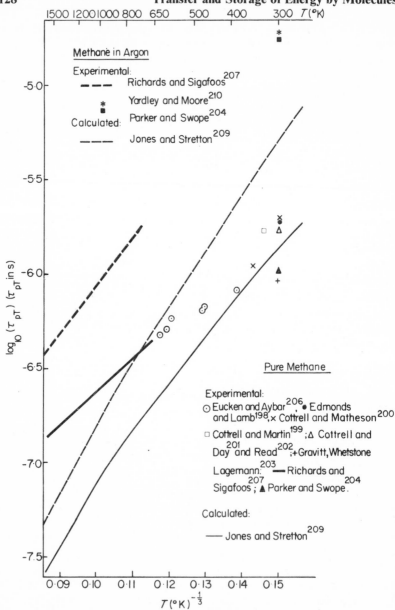

Figure 2.15 Experimental and calculated relaxation times for pure methane (CH_4) and methane in argon

showed the earlier work to be in error. They found that the relaxation time of v_3 thus measured is very close to the acoustic one for CH_4 implying that energy is rapidly transferred from v_3 to v_4, deactivation of which is the rate-determining step. This was confirmed by the laser experiment of Yardley and Moore[210] (see Section 1.4.4).

In Figure 2.16 are shown the calculated[55] probabilities for the transitions between the four modes of CH_4. It can be seen that the two upper

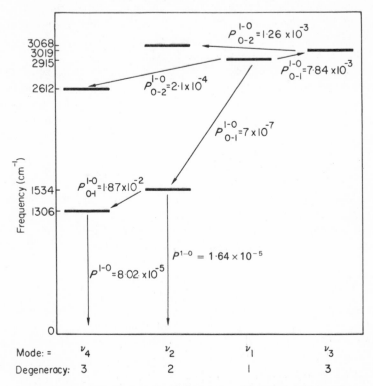

Figure 2.16 The frequencies of the modes of methane (CH_4) and the calculated[55] probabilities for transitions between them

modes, v_1 and v_3, exchange energy most rapidly with the first harmonics of v_4 and v_2 and that the quickest route for deactivating v_3 is via v_2. Yardley and Moore[210] using their laser technique measured the relaxation time for the transfer of energy from v_3 to v_4 and obtained a value of 6 ± 2 ns;

this is nearly an order of magnitude shorter than the calculated[55] value of 56 ns[†]. The disagreement could again be due, *inter alia*, to the SSH theory underestimating the probabilities of multi-quantum jumps. Yardley and Moore further found that the rates of deactivation of v_3 at CH_4/Ne and CH_4/Ar collisions were at least an order of magnitude slower than at CH_4/CH_4 collisions; this is additional evidence for complex processes occurring *inter-* rather than *intramolecularly* since the effect of the inert gases on the simple deactivation of v_4 is much less marked[210].

Gravitt, Whetstone and Lagemann[203] have determined the relaxation times of the four deuterated derivatives of CH_4 from ultrasonic absorption measurements at 26°C. They obtained values of 1·27, 1·72, 2·57 and 4·67 μs for the relaxation times of CH_3D, CH_2D_2, CHD_3 and CD_4 respectively. The result for CD_4 is quite close to the value of 3·9 μs of Cottrell and Matheson[200]. Thus the relaxation time for CD_4 is about *double* that for CH_4: theoretical SSH calculations[55] predict that it should be *half* because, although CD_4 is a little heavier, the frequency of its lowest mode is 996 cm^{-1} compared with 1306 cm^{-1} for CH_4[‡]. Cottrell and Matheson have suggested that this discrepancy is due to energy exchange between vibration and rotation; this is discussed in Section 2.5.1. They have also measured the relaxation times of the two monosilanes SiH_4 and SiD_4 and found a similar state of affairs: the lowest mode of SiH_4 has a frequency of 914 cm^{-1} while for SiD_4 it is 675 cm^{-1}, yet their relaxation times at 298°K are 0·11 and 0·20 μs respectively[200].

2.4.1.4.2 *Methyl halides.* Measurements have been made on three methyl halides (CH_3F, CH_3Cl, CH_3Br) and their fully deuterated derivatives. The experimental relaxation times have been tabulated by Stretton[55] who compared them with values calculated using the SSH theory. For the bromides and chlorides slight multiple relaxation is predicted (similar to that for C_2H_4, Section 2.4.1.5.1) but sufficiently accurate experimental data are not available to test these predictions. For all the halides except CD_3F (see Figure 2.6), v_3 is the mode with the lowest frequency. This mode is essentially a stretching of the carbon–halogen bond and has only a small vibrational factor compared with that of v_6, the next mode up, which involves extensive movement of the lighter hydrogen or deuterium atoms (see Section 2.2.5). For two of the halides, CH_3F and CD_3Cl, this

[†] The second half of Table 3 of reference 55 contains an error; in the column headed 'Process' the second and third rows should be interchanged.

[‡] Note added in proof: it is usually taken that the intermolecular potential is the same for a molecule and its isotopic analogues. For an examination of this postulate based upon the properties of liquid CH_4 and CD_4 see the recent papers by Grigor and Steele[279].

effect outweighs that of v_3 having a lower frequency and thus, similar to C_2D_2, the quickest route for the relaxation of the vibrational energy is *via* the second lowest mode.

The intermolecular force constants are most reliable for CH_3Cl and the relaxation times calculated[55] using the Monchick and Mason[49] '12–6–3' potential are in good agreement with the experimental results†. A slight decrease of the relaxation time with temperature is predicted due to the effect of the dipole moment, whereas the experimental results indicate that the relaxation time is almost independent of temperature at about $0.2 \mu s$. Blythe, Cottrell and Read[134] carried out a semiclassical calculation for CH_3Cl using a Morse intermolecular potential (Equation 2.30); they obtained good agreement with experiment, though an error has since been noted in their analysis[36]. The intermolecular force constants are not so reliable for the fluorides and bromides. However one main point does arise from a comparison of the calculated and experimental results, namely that the ratio of the relaxation times of the deuterated and the corresponding non-deuterated molecules is always wrongly predicted in a sense similar to that for CD_4 and CH_4 as discussed in the previous sub-section.

2.4.1.4.3 *Methylene halides.* The methylene halides are an important class of compounds because several of them show multiple rather than the usual single relaxation. The criterion for this appears to be the ratio of the frequencies of the lowest two modes, and the patterns of the fundamental frequencies of the methylene halides are shown in Figure 2.17.

CH_2F_2 has been found to have only one relaxation time; it has a value of $7.5 \mu s$ at $300°K$ and the experimental work has been reviewed by Cottrell and McCoubrey[4]. With CH_2F_2 the frequency of the first harmonic of the lowest mode is very close to that of the next lowest mode.

CH_2Cl_2 is one of the best examples of double relaxation. Sette, Busala and Hubbard[212] deduced two relaxation times (9.46 and 0.195 μs) from their dispersion measurements at $303°K$. Tanczos[58] carried out SSH calculations on CH_2Cl_2 and showed that the multiple relaxation was due to the slow exchange of energy between the lowest mode (283 cm^{-1}) and the next up (704 cm^{-1}) using a process of the type P_{0-1}^{2-0}. Both the predicted values for the two relaxation times and the gap between them are greater than those found experimentally. The calculations were repeated by Stretton[62] using Equation (2.38) with the Monchick and Mason[49] '12–6–3'

---

† Similar agreement was obtained by Tanczos[58] at one temperature using the Krieger potential which was later discounted[49].

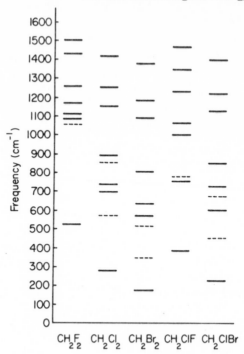

Figure 2.17 Fundamental frequencies of the
methylene halides

potential and correcting minor errors of Tanczos; this resulted in a wider
gap, six times greater than the experimental value. Double relaxation of
the vibrational energy has also been observed in liquid CH_2Cl_2. Hunter
and Dardy[213] found that the upper modes relaxed at frequencies around
150 MHz whilst relaxation of the lowest mode has been detected at
hypersonic frequencies around 5 GHz by Hanes, Turner and Piercy[214]
and by Clark, Moeller, Bucaro and Carome[215], both using the stimulated
Brillouin scattering of a laser beam.

Meyer[216] made measurements with gaseous CH_2Br_2 and there appears
to be multiple relaxation but the absorption and dispersion results are not
consistent with each other†. Dickens and Schofield[217] carried out SSH
calculations on CH_2Br_2 using Equation (2.38) and predicted that there
should be two widely separated relaxation zones because of slow energy

† Note added in proof: further results on gaseous CH_2Br_2 have recently been reported
by Hageseth[280].

exchange between the lowest mode $(174\,\mathrm{cm}^{-1})$ and the second one $(576\,\mathrm{cm}^{-1})$ using a complex process of the type P_{0-1}^{3-0}, a rare example of a triple jump. In liquid CH_2Br_2, Hunter and Dardy[218] have found that all the modes, *except* the lowest, relax together at frequencies around 400 MHz; thus it would appear that there is double relaxation with CH_2Br_2 in accordance with the predictions of Dickens and Schofield but at variance with the earlier measurements by Meyer on gaseous CH_2Br_2, for which further experimental results are required.

Only one 'mixed' methylene halide has been investigated experimentally, namely CH_2ClF. Rossing and Legvold[219] made dispersion measurements and, though they were not able to apply corrections for the non-ideality of the gas, all the vibrational energy appears to be relaxing with a single relaxation time. There is presumably rapid energy exchange between the lowest mode of CH_2ClF and the second one, because the latter has a similar frequency to the first harmonic of the lowest mode (see Figure 2.17). Such a situation does not exist with CH_2ClBr and this would be expected to show double relaxation, but has yet to be investigated.

2.4.1.4.4. *Other halo-derivatives of the Group IV hydrides XH_4.* A large number of halo-derivatives have been investigated and the experimental results have been reviewed by Cottrell and McCoubrey[4] and by Read[6]. Only new results or those where a comparison can be made with theory are discussed here.

The simplest derivatives are those where all four hydrogen atoms have been replaced by the same halogen atom. Of those based on methane, CF_4 is the case with the highest frequency for its lowest mode $(435\,\mathrm{cm}^{-1})$ and has the longest relaxation time, about $0.8\,\mu s$ at $300°K$. It has been fully studied by experiments over the temperature range $290–600°K$[4,6] but no theoretical investigations have yet been made. Calculations have been carried out for CCl_4 by Tanczos[58]; his results indicated that the three lowest modes $(218, 314, 458\,\mathrm{cm}^{-1})$ relaxed together but the top one $(776\,\mathrm{cm}^{-1})$ exchanged energy very slowly with the rest and so relaxed separately. Experimental measurements have been made on CCl_4 by Sette, Busala and Hubbard[212] and by Hinsch[220,221]; the former results show a possible indication of a second relaxation zone at low effective frequencies but with both the absorption and dispersion results of Hinsch[220] there are no signs of multiple relaxation. The relaxation time calculated by Tanczos for the lowest mode is about ten times larger than the experimental values of $19\,\mathrm{ns}$[220] and $21\,\mathrm{ns}$[212] at room temperature; a value of $4.15\,\text{Å}^{-1}$ was obtained for α^*, a higher value near the $5\,\text{Å}^{-1}$

found for diatomic molecules would give better agreement. Hinsch[220] has made absorption and dispersion measurements on $SiCl_4$, $GeCl_4$ and $SnCl_4$; all show single relaxation. The frequencies of the lowest modes are 150, 134 and 104 cm^{-1} and the relaxation times have values of 5·9, 3·4 and 2·5 ns respectively at 22°C.

Tanczos[58] also carried out calculations for gaseous chloroform, $CHCl_3$; he again predicted double relaxation due to slow energy exchange of the upper modes. The dispersion results of Sette, Busala and Hubbard[212] and the later absorption and dispersion measurements of Haebel[222] show only single relaxation with relaxation times of 13·5 and 15·0 ns at 30 and 22°C respectively. Haebel[222] also measured the relaxation times of $CHCl_2Br$ and $CHClBr_2$; at 22°C he deduced relaxation times of 9·8 and 6·3 ns for these two molecules whose lowest modes have frequencies of 217 and 185 cm^{-1} respectively. No other measurements have been reported for pure halo-methanes since the review by Read[6].

Olsen and Legvold[223] and Miller, Boade and Legvold[224] have made dispersion and absorption measurements on several mixtures of the halo-methanes with inert gases. Boade[225] has shown that the relative efficiencies of the inert gas atoms as collision partners are in accordance with the SSH theory.

2.4.1.5   *Hydrocarbons*

The simplest hydrocarbons, methane and acetylene, have been dealt with already; this sub-section is concerned with ethylene and ethane and their higher homologues.

2.4.1.5.1   *Ethylenes.*   Experimental measurements and theoretical calculations have been reported for both C_2H_4 and C_2D_4 and the results are set out in table 2.3, taken from Stretton[55]. For both molecules v_{10} is the mode with the lowest frequency (810 and 589 cm^{-1} respectively).

Holmes and Tempest[226] measured the absorption in C_2H_4 at 298°K and obtained a peak which is 4% lower than the one expected for a single relaxation time; they interpreted this as being due to slight double relaxation. The calculated[55] relaxation times (τ_{vs}) associated with the three largest relaxation constants (see Section 2.3.1.2) are given in Table 2.3. One relaxation time governs 97% of the absorption or dispersion (see Equations 2.65 and 2.66), the remaining 3% being associated with two much shorter relaxation times. This would fit their experimental results but differs from the interpretation of Holmes and Tempest who postulated two relaxation times of similar magnitude. At higher temperatures it is

Table 2.3

Calculated and experimental relaxation times of the ethylenes (from Stretton[55])

					$\tau_{pT}(\mu s)$	
	$T(°K)$	$Z(\text{ns}^{-1})$	$\alpha^*(\text{Å}^{-1})$	$P^{1-0}(v_{10})\times 10^4$	calc.	expt.
C_2H_4	263	9.89	4.74	1.61	1.35	
	298	9.29	4.72	2.46	1.01	0.287[226] 0.39[184] 0.26[227]
	353	8.54	4.69	4.43	0.670	0.226[226] 0.126[184]
	573	6.70	4.62	24.7	0.204	0.183[226]
C_2D_4	298	8.69	4.76	15.1	0.211	0.14[227]

Effective relaxation times (τ_{vs}) and constants for C_2H_4

$T = 298°K$		$T = 353°K$	
$\tau_{vs_i}(\mu s)$	D_i	$\tau_{vs_i}(\mu s)$	D_i
0.726	7.27×10^{-2}	0.418	9.97×10^{-2}
0.018	9.4×10^{-4}	0.037	1.7×10^{-3}
0.039	8.1×10^{-4}	0.016	1.57×10^{-3}
	$\sum_i D_i = 7.50 \times 10^{-2}$		$\sum_i D_i = 1.040 \times 10^{-1}$

predicted that less of the total absorption is associated with the major relaxation time, in agreement with experiments. The small fraction of the absorption associated with the two minor relaxation times will be found at effective frequencies above 10 MHz atm^{-1}, where it may be obscured by the rotational and classical absorption (see Chapter 1). The slight double relaxation predicted by C_2H_4 is due to the exchange of energy between the closely spaced upper modes being very much *faster* than the simple deactivation of the lowest mode. The opposite is true for cases such as CH_2Cl_2 or SO_2. With C_2H_4 the slight double relaxation is barely detectable by dispersion measurements; deactivation of the lowest mode is the rate-determining step for C_2H_4 and C_2D_4 and details for this transition are given in Table 2.3.

The predicted relaxation time for C_2H_4 is 3–4 times too long at room temperature and the temperature dependence is 50% steeper than that found experimentally; both of these disagreements could be due to α^* being too small. Holmes and Tempest deduced a value of α^* from the temperature dependence of their measurements; they assumed that the temperature dependence of the reciprocal relaxation time for C_2H_4 was directly proportional to the transition rate for the simple deactivation of the lowest mode $k^{1-0}(v_{10})$ (see Section 1.3.1.2). This is probably incorrect: between 263 and 353°K the calculations presented in Table 2.3 show that $(1/\tau_{pT})$ increases 2·02 times whilst k^{1-0} (v_{10}) increases by a factor of 2·67. Using the calculations to give the relation between $(1/\tau_{pT})$ and k^{1-0} (v_{10}), one obtains $\alpha^* = 6·1$ Å^{-1} from the results of Holmes and Tempest as against their value of 6·72 Å^{-1}. This demonstrates the dangers of trying to deduce values of α^* for polyatomic molecules from temperature-dependence measurements.

The relaxation time of C_2D_4 has been measured by Hudson, McCoubrey and Ubbelohde[227] whose measurements covered half the dispersion zone. The calculations predict that the relaxation time of C_2D_4 should be shorter than that of C_2H_4 by a margin which is much greater than that found experimentally.

2.4.1.5.2 *Propylene.* The next higher homologue after ethylene is propylene, $CH_3—CH=CH_2$. This has been investigated by Holmes, Jones and Pusat[228], who made absorption and dispersion measurements up to an effective frequency of 1 GHz atm^{-1}. The experimental data had to be carefully corrected for classical absorption and translational dispersion (see Chapter 1) and they deduced a relaxation time (τ_{ps}) of $14·8 \pm 0·8$ ns at 30°C.

2.4.1.5.3 *Ethane*. Lambert and Salter[229] discovered double relaxation in ethane (CH_3CH_3) but were only able to define the low frequency part of the dispersion curve. Valley and Legvold[230] extended the measurements to higher effective frequencies but they omitted the correction for translational dispersion; their results have been recalculated[96,228]. Holmes, Jones and Pusat[228] made absorption and dispersion measurements up to an effective frequency of 1 GHz atm^{-1} which was sufficient to just cover both relaxation zones. They obtained relaxation times (τ_{pT}) of 15.0 ± 0.03 ns and 0.96 ± 0.25 ns at 30°C. The shorter relaxation time is associated with the specific heat of the lowest mode, a hindered torsional rotation at 290 cm^{-1}† which does not interact with the other modes[231]. All the other modes relax together, the second lowest mode has a frequency of 821·5 cm^{-1}; this mode relaxes much faster than the lowest mode of C_2H_4 (810 cm^{-1}). Since these two molecules have similar structures and molecular weights, it suggests that the 821 cm^{-1} mode of C_2H_6 relaxes *via* a series mechanism involving a triple quantum jump for the torsional mode. No theoretical calculations have been carried out yet for C_2H_6 as the lowest mode is not a harmonic oscillator.

2.4.1.5.4 *Propane*. Holmes, Jones and Pusat[228] obtained a relaxation time (τ_{pS}) of 15.6 ± 0.8 ns; only single relaxation was observed. It should be noted that propane $(CH_3CH_2CH_3)$ with a lowest mode at 202 cm^{-1} and a hindered rotation at 283 cm^{-1} relaxes much slower than the lowest mode of ethane; notwithstanding the fact that propane is the heavier molecule, this is rather surprising.

2.4.1.5.5 *Butane*. The relaxation time of n-butane has recently been measured by Holmes, Jones and Lawrence[232]. They obtained a value of 1·3 ns at 298°K; the lowest mode is a torsional one with a calculated[232] frequency of 102 cm^{-1}.

2.4.1.5.6 *Pentanes*. Holmes, Jones and Lawrence[232] have investigated two isomers and the ring form of this five carbon paraffin. The straight chain isomer has a relaxation time at 298°K of 1·4 ns; for neo-pentane, $(CH_3)_4C$, the value is 4·35 ns and for cyclopentane, $(CH_2)_5$, it is 1·93 ns. n-pentane has a lowest frequency of 88 cm^{-1}, for the other two molecules the values are uncertain but are thought to lie between 200 and 300 cm^{-1} [232].

2.4.1.5.7 *Hexane*. At 298°K the relaxation time of the straight chain

---

† Note added in proof: using neutron scattering, Strong and Brugger[281] have observed the harmonic frequency of the torsional mode to be 268 cm^{-1} as against the previously accepted value of 290 cm^{-1}. However the latter value has just received support from Weiss and Leroi[282].

isomer was found by Holmes, Jones and Lawrence[232] to be 1·23 ns; the lowest vibrational mode of this molecule is a torsional one with a calculated frequency of 61 cm^{-1}.

2.4.1.6   Sulphur Hexafluoride and Tetrafluoroethylene

Since the review of experimental results for these two gases by Cottrell and McCoubrey[4], further measurements of their relaxation times have been reported in papers[96,233] dealing with their mixtures with other polyatomic gases. Both molecules have several low modes and hence large vibrational specific heats. SF_6 with its high symmetry mostly has degenerate modes, the lowest has a frequency of 344 cm^{-1} and the room temperature relaxation time is just under a microsecond†. The lowest mode of C_2F_4 is at only 190 cm^{-1} and its relaxation time is about 10 ns. Stretton[55] has carried out SSH calculations for SF_6 (see Figure 2.6); the predicted relaxation time was much too long and this was ascribed to the small value of α^* (3·6 Å^{-1}) derived from viscosity data.

2.4.1.7   Empirical Correlations

Two empirical correlations have been put forward for experimentally determined vibrational relaxation times. The more recent one suggested by Losev and Osipov[234] and developed by Millikan and White[136] applies mainly to diatomic molecules. It is found that the relaxation times of these molecules and their mixtures with the inert gases are given over wide ranges of temperature by the formula:

$$\ln(\tau_{pT}) = 1·16 \times 10^{-3} \mu^{1/2} (h\nu/k)^{4/3} (T^{-1/3} - 0·015\mu^{1/4}) - 18·42 \quad (2.81)$$

This equation contains no terms relating to the intermolecular potential: this is reasonable since the most important parameter (α^*) is similar for the diatomic molecules (see Table 2.1). With the SSH theory τ_{pT} can be expressed as a function of μ and ν using Equations (2.41) and (2.58), but, because of the complexity of the resulting expression, it is not possible to reduce it to the form of say Equation (2.81). If one assumes that the 3ξ is the dominant term in Equation (2.41), as was done by Losev and Osipov[234], one would expect $\ln(\tau_{pT})$ to vary as $\mu^{1/3}$ and $(h\nu/k)^{2/3}$. However this does not mean that Equation (2.81) is necessarily at variance with the SSH theory because, as has been seen, predictions of the latter are in good agreement with experiment for diatomic molecules. Equation (2.81)

---

† Note added in proof: further results have been reported for SF_6 over a temperature range[283,284] and for its analogue WF_6[284].

does not work for H_2, D_2 and the hydrogen halides[7] nor for polyatomic molecules generally[225].

The second empirical correlation concerns polyatomic molecules. It is due in its final form to Lambert and Salter[229] and is shown in Figure 2.18; it consists of a plot of the collision number for the lowest mode of a molecule against its frequency (ν_{min}). The collision number (z_{10}) is the average number of collisions required to deactivate the lowest mode; it is thus the inverse of P^{1-0} (ν_{min}). z_{10} is calculated from the experimental relaxation time using first the series formula, Equation (2.79)†, and then

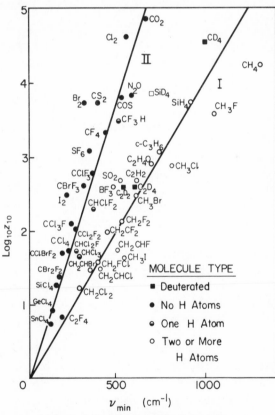

Figure 2.18 The Lambert Salter plot

† The probable error involved in using this equation is small enough to be discounted for the purposes of this exercise; for molecules with two relaxation times, that for the lowest mode is used directly.

equation (2.58). The plot shows that there are two broad classes of molecules: Class I with two or more hydrogen atoms and Class II with none. In between lie a few molecules with only one hydrogen atom and the deuterated molecules. There are a few anomalies, such as H_2O, but the plot is useful for predicting the relaxation time of a polyatomic molecule. Reasons behind these two correlations are considered in Section 2.5.

2.4.2 Binary Mixtures Containing Two Relaxing Components

Most of the information just discussed relates only to simple processes, i.e. the complete interconversion of vibrational and translational energy. The few cases of multiple relaxation in pure gases give some information about complex processes at *homomolecular* collisions and the reason for studying binary mixtures is to investigate complex processes at *hetero-molecular* collisions. A general review of vibration–vibration transfer has recently been made by Lambert[235].

The mixtures studied can be divided into two groups; those whose components are both comparatively simple molecules (which each have only one or two effective modes) and mixtures of more complicated polyatomic gases. The former have been investigated experimentally using both acoustic and optical techniques and, because there are only a few modes to consider, they are amenable to a full theoretical treatment for the calculation of the relaxation times. However, the mixtures in the second group have so many modes that their behaviour can only be interpreted in terms of simplified models and only acoustical measurements have been made on such mixtures. The optical experiments, although of limited application, are simpler to interpret and can lead directly to values for the rates of complex processes. In contrast the acoustic methods need to be carefully analysed in order to determine the relaxation mechanism and this requires that measurements be made over the whole concentration range.

In this review the general principles for mixtures will first be established and then applied to acoustic investigations of mixtures of diatomic molecules. Next the acoustic results for other simple mixtures will be described. Then the experimental results for this group will be discussed as a whole (including the optical measurements) and compared with the theoretical calculations for complex processes. Finally the polyatomic mixtures will be dealt with. Historically, this group was investigated first and most of the work on the simple mixtures has appeared only recently.

2.4.2.1   *General Principles and Acoustic Investigations of Mixtures of Diatomic Gases*

Consider a mixture of two diatomic gases A and B (e.g. N_2 and O_2) where, by convention, pure gas B relaxes faster than pure A. A molecule of A with vibrational energy (denoted by A*) can be deactivated either at a *homomolecular* collision with another A molecule, as in the pure gas, or at a *heteromolecular* collision with a B molecule, at which the latter acts simply as a collision partner and could just as well be an inert gas atom. These are processes I and II respectively of the energy-exchange scheme set out in Table 2.4; processes I and II are both *simple*, involving complete conversions of a vibrational quantum into translational energy. Similar considerations apply to the deactivation of an excited B molecule, processes IV and V. The remaining one is a *complex* process involving exchange of vibrational energy between A and B. If this complex process III is fast, the modes will be strongly coupled and will relax together with a single relaxation time for both modes just like a pure polyatomic gas whose vibrational modes exchange energy rapidly with each other. If process III is slow, coupling is weak, and there will be double relaxation if the rates of processes I and II differ greatly from those of processes IV and V (little can be learnt from mixtures of gases whose relaxation times are similar, though see results for SF_6/C_2H_4 mixtures[233]).

Table 2.4

Energy-exchange scheme for a binary mixture of
diatomic gases

Process		
I	$A* + A \rightleftharpoons A + A$	(vib. $\rightleftharpoons$ trans. : simple process)
II	$A* + B \rightleftharpoons A + B$	(vib. $\rightleftharpoons$ trans. : simple process)
III	$A* + B \rightleftharpoons A + B*$	(vib. $\rightleftharpoons$ vib. : complex process)
IV	$B* + B \rightleftharpoons B + B$	(vib. $\rightleftharpoons$ trans. : simple process)
V	$B* + A \rightleftharpoons B + A$	(vib. $\rightleftharpoons$ trans. : simple process)

* denotes vibrational excitation.

All the five processes are reversible, the forward and backward rates being related by the principle of detailed balancing. The rates of the processes are functions of the concentration because the hetero- or homomolecular collision rates depend upon the mole fraction of each gas (see Section 2.3.1.2), as do the specific heat contributions. Thus the relaxation times are functions of the concentration, their behaviour depending upon the relative rates of the five processes. In order to determine whether there is single or multiple relaxation, experiments have to be carried out at concentrations where the vibrational specific heats of each component are comparable.

Bauer and Roesler[100] have investigated the relaxation of mixtures of oxygen and nitrogen over the whole concentration range. They used their acoustic-resonator technique to measure the long relaxation times. Pure N_2 relaxes very much slower than pure O_2 but in the mixtures it was found that there was only a single relaxation time and its concentration dependence is shown in Figure 2.19. Bauer[90] has shown that the experimental points should lie on a curve which is a branch of a hyperbola. Since the mixture has two modes there are nominally two relaxation times for each concentration; in this case the second one is only 'imaginary' in the sense that it has negligible relaxing specific heat associated with it (see Section 2.3). The concentration dependence of this second relaxation time is given by the upper curve which is the other branch of the same hyperbola; the ends of the two branches are tangential to a pair of intersecting straight lines as shown in Figure 2.19. From the properties of these two lines the rates of the five processes can be deduced[90]; the values obtained from the experimental results are set out in Table 2.5, together with theoretical rates calculated by Jones[98] using the SSH theory (Equations 2.38, 2.67 and 2.68). The rate for the simple deactivation of N_2 (process I) is too slow to be determined experimentally, as is that for process II. The complex exchange from N_2 to O_2 has a rate comparable to that for process IV or V; there is thus strong coupling between the two modes leading to a single relaxation time and the nitrogen molecules only exchange energy with translation *via* the oxygen molecules.

In Section 2.3 two methods for calculating relaxation times were discussed: the Bauer method which results in values of isothermal relaxation times (τ_{pT}), and the SSH method which gives values of the adiabatic relaxation times (τ_{vS}). The differences between the values for these types of relaxation time depend upon the magnitudes of the relaxing specific heats (see Chapter 1). With diatomic mixtures this is very small ($< 0.1 \, cal \, (°K)^{-1} \, mole^{-1}$) and the two types of relaxation time are numerically indistinguishable, so such mixtures afford a good opportunity for comparing the two methods. Jones[98] has carried out calculations for the O_2/N_2 mixtures and shown that, starting with the same transition rates, similar results are obtained from both methods. Bauer uses a

Table 2.5

Experimental and calculated rates for the five energy-exchange processes in mixtures of oxygen and nitrogen
$T = 300°K$

	Process	Collision rate (ns^{-1})	Probability	Calculated[98] Transition rate (s^{-1})	Experimental[100] Transition rate (s^{-1})
I	$N_2^* + N_2 \rightarrow N_2 + N_2$	7·30	$P^{1-0} = 1·16 \times 10^{-11}$	$k^{1-0}/m_{N_2} = 8·48 \times 10^{-2}$	$k^{1-0}/m_{N_2} = <5$
II	$N_2^* + O_2 \rightarrow N_2 + O_2^*$	6·68	$P^{1-0} = 1·19 \times 10^{-11}$	$k^{1-0}/m_{O_2} = 7·96 \times 10^{-2}$	$k^{1-0}/m_{O_2} = ?$
III	$\{ N_2^* + O_2 \rightarrow N_2 + O_2^*$	6·68	$P_{0-1} = 4·8 \times 10^{-8}$	$k_{0-1}/m_{O_2} = 321$	$k_{0-1}/m_{O_2} = 150$
	$\{ O_2^* + N_2 \rightarrow O_2 + N_2^*$	6·68	$P_{0-1} = 1·12 \times 10^{-9}$	$k_{0-1}/m_{N_2} = 7·5$	$k_{0-1}/m_{N_2} = 3·5$
IV	$O_2^* + O_2 \rightarrow O_2 + O_2$	6·11	$P^{1-0} = 1·20 \times 10^{-8}$	$k^{1-0}/m_{O_2} = 73·5$	$k^{1-0}/m_{O_2} = 52$
V	$O_2^* + N_2 \rightarrow O_2 + N_2$	6·68	$P^{1-0} = 1·22 \times 10^{-8}$	$k^{1-0}/m_{N_2} = 81·3$	$k^{1-0}/m_{N_2} = 40$

m_{N_2} and m_{O_2} are the mole fractions of nitrogen and oxygen respectively. In order to obtain the transition rate for a given concentration, the above values should be multiplied by the appropriate value of m_{N_2} or m_{O_2} as the case may be.

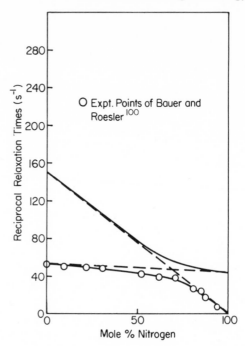

Figure 2.19 Relaxation times[100] for mixtures
of oxygen and nitrogen at 298°K

two-state model for each vibration but he asserts that the same results
would be obtained with harmonic oscillators[236].

With either method the eigenvalue equation for the two relaxation
times reduces to a quadratic and the hyperbolic concentration dependence
of the relaxation times is a direct consequence of their being governed by
a quadratic equation. It can be solved analytically for either end of the
concentration range; for say $100\% \, O_2/0\% \, N_2$, one relaxation time is
that for pure O_2 and has all the specific heat associated with it. The
second solution is for one N_2 molecule in otherwise pure oxygen; it is
found that the inverse of this relaxation time equals the sum of the transition rates of the two heteromolecular processes II and III, which are
the only two open to the lone N_2 molecule. This is illustrated by Figure 2.20
in which the curves were calculated[98] from the theoretical rates given in
Table 2.5. Figure 2.19 differs in appearance from Figure 2.20 mainly
because the theoretical SSH calculations have overestimated the rates

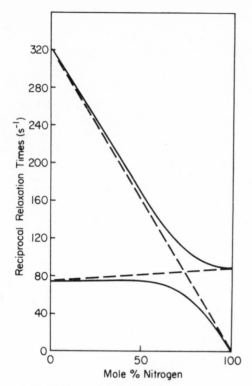

Figure 2.20 Relaxation times for mixtures of oxygen and nitrogen calculated from theoretical transition rates[98]

for processes III and IV; all the relaxing specific heat is still associated with the relaxation time whose concentration dependence is given by the lower curve. Comparison of Figure 2.19 with Figure 2.20 shows how sensitive the concentration dependences of the relaxation times are to the rates of the various processes.

Bauer and Roesler[100] also made measurements on a few mixtures of oxygen and carbon monoxide; unfortunately the CO was contaminated with a little hydrogen and so the relaxation time of the CO sample was shorter than that for their pure oxygen; this is contrary to the accepted values for the pure gases (see Section 2.4.1.1.2) and casts doubt on the results for the mixtures. There is also the possibility of radiation from the CO affecting the issue, a point discussed separately by Bauer[237].

2.4.2.2   *Acoustic Investigations of Other Simple Mixtures*

2.4.2.2.1   *Mixtures of oxygen with water vapour.* A considerable amount of attention has been paid to these mixtures. Tuesday and Boudart[178], using the impact tube, made measurements on both O_2/H_2O and O_2/D_2O mixtures containing up to 2% of water vapour. They established that both forms of water considerably shortened the relaxation time of oxygen, but whereas the relaxation time varied *linearly* with the D_2O concentration, the relationship was *quadratic* with H_2O. The latest experimental work has been carried out by Henderson, Clark and Lintz[179] and they have also reviewed the earlier work. They used the humidity control technique to obtain varying concentrations of water vapour; with this method the oxygen is under high pressure and there is excess liquid water in the apparatus so that there are no errors due to the absorption of water vapour on the walls which could have affected earlier work†. The measurements were made using the acoustic-resonator technique and the lines for the concentration dependences of the relaxation times which they found to be the 'best fits' to their own and other results are shown in Figure 2.21.

These mixtures have been considered theoretically by Tuesday and Boudart[178], Henderson and Herzfeld[238] and by Jones, Lambert and Stretton[62,98,180]. Both H_2O and D_2O have effectively only one vibration (1596 and 1178 cm^{-1} respectively); the frequency for oxygen is 1554 cm^{-1} and so conditions are favourable for complex transfer between O_2 and H_2O (process III). Jones, Lambert and Stretton[180] calculated the collision number (inverse probability) for this and the results are given in Table 2.6. Process III is predicted to be much slower than process IV (the calculation for the latter is unreliable because the probability is so large) which leads to double relaxation over the main concentration range. But for H_2O, process V is several orders of magnitude slower than process III or IV and plays no significant rôle. Thus while process III is normally rate determining for the relaxation of the oxygen, at low H_2O concentrations ($<2\%$) the H_2O-H_2O collision rate is so small that process IV becomes rate determining giving rise to single relaxation and the quadratic concentration dependence because the H_2O-H_2O collision rate is proportional to the *square* of the water concentration. The collision number for process V is predicted to be much greater than that for process IV because only with H_2O-H_2O collisions is there the strong dipole interaction which aids

† Further work by Harlow and Nolan[239] has confirmed that this is indeed the case.

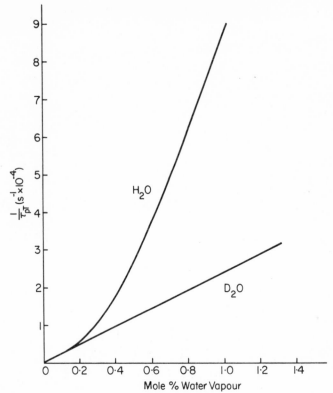

Figure 2.21 Concentration dependences for the relaxation times of
mixtures of water vapour and oxygen

energy transfer. Henderson and Herzfeld[238] have also analysed the
experimental results for H_2O/O_2 mixtures and obtained a similar explanation for the quadratic dependence; they disproved a suggestion by
Calvert[240] that this is due to a dimer $(H_2O)_2$ being the active species for
the deactivation of H_2O.

Since process IV is rate determining, the relaxation time for pure H_2O
can be deduced from the results of the H_2O/O_2 mixtures, see Section
2.3.1.2.7. Jones[98] has shown that the expressions derived by Henderson
and Herzfeld[238] for the relaxation times of the mixtures following the
approach of Tuesday and Boudart (see reference 3, p. 212) are identical to
those obtained from the SSH–Tanczos method. If the experimental

Table 2.6

Collision numbers (z) for energy transfer in oxygen mixtures at 296°K (from Jones, Lambert and Stretton[180])

Molecule B	ν_{min}(cm^{-1})	'Combined' force constants		Process (II)	ν_{ex}(cm^{-1})	Process (III)		Process (IV)	Process (V)
		ε/k(°K)	r_0(Å)	z		$\Delta\nu$(cm^{-1})	z	z	z
CH$_4$	1306	114	3·65	$5\cdot61 \times 10^6$	1534	20	173	11600	49650
CD$_4$	996	114	3·65	$1\cdot50 \times 10^7$	1092	462	15940	4957	11910
C$_2$H$_4$	810·3	134	3·89	$1\cdot49 \times 10^8$	1444	110	487	4054	3500
C$_2$H$_2$	611·8	128	3·88	$1\cdot10 \times 10^8$	2×729	96	1841	488	488
H$_2$O	1596	234	3·10	$3\cdot21 \times 10^5$	1596	-42	76	~1	21690
D$_2$O	1178	234	3·10	$4\cdot71 \times 10^5$	1178	376	1159	~1	2726
HDO	1402	234	3·10	$3\cdot91 \times 10^5$	1402	152	137	~1	8652

O_2, $\varepsilon/k = 88°K$, $r_0 = 3\cdot54$ Å, $\nu = 1554$ cm^{-1}, process (I) $z = 8\cdot31 \times 10^7$
H_2O, $\varepsilon/k = 506°K$, $r_0 = 2\cdot71$ Å, $\delta^* = 1\cdot2$

For full details of data used for other molecules see Stretton[55].

measurements were carried out with up to 10 mole % H_2O, it is predicted that the concentration dependence would become linear from which the rate of process III could be deduced[98].

With mixtures of D_2O and O_2, process III is almost as slow as process IV (because of the larger energy gap) and remains rate controlling down to a much lower water concentration: hence the linear concentration dependence and the corollary that, while no firm deductions can be made regarding the relaxation time of pure D_2O (process IV), the rate of the complex transfer (process III) can be obtained. A collision number of 2200 (D_2O–O_2 collision rate $5.32 \, ns^{-1}$) has been calculated[98] from the experimental results[179]; this is in good agreement with the value calculated theoretically using the SSH method, see Table 2.6.

Henderson, Clark and Lintz[179] also studied the effect of HDO on the relaxation of oxygen; the position is complex because isotopic substitution leads to the formation of an equilibrium mixture of H_2O, HDO and D_2O. Their assertion that HDO is much more effective than H_2O in shortening the relaxation time of oxygen is based upon the unjustifiable assumption that there is no energy exchange between the various isotopic forms of water. If it could be obtained pure, HDO would be expected to resemble H_2O rather than D_2O, see Table 2.6. Measurements have been made of the relaxation of air–water vapour mixtures[179]; these have also been analysed by Henderson and Herzfeld[238]. The relaxation of such mixtures means that the degree of humidity affects the absorption of sound by the atmosphere; this has been discussed by Kneser[5] and data concerning the effect have been tabulated by Harris[241].

2.4.2.2.2 *Mixtures of oxygen with simple hydrocarbons.* The first measurements were made on oxygen/methane mixtures and Schnaus[242] reported that traces of either CH_4 or CD_4 considerably reduced the relaxation time of oxygen at room temperature, see Figure 2.22. White[243] drew similar conclusions from his shock-tube experiments at high temperatures and pointed out that linear extrapolation of the results to zero oxygen concentration suggested that oxygen was more effective in deactivating methane (process V) than methane itself (process IV). As this appeared contrary to the simple theory that the probability of energy exchange varies inversely with the reduced mass, Jones, Lambert and Stretton[180] carried out a full calculation for methane/oxygen mixtures. Three modes were considered, the one for O_2 and v_2 and v_4 of CH_4; process III for these mixtures is a rapid exchange process between the oxygen mode and v_2 of CH_4 (see the transition scheme in Figure 2.23).

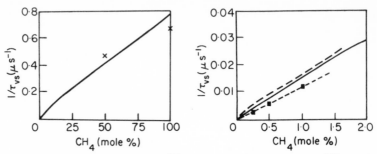

× Expt. results of Cottrell and Day[201]; ■ Expt. results of Schnaus[242]; – – Expt. curve of Parker and Swope[204]; —Calculated curve of Jones, Lambert and Stretton[180].

Figure 2.22   Relaxation times for mixtures of oxygen and methane

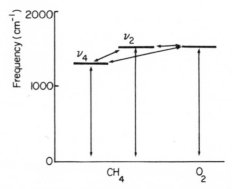

Figure 2.23   Transition scheme for methane/oxygen mixtures

From the calculated transition probabilities, see Table 2.6, the relaxation times were calculated for the whole concentration range. It was found that all the relaxing specific heat was associated with only one relaxation time whose concentration dependence was curved (see Figure 2.22), though at low methane concentrations it would appear to be linear. This demonstrates how erroneous conclusions can be obtained for mixtures where vibration–vibration exchange is likely to occur if results are obtained over only a small concentration range and then extrapolated. Unfortunately the relaxation behaviour is insensitive to the rate of the complex process and no rate can be deduced for this near resonant transfer from the experimental results[98]. Confirmation of the predicted relaxation behaviour was provided by the measurements made on a 50:50 methane/

oxygen mixture by Cottrell and Day[201]†. Parker and Swope[204] have since reported further measurements at low methane concentrations which agree very well with the calculations. For methane concentrations <0·5%, they found that the relaxing specific heat of the mixture was less than the sum of the vibrational specific heats of the two components; they suggested that it was due to radiative losses from the v_4 mode of CH_4, but this has a long radiative lifetime of 0·4 s [210].

White[243] made further shock-tube measurements on mixtures of oxygen with two other hydrocarbons, ethylene and acetylene. Both considerably shorten the relaxation time of oxygen and the rates of the various relevant processes have been calculated by Jones, Lambert and Stretton[180], see Table 2.6. For these calculations the exchanging mode of the hydrocarbon molecules (v_{ex}) is taken to be the one whose frequency is closest to that of the oxygen mode.

2.4.2.2.3 *Carbon dioxide mixtures.* It has long been known that addition of water vapour to CO_2 considerably shortens the relaxation time. The earlier work has been listed by Lewis and Lee[245] who made recent measurements with CO_2 containing 0·2, 0·3 and 0·4 mole % H_2O. Over this narrow range they found a linear concentration dependence and that with inceasing temperature the efficiency of the water vapour decreases. They extrapolated their results to 100% H_2O concentration which could be erroneous, as was pointed out with the CH_4/O_2 mixtures, because there is likely to be a favourable complex transfer between H_2O and the middle mode of CO_2, see Figure 2.24. An abstract[246] of results obtained

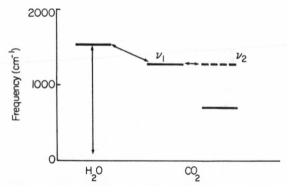

Figure 2.24 Possible transition scheme for carbon dioxide/water mixtures

† Their theoretical calculations have since been found to be incorrect[244].

for higher water concentrations has been published and further details are awaited[†].

Water vapour is an important additive in a CO_2 gas laser[164,247]; nitrogen molecules are excited by an electric discharge to the first vibrational level and exchange energy with the v_3 mode of CO_2, a near resonant transfer. The laser action emits a quantum of wavelength $10\cdot6$ μm and simultaneously the v_3 quantum is converted to a v_1 quantum. In order to maintain an inverted population for the laser action, the v_1 mode must be rapidly de-excited and a little water vapour is added to aid this.

Widom and Bauer[248] carried out semiclassical calculations for the deactivation of a CO_2 molecule in collision with an H_2O molecule. In order to explain the high efficiency of such collisions they postulated that there is a very large attractive term in the intermolecular potential due to a 'chemical interaction'; however there is no independent evidence for this interaction. Marriott[72] has reported calculations on CO_2/H_2O mixtures using his quantum-mechanical method. At low concentrations he predicted double relaxation, as has been found by Pielemeir, Saxton and Telfair[249], but with 10 mole % of H_2O or more there is only single relaxation with a temperature dependence similar to that measured by Eucken and Nümann[250]. In obtaining these results Marriott did not have recourse to an intermolecular potential with a very deep well and he attributes the success of the calculations to his taking account of coupling between the vibrational states, see Section 2.2.6. No account was taken of heteromolecular energy exchange between the H_2O and CO_2 molecules.

Cottrell, Macfarlane and Read[160] have investigated CO_2/N_2 mixtures using the spectrophone[‡]. They obtained evidence of rapid energy exchange between the two molecules and found that the concentration dependence of the relaxation of the v_3 mode of CO_2 is non-linear in consequence. Similar results were obtained with N_2O/N_2 mixtures though the curvature of the concentration dependence is less marked.

2.4.2.3   *Results Obtained from Simple Mixtures and Comparison with Theoretical Calculations*

Optical methods afford an almost direct method of measuring the rates of complex processes and one such is the flash technique used by Callear

---

† Note added in proof: Higgs and Torborg have since published a fuller account of their work[105]. Shields and Burks[285] have just published results for mixtures of CO_2 with D_2O.

‡ Note added in proof: further experimental and theoretical results on CO_2/N_2 (and on CO_2/O_2) mixtures have recently been reported by Herzfeld[277].

and his colleagues[93,147,251,252,253]. A diatomic molecule A diluted considerably in an inert gas is excited optically and the relaxation of its vibrational energy is followed spectroscopically. This is very slow unless a small quantity of another gas B is added to which the vibrational energy can be transferred. This gives a direct measure of the rate of process III providing the other processes are all comparatively slow and transfer back to A is not a factor; Callear and Williams have considered the latter point and concluded that the effect is negligible[147]. The other processes are slow because they involve vibration–translation transfer.

The first experiments[252] concerned transfer from various electronic states of NO to other diatomic molecules. The result that about a thousand collisions were necessary, on average, for the resonant transfer from $NO(A^2 \Sigma^+)$, $v = 2342$ cm^{-1}, to N_2, $v = 2331$ cm^{-1}, was the first confirmation of equation (2.39) and demonstrated that not even resonant processes took place at every collision.

Millikan has made similar measurements for complex processes using his fluorescence technique for the rates of vibrational-energy exchange between CO and O_2[141] and between CO and CH_4[254]†. As suggested by Callear[252], all these results can be conveniently presented graphically as a plot of the collision number (z) against Δv for the complex exchange. This is done in Figure 2.25 for all the results obtained from optical and acoustic measurements and it can be seen that the results fall into three groups A, B and C.

Those about line 'A' are all for pairs of diatomic molecules and many theoretical calculations have been carried out for these complex transitions. Rapp[68] showed that the semiclassical treatment gave good agreement with experiment for the three cases considered: $NO(A^2 \Sigma^+)/N_2$, $NO(X^2\Pi)/CO$ and $NO(X^2\Pi)/N_2$. The SSH calculation[98] for N_2/O_2 is within a factor of two, see Table 2.5. For CO/O_2 the result of Millikan[141] is to be preferred to that of Bauer and Roesler[100], for reasons already discussed, and it is more in line with the other results. No specific theoretical calculations have been carried out for this process. Hooker and Millikan[125] and Millikan and White[254a] have found that, behind a shock wave, the relaxation times of N_2 and CO are the same in a mixture containing 5% CO; this demonstrates that there is a rapid energy exchange between them but no actual value of the rate could be obtained‡. It has

† Note added in proof: further results for mixtures of CH_4 with O_2 and NO have just been reported by Yardley and Moore[286] who used their laser technique.

‡ Note added in proof: fresh results for simple mixtures at high temperatures have recently been reported[287,288,289].

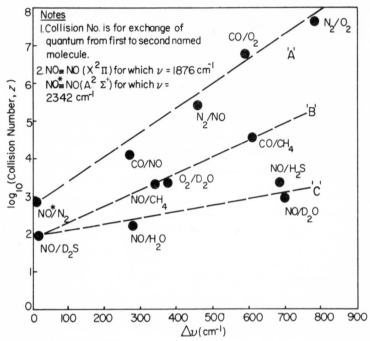

Figure 2.25 Experimental collision numbers for vibration–vibration energy exchange between simple molecules at room temperature

been calculated[255] using the SSH theory that the calculated collision probability for exchange from N_2 to CO ($\Delta E = 188 \text{ cm}^{-1}$) is $8 \cdot 75 \times 10^{-5}$ at 500°K (collision rate = $5 \cdot 57 \text{ ns}^{-1}$). The calculation was carried out because this process is a factor in the heat balance for the Martian atmosphere; the CO is infrared active and so could radiate energy from the planet[256]. Callear has shown that the slope of the line drawn in Figure 2.25 through the 'A' group is in general agreement with the SSH theory.

    The line through the 'B' group falls well below the 'A' line because with the former the average size of quantum exchanged is smaller, the colliding molecules have a smaller reduced mass and the hydride molecules have larger vibrational factors than the diatomic molecules. Theoretical calculations have been carried out for specific cases in the 'B' group: for the transfer from O_2 to D_2O the SSH calculation (see Table 2.6) underestimates the collision number by a factor of two. Jones, Lambert and

Stretton[257] carried out an SSH calculation for transfer from CO to CH_4; they showed that the most favourable transition was to the v_2 mode of CH_4 involving an energy discrepancy (Δv) of 617 cm^{-1}. Theoretical values for the collision numbers for this process are 21,000 at 303°K dropping to 13,800 at 363°K and are within 50% of Millikan's[254] experimental values of 33,000 and 24,000 respectively and have almost the correct temperature dependence. Because of closer correspondence between the vibrational levels, exchange from CD_4 to CO is expected[257] to be much faster, but no experimental results are available yet.

Jones[98] has carried out SSH calculations for vibration–vibration transfer from $NO(X^2\Pi)$ and the results are set out in Table 2.7. Agreement with experiment is generally good except for NO/H_2S and NO/D_2O when the calculated collision numbers are an order of magnitude greater than the experimental ones. It would appear that experiment and theory diverge for exchanges between NO and the polar hydrides by an amount depending upon Δv, since there is accord for NO/D_2S for which $\Delta v \approx 0$. Callear and Williams[147] drew similar conclusions from less elaborate calculations and suggested that the hydrogen bonding between NO and polar hydrides enhanced the conversion of the vibrational discrepancy (Δv) to translational energy. In such cases z varies only slowly with Δv as shown by the examples in the 'C' group of Figure 2.25.

There is thus substantial agreement between theory and experiment for the complex exchange of vibrational energy between simple molecules at heteromolecular collisions. This reflects not only upon the success of the theory but also upon the efficiency of the simple combining rules[50] used to obtain the required heteromolecular force constants.

2.4.2.4   *Mixtures of Polyatomic Molecules*

Binary mixtures of polyatomic gases were investigated before their counterparts with simple molecules and to date only acoustical methods have been used. The first results were reported by Amme and Legvold[258]. They found that mixtures of halomethane gases showed only single relaxation over the whole concentration range whereas Lambert, Edwards, Pemberton and Stretton[233] later found double relaxation in mixtures of SF_6 and some halomethanes with C_2F_4 and CH_3OCH_3. The first quantitative explanation of the behaviour in terms of vibration–vibration transfer was given by Lambert, Parks-Smith and Stretton[96].

They reported measurements made on mixtures of SF_6 and $CHClF_2$; single relaxation was observed for each of the mixtures with the relaxation

Table 2.7

Calculated and experiment collision numbers (z) for complex processes between $NO(X^2\Pi)$ and other simple molecules at 298°K (from Jones[98])

Molecule B^b	ν_B (cm⁻¹)	$\Delta\nu$ (cm⁻¹)	Heteromolecular force constants^c ε/k(°K)	r_0(Å)	δ^*	Collision rate (Z)(ns⁻¹)	Calculated^d z	Experimental^{a,d} z	Expt. Ref.
H_2O	1596	280	240	3·11	0·14	5·60	2·4, 2	1·4, 2	147
D_2O	1178	698	240	3·11	0·14	5·42	1·33, 4	8·9, 2	147
H_2S	1188	687	197	3·50	0·06	5·99	6·99, 4	2·3, 3	147
D_2S	1892	−16	197	3·50	0·06	5·88	1·55, 2	8·7, 1	147
CH_4	1534	342	133	3·62	0	7·86	2·75, 3	2·06, 3	147
N_2	2331	−455	97	3·61	0	6·65	3·77, 6	2·29, 6	253
CO	2143	−267	102	3·59	0	6·57	1·31, 5	4·39, 4	253

^{a} Calculated from experimental rate coefficient using collision rate quoted in this table.

^{b} For this Table *only*, molecule A is $NO(X^2\Pi)$ ($\nu = 1876$ cm⁻¹); collision numbers for transfer from A → B.

^{c} Calculated using the usual combining rules[50]; for homomolecular force constants see Section 2.4.1.

^{d} Values in floating style, e.g. 2·4, 2 ≡ 24×10^2.

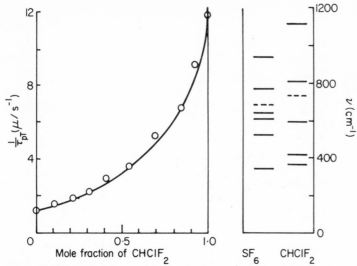

Figure 2.26 Reciprocal relaxation times and energy-level diagram for $SF_6 + CHClF_2$ mixtures (from Lambert, Parks-Smith and Stretton[96])

times having the concentration dependence shown in Figure 2.26. Although the relaxation times of the two pure gases differ by nearly an order of magnitude, the lowest vibrational levels of the two molecules have similar frequencies (see energy-level diagram in Figure 2.26) so that one would expect a rapid complex transfer between them in the mixtures. In order to elucidate the relaxation mechanism, calculations of the relaxation behaviour for the whole concentration range were carried out based upon the energy-exchange scheme of Table 2.4, SF_6 being component A. To simplify the calculation, the upper modes of each molecule were represented by a single composite mode having the same specific heat as the modes which it replaced and exchanging energy rapidly with the lowest mode[62]. Using this model for a polyatomic molecule, whose vibrational energy relaxes all together, reduces the problem for a binary mixture to one of four modes with complex transfer (process III) assumed to take place between only the lowest modes, see Figure 2.27.

The rates of processes I and IV are known from the results with the pure gases and plausible values were inserted for the other three; then the relaxation equations were solved numerically for the whole concentration

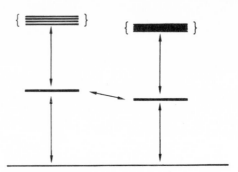

Figure 2.27 Model used by Lambert, Parks-Smith and Stretton[96] for calculations on binary mixtures of polyatomic molecules

range (see Section 2.3.1.2). It was found that the best fit to the experimental results was with process II twice as fast as process I, process V half as fast as process IV and the collision number (z) for the complex process III equalling 50. Process III is the fastest one and leads to the observed single relaxation involving the total vibrational specific heat of the mixture at all concentrations; the sensitivity of the fit is such that the probability of the complex process can be estimated to within 30%. The non-linear concentration dependence is mainly due to the rate of process I exceeding that for process II.

A similar treatment was applied to the experimental results reported by Valley and Legvold[230] for mixtures of $C_2H_4(A)$ and $C_2H_6(B)$. Pure ethane has two relaxation times (see section 2.4.1.5.3), the low frequency torsional mode relaxing faster than the rest; in the mixtures the vibrational specific heat of the C_2H_4 and the upper modes of C_2H_6 relax together and there is a second zone for the torsional specific heat. Making certain assumptions[96] it was found that the probability for transfer from the lowest mode of C_2H_4 ($v = 810 \, cm^{-1}$) to the second one of C_2H_6 ($v = 821.5 \, cm^{-1}$) was $\frac{1}{40}$. Another case of single relaxation in binary polyatomic mixtures has been reported for CO_2 and ethylene oxide (C_2H_4O) by Seshagiri Rao and Srinivasachari[259]; a non-linear concentration dependence of the relaxation time was found. Parks-Smith[192] has investigated mixtures of SO_2 and CH_2F_2: the energy levels are shown in Figure 2.28 for the two molecules. It can be seen that favourable transitions are available between each of the three SO_2 levels and those of CH_2F_2, and it

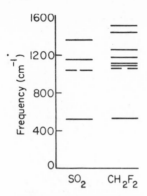

Figure 2.28 Energy levels of SO_2 and CH_2F_2

was found that mixtures of the two gases had only one relaxation time, though pure SO_2 has two relaxation zones (see Section 2.4.1.2.2). This implies that the upper two modes of SO_2 exchange energy with translation via adjacent CH_2F_2 levels and that the lowest mode of SO_2 is also strongly coupled to that of CH_2F_2. This is a direct demonstration of complex processes occurring between the upper modes of polyatomic molecules.

Many binary mixtures of halo-methanes have been investigated and in all cases only a single relaxation time has been observed. Amme and Legvold[258] made measurements on $CHClF_2/CF_4$ and $CHCl_2F/CClF_3$ mixtures and Boade and Legvold[260]† studied all ten binary combinations of the five gases: CH_2F_2, CHF_3, CF_4, CCl_2F_2 and $CHCl_2F$. Some of the results of the latter work are plotted in Figure 2.29. Calvert and Amme[261] interpreted the results in terms of parallel excitation for each molecule, i.e. they wrote Equation (2.80) for each component and then combined them, the relaxation time of the mixture being the mean of these two relaxation times weighted according to the concentrations. It was found[260] that this agreed with the experimental results so long as both the specific heats and relaxation times of the pure components were similar, for example as for $CCl_2F_2/CHCl_2F$ but not for $CF_4/CHCl_2F$. This approach is erroneous because it takes no account of vibration–vibration transfer; in all cases the vibrational levels in the pair of molecules are close enough for complex transfer to be favourable. In Figure 2.29, the behaviour of three of the

† These results were from dispersion measurements, some similar results were obtained from absorption measurements[262].

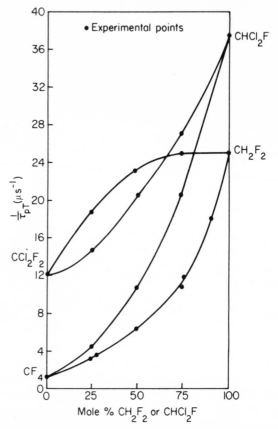

Figure 2.29 Experimental results of Boade and
Legvold[260] for binary halomethane mixtures

series of mixtures resemble $SF_6/CHClF_2$ whilst the concentration dependence of the single relaxation time for the CCl_2F_2/CH_2F_2 mixtures curves the other way similar to CH_4/O_2. No quantitative interpretation of the behaviour of the mixtures in terms of vibration–vibration transfer has been carried out.

Lambert, Edwards, Pemberton and Stretton[233] investigated several series of binary polyatomic mixtures and found that for all of them there was double relaxation at all concentrations, each component having a

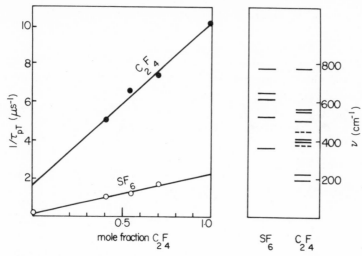

Figure 2.30 Reciprocal relaxation times and energy diagram for $SF_6 + C_2F_4$ mixtures (from Lambert, Edwards, Pemberton and Stretton[233])

relaxation time associated with its share of the vibrational specific heat. The pure components have widely different relaxation times and an example is the series of mixtures of SF_6(A) and C_2F_4(B) for which the results are summarized in Figure 2.30. The behaviour of these mixtures was misinterpreted at first[233] but this was corrected by Lambert, Parks-Smith and Stretton[96]. The relaxation time of the A component is considerably shortened by addition of B; this could be because either process II or III is fast. Since the reductions are large and more than one would expect from theory for process II, it is likely that process III is responsible, for which the vibrational levels of the molecules are favourably disposed. However, with these mixtures the complex exchange is intermediate between the rates for process I (or II) and process IV (or V) and so is not fast enough to lead to single relaxation. The mixtures are comparable in behaviour to pure gases such as CH_2Cl_2 which have two relaxation times; the energy of the A component is exchanged with translation via component B. This mechanism was confirmed by calculating the relaxation times for a series of mixtures using a model similar to that described above for SF_6 and $CHClF_2$.

The linear concentration dependence of the relaxation time for the B component gives information about the relative rates of processes IV and V, whereas that for A can yield the rate of the complex process in a way similar to that discussed for mixtures of N_2 and O_2 in Section 2.4.2.1. The inverse of the relaxation time, obtained by extrapolation, for a single molecule of A in otherwise pure B is the sum of the rates of processes II and III. The rate of process II is taken to be comparatively slow so that application of the series formula (Equation 2.79), as a rough correction for the effect of the upper modes, gives the approximate rate at which the lowest mode of A exchanges energy with the nearest mode of B (see Figure 2.30 for the energy-level diagram for SF_6 and C_2F_4). The results thus obtained from several mixtures are summarized in Table 2.8 together with values deduced from the mixtures having only one relaxation time. z_{AA} and z_{BB} are the collision numbers for processes I and IV and z_{AB} is for the complex process. Note that whereas with mixtures having only *one* relaxation time z_{AB} is smaller than either z_{AA} or z_{BB}, it is intermediate between the two for mixtures with *two* relaxation times.

The values of Z_{AB} given in Table 2.8 are in general accord with the SSH formula for resonant complex processes (Equation 2.39). Thus the lowest collision number, five, is for exchange of the smallest quantum ($250 \ cm^{-1}$) and this is about a hundred times smaller than the collision number for a quantum ten times bigger, for example the exchange between NO ($A^2\Sigma$) and N_2, see Figure 2.25. This agrees with Equation (2.39) that, all other factors being equal, the collision number varies as the square of the size of the quantum exchanged. Values of $z_{AB} \sim 50$ are generally obtained from mixtures involving a single quantum transfer; for C_2H_6/C_2H_4 the fact that this is the largest quantum exchanged is counteracted by the molecules having a low reduced mass. For exchanges involving a double-quantum jump the collision numbers are around a hundred.

In general the mechanism of energy exchange in binary mixtures of simple or polyatomic molecules is now understood and the criteria have been established for single or multiple relaxation. Early work was concerned with mixtures containing at most only a few per cent of one component; this generally does not allow one to decide the energy-exchange mechanism and so such work has not been considered here. However, much information has been obtained about vibration–vibration transfer from those mixtures for which the relaxation mechanism can be elucidated and this is usually in accord with the theory.

Table 2.8

Collision numbers for energy transfer between polyatomic molecules
(from Lambert, Parks-Smith and Stretton[96])

A	B	v_A (cm^{-1})	v_B (cm^{-1})	i	Δv (cm^{-1})	z_{AB}	z_{AA}	z_{BB}
			Singly-relaxing mixtures					
SF$_6$	CHClF$_2$	344	369	1	+25	50	1005	122
C$_2$H$_4$	C$_2$H$_6$	810	821·5	1	+11·5	40	970	74
			Doubly-relaxing mixtures					
CCl$_2$F$_2$	CH$_3$OCH$_3$	260	250	1	−10	5	73	<3
CH$_3$Cl	CH$_3$OCH$_3$	732	250	3	+18	70	421	<3
SF$_6$	CH$_3$OCH$_3$	344	164	2	−16	80	1005	<3
CHF$_3$	C$_2$F$_4$	507	507	1	0	50	1500	5·5
SF$_6$	C$_2$F$_4$	344	190	2	+36	70	1005	5·5
CF$_4$	C$_2$F$_4$	435	220	2	+5	110	2330	5·5

z_{AB} is the collision number for transfer between one quantum of mode v_A of molecule A and i quanta of mode v_B of molecule B.

2.5 GENERAL DISCUSSION

2.5.1 The Effect of Rotation

So far the theory of the excitation and de-excitation of vibrational energy has been based upon the assumption that the vibrational energy is coupled directly with the translational energy. On the whole this has worked well but it has ignored the possible effects of molecular rotation. The first suggestion that rotation might be involved came with the study of the relaxation of deuterated molecules. Cottrell and Matheson[200] discovered that CD_4 had a *longer* relaxation time than CH_4 even though the lowest mode of CD_4 has much the lower frequency so that, according to the simple theory, the relaxation time of CD_4 should be *shorter* in spite of the small difference in molecular weights. A similar effect was also noted with the silanes and the Group V hydrides as has already been pointed out in Sections 2.4.1.4.1 and 2.4.1.3.2. Stretton[55] showed that a similar discrepancy existed between theory and experiment with the methyl halides and the ethylenes. Generally, if Q is the relaxation time of the hydrogenated molecule divided by that for the corresponding deuterated one, the predicted values of Q according to the SSH theory are about four times greater than the experimental ones; only with the acetylenes is there anything like agreement between the theoretical and experimental values of Q [55].

Cottrell and Matheson[200] suggested that since the moments of inertia of hydrocarbon molecules are generally very small the peripheral velocities are correspondingly high and this aids vibrational-energy exchange. Thus comparing, say, CH_4 and CD_4, this would favour CH_4 which has a rotational velocity $\sqrt{2}$ times faster than its deuterated analogue. As pointed out by Millikan[7], the average classical peripheral velocity of the rotating hydrogen atoms in CH_4 is greater than their translational velocity. A striking demonstration that rotation can indeed be involved in vibrational-energy exchange is provided by the elegant studies of Millikan and Osberg[263] on the separate effects of *ortho-* and *para*-hydrogen on the relaxation of carbon monoxide. These two forms of hydrogen differ only in the relative alignment of their nuclear spins; this affects the occupation of the rotational states so that with the *ortho* form only the *odd*-numbered levels ($J = 1, 3, 5 ...$) can be filled whereas with the *para* form only the *even* ones ($J = 0, 2, 4, ...$) are used. Employing the fluorescence technique (see Section 1.4.2) it was found that *para*-hydrogen was more than twice as efficient as the *ortho* form in deactivating CO at $288°K$ [263]. This result is contrary to the idea that a high peripheral velocity of rotation aids vibrational-energy exchange since, at 288°K, only 30% of the *para* form is in the $J = 2$ state, the rest is in the non-rotating $J = 0$ state. Another possibility is that vibration–rotation exchange takes place because the $J = 2$ to $J = 6$ transition is nearly resonant with a vibrational quantum of CO. However this explanation can be discounted because at 110°K, when the $J = 2$ state of *para*-H_2 is virtually unpopulated, the *para* form is about twice as effective[7] in deactivating CO, but above 450°K the difference between the two forms is negligible[263]. Even so there is a clear implication that rotation is somehow involved.

Cottrell and Matheson[200] put forward a semiquantitative theory of vibration–rotation energy exchange and this was developed by Cottrell, Dobbie, McLain and Read[197]. They used the semiclassical vibration–translation theory (Section 2.2.1.2) with the peripheral velocity of rotation replacing the translational velocity. The molecular rotation is thus treated classically and the resulting expression is integrated over the thermal distribution of rotational velocities. This method was used to compute ratios of relaxation times, rather than for calculating absolute values, and correctly predicted the values of Q for the methanes and silanes[200]. The predicted temperature dependence for CH_4, though agreeing with the two experimental values of Cottrell and Matheson[200], is much steeper than that shown by the consensus of experimental results, see Figure 2.15. However the theory also gives the correct values of Q for the relaxation times of the group V hydrides[197], see Table 2.2, and predicts that the relaxation time of phosphine will be close to that of arsine, because the moments of inertia are similar; this is borne out by experiment[197]. On the other hand vibration–translation theory would suggest[197] that the relaxation time of arsine would be two orders of magnitude longer than that of phosphine because of its larger molecular weight (78 as against 34)†.

This theory of vibration–rotation transfer has been extended and carefully examined by Moore[264]. From a semiclassical treatment averaged over the thermal distribution of angular velocities he obtained the expression:

$$P^{1-0} = P_0 \frac{32\pi^4(2\pi)^{1/6}}{3^{1/2}} \frac{I^{13/6}v^{4/3}}{d^{13/3}(kT)^{1/6}\hbar M\alpha^{7/3}} \exp\left[-3\left(\frac{2\pi^4 Iv^2}{d^2\alpha^2 kT}\right)^{1/3} + \frac{hv}{kT} \right]$$

(2.82)

† Note added in proof: in contrast, the ratio of the relaxation times of SF_6 and WF_6 molecular weights (146 and 297·4) is correctly predicted by the SSH theory and with these large moments of inertia vibration–rotation transfer is unimportant[284].

The symbols are as used in Section 2.2; I is the moment of inertia and d the distance between the centre of rotation and peripheral atoms. The leading exponential term is the same as that obtained with the vibration–translation treatment (e.g. see Equation 2.41) except that the reduced mass of the colliding pair of molecules (μ) is replaced by I/d^2. Moore directed his attention to molecules whose peripheral velocity is greater than the translational ($I/d^2 \ll M$ where M is the reduced mass of the vibrational normal coordinate) and considered some twenty-five such cases. Equation (2.82) has two effectively unknown parameters, P_0 and α, and Moore found that the experimental data for these molecules could be correlated according to Equation (2.82) by putting $\alpha = 2.94\,\text{Å}^{-1}$ and $P_0 = 0.2$, the average deviation being a factor of $\frac{3}{2}$ and the maximum $\frac{5}{2}$. He concluded that this was strong evidence for the vibration–rotation model though the temperature dependence was incorrectly predicted. The calculations showed that the most favourable peripheral velocity for energy transfer exceeded the average translational velocity by factors varying from twenty for HBr to three for CD_4, the average being around five; thus in these cases the translational energy will be unimportant. It will have an effect if the moment of inertia is large, or, paradoxically, when it is very small because then the spacing between the rotational energy levels will be large and translational energy will be needed to make up any discrepancy. For such molecules, for example HBr, the rotation will have to be treated quantum mechanically and exchange between vibration, rotation and translation considered together.

Benson and Berend[46] have considered classically the vibrational-energy exchange of a rotating molecule. They carried out calculations in two dimensions for collisions between oxygen molecules and argon atoms. Providing the rotational energy was small compared with the translational, it was found that the rotation had little or no effect upon the efficiency of vibrational-energy exchange but at high rotational energies the efficiency dropped away considerably. Viewed classically, the reason is that with a rapidly rotating oxygen molecule the chances are high that the collision will be a glancing one with only a small component of the force at impact acting along the axis of the diatomic molecule. This is an unfavourable situation for affecting the vibration, but with increasing translational energies the axial component of the impact force will increase and with it the probability of vibrational-energy exchange. Benson and Berend[46] found that the critical ratio is $E(\text{trans.}) \approx 0.7\,E(\text{rot.})$; for low translational energies the chances of vibrational-energy exchange are very small but above the critical value translation and rotation are equally effective. Quantitative calculations in two dimensions were carried out for the relaxation of oxygen by argon atoms over a temperature range and compared with previous one-dimensional calculations[15]. The effect of the rotation is to lower the transition probability by about a factor of three without affecting the temperature dependence; this was compensated for by increasing the repulsive potential parameter $(\alpha)_{BB}$ from 3·8 to 4·1 Å^{-1}.

This work furnishes an explanation of Millikan and Osberg's work on CO and *ortho-* and *para-*hydrogen. Rotation can decrease the vibrational-energy exchange of diatomic molecules and thus the non-rotating form of hydrogen, the *para* form with $J = 0$, will be the most efficient for bringing about vibrational transitions. The effect of the *para* form will be most marked at low temperatures when the $J = 0$ state is prevalent; with increasing temperature the higher rotational states will be occupied and the superiority of the *para* form over the *ortho* will become progressively less; this is in accord with the experimental results. Thus the contention of Benson and Berend that rotation can hinder the vibrational-energy exchange of diatomic molecules is borne out—but why does rotation appear to help with cases such as CH_4? Considered classically, rotation favours glancing molecular encounters; as has been seen this is of little use for diatomic molecules for whom the displacement of the vibrating atoms is perpendicular to their direction of rotation, but with the v_4 mode of CH_4 the displacement is parallel to the direction of rotation (or the molecular 'surface') and so glancing blows will be advantageous. Thus any treatment of the simultaneous exchange of vibrational and rotational energy will have to take account of the direction of the atomic displacements in a vibration and so the 'breathing sphere' model, which only takes account of their magnitude, will not suffice. A steric factor to allow for random orientation of the molecules will still be needed. In a preliminary communication Kelley and Wolfsburg[47] have examined classically in two dimensions the efficiencies of vibrational- and rotational-energy exchange as a function of the collision angle and have shown that a 'head-on' collision is not necessarily the best for vibrational-energy exchange when rotation is also considered†. Takayanagi[11] has given some consideration to a quantum-mechanical treatment of vibration and rotation together and produced a complex formula but no numerical results are available yet.

There are several cases where the vibrational and rotational relaxation zones overlap: examples are ammonia, water and the large hydrocarbons.

† Note added in proof: a related study by Shin has just appeared[290].

These could be cases where there is simultaneous rotation and vibration energy exchange at one collision. Evidence has also been adduced from work on mixtures of carbon dioxide with He^{171}, H_2 and $D_2{}^{265}$: the total relaxing specific heat (after allowance for viscothermal and diffusion effects, see Chapter 7) exceeds the vibrational specific heat and is taken as being part of the rotational specific heat. This was attributed to the influence of rotational transitions simultaneous with, and in the opposite direction to, the vibrational transition.

It has been suggested[200] that the explanation of the classes of the Lambert–Salter Plot is that the molecules containing two or more hydrogen atoms (Class I) exchange vibrational energy with rotation whilst the others exchange with translation. While this merits attention it is unlikely that the issue is as clear cut because Class I contains molecules such as cyclopropanes and ethylene oxide which have comparatively large moments of inertia[55,264]. It is probable that the two classes result from a variety of reasons: comparing molecules in the two classes which have the *same* lowest frequency, those in Class I have smaller masses and large vibrational factors or else they have large dipole moments, all of which enhance the interconversion of vibrational and translational energy.

2.5.2 Conclusion

Further understanding of the exchange of vibrational energy at molecular collisions will be brought about in three ways: by obtaining more precise and specific experimental results for the various energy exchange processes, refining the methods of calculating transition probabilities and the use of more realistic potentials. What needs to be done in these three directions and what are the prospects?

The importance of temperature-dependence measurements has been stressed already and the relaxation times of the diatomic gases need to be reliably established over as wide a temperature range as possible as has been done for oxygen; in particular, data are lacking for the hydrogen halides. With polyatomic gases there is a general dearth of temperature-dependence results. One looks to the shock tube to provide these, as has recently been done for CO_2 and CH_4, with measurements being made down to near room temperature so as to overlap with the acoustic measurements. Acoustic techniques are good for determining accurately the relaxing specific heat and for detecting multiple relaxation; with careful design and the use of transducers able to resist high temperatures, acoustic measurements could be carried out over wider temperature ranges.

The simple, more symmetrical molecules should be examined first as these are more amenable to theoretical treatment. In general the shock tube and the acoustic experiments provide information about simple processes whereas optical methods can measure the rates of complex ones. The advent of the laser has stimulated the development of the latter, as is evident in Chapter 1, but some thought needs to be given as to what exactly is being measured and how the observed relaxation times relate to the various transition rates, i.e. one requires a treatment analogous to the Tanczos or Bauer ones for acoustic work.

Ideally one would like to be able to determine all the independent elements of the transition rate **k** matrix (equation 2.63) for a polyatomic molecule. However, even with a triatomic molecule (the simplest case) insufficient information is obtained from acoustic measurements even when there is more than one relaxation (see Shields' work on SO_2). The problem is analogous to determining the intramolecular force field for a molecule's vibrations by using the **GF** matrix method[60]; knowing the **G** matrix from the molecular geometry one attempts to determine the **F** matrix knowing the molecule's measured fundamental frequencies which are directly related to the eigenvalues (λ) of

$$|\mathbf{GF} - \lambda \mathbf{I}| = 0$$

Compare this situation with Equation (2.64): although the elements of the **b** matrix are known, one has only one or perhaps two measurable relaxation times (i.e. known eigenvalues) and the problem is indeterminate for **k**. Thus additional experimental information is required about the complex processes which couple the upper modes and the optical methods should be able to provide this.

Another technique, which is still in its infancy, is that of molecular beam scattering. This has yet to be applied to measuring vibrational inelastic cross-sections, though the method appears to be feasible providing that the inelastic cross-sections are reasonably large. There would be the added advantage that polar molecules, such as H_2O or NH_3, could be aligned in an electric field prior to collision so that the angular dependence of the inelastic cross-section could be studied, see for example the work of Toennies, Bennewitz and Kramer[266].

Molecular beams also provide a good method for studying intermolecular potentials; this has recently been reviewed by Pauly and Toennies[267]. Once again angular dependence could be studied rather than obtaining angularly independent values from transport data. One general point concerning the intermolecular potential which has emerged from comparing theoretical and experimental results is that, when there is agreement between theory and experiment, the value of the exponential repulsion parameter (α^*) is about 5–5·5 Å^{-1}. This applies to both diatomic and polyatomic molecules[55] and it should be remembered that there is no term for the intermolecular potential in Millikan and White's correlation (Equation 2.81). This is not to say that all molecules have the same intermolecular repulsive potential, rather that the values for the degree of steepness of the repulsive forces lie in a comparatively narrow range— indeed the value of α is, according to the SSH theory, a function of the collision energy. There is as yet no evidence that a particular group of atoms or molecules, e.g. peripheral hydrogen atoms, have steeper potentials than the rest. With more detailed intermolecular potentials available it should be possible to treat collisions as encounters between pairs of peripheral atoms rather than between two molecules.

There are many refinements which can be made to the theoretical calculations; several have already been discussed in section 2.2.6 and generally they will necessitate the use of numerical rather than analytical methods for solution. A full treatment is now required for the simultaneous interchange of vibrational, rotational and translational energy. Some such studies have already been made using a classical approach and the problem presented by polyatomic molecules needs to be tackled because many experimental results have now been obtained for molecules with small moments of inertia. So far theoretical predictions of temperature dependence for polyatomic molecules have been steeper than the measured values, even if some account is taken of rotation[264], yet there is agreement for diatomic molecules; a combined vibration–rotation–translation theory might account for this. Multi-quantum jumps are not yet well understood: as has been seen in Section 2.4, theoretical methods such as the SSH one appear to underestimate their importance and Sharp and Rapp's method of calculation needs to be applied to a case where a direct comparison can be made with experiment†. Molecules which exhibit double relaxation would give a good test of such calculations, as would recent studies of double quantum jumps in iodine[268]. So far complex processes have been assumed to involve only two modes, yet with say CCl_4 the top mode (776 cm^{-1}) could convert one of its quanta into two quanta, one drawn from each of the two lower vibrations (314 and 458 cm^{-1}). Such a

† Note added in proof: Thiele and Weare[291] have just published a paper dealing with higher order terms of the distorted treatment for n quanta transitions.

process would be near resonant but might involve an unfavourable steric factor. With multi-quantum jumps the effects of the anharmonicity of the molecular vibrations become more important. A wealth of information concerning molecular vibrations, including the degrees of anharmonicity, has been obtained spectroscopically, yet this is only beginning to be drawn upon[164,168]†.

The emphasis of this review has been the comparison between theory and experiment in order to try to indicate the areas where there is disagreement or, say, a lack of experimental results to test a theoretical prediction (or *vice versa*). The present overall state of agreement can only be considered to be fair. The major factors influencing the exchange of vibrational energy for simple molecules are understood but much refinement is necessary. Only by perturbing molecules can one investigate their nature; this review has been concerned with the perturbation of molecular vibrations at collisions in the gaseous state. The extent of agreement between theory and experiment for such processes is one measure of the state of knowledge of molecular properties.

Acknowledgements

The author would like to thank Dr D. G. Jones for many valuable discussions and suggestions and the many authors who allowed him access to their results prior to publication. He also wishes to acknowledge the support of the Central Electricity Generating Board for the period during which the major part of his research in this field was carried out.

REFERENCES

(As far as possible the literature has been surveyed up to the end of June 1967)
1. S. S. Penner, *Quantitative Molecular Spectroscopy and Gas Emissivities*, Pergamon Press, London, 1959.
2. W. M. Madigosky, *J. Chem. Phys.*, **39**, 2704 (1963).
3. K. F. Herzfeld and T. A. Litovitz, *Absorption and Dispersion of Ultrasonic Waves*, Academic Press, London, 1959.
4. T. L. Cottrell and J. C. McCoubrey, *Molecular Energy Transfer in Gases*, Butterworths, London, 1961.
5. H. O. Kneser, 'Relaxation Processes in Gases', in *Physical Acoustics* (Ed. W. P. Mason), Volume 2, Part A, Academic Press, London, 1965, Chapter 3, pp. 133–202.

† Note added in proof: the vibrational relaxation of anharmonic oscillators has recently been discussed by Treanor, Rich and Rehm[292].

6. A. W. Read, 'Vibrational Relaxation in Gases', in *Progress in Reaction Kinetics* (Ed. G. Porter). Volume 3, Pergamon Press, Oxford, 1965, Chapter 5, pp. 203–236.
7. R. C. Millikan, 'Relaxation Processes for Vibrational Energy in Gases', in *Molecular Relaxation Processes* (Chemical Society Special Publication No. 20) Chemical Society and Academic Press, London, 1966, pp. 219–234.
8. K. Takayanagi, *Advan. At. Mol. Phys.*, **1**, 149 (1965).
9. J. F. Clarke and M. McChesney, *The Dynamics of Real Gases*, Butterworths, London, 1964, Chapter 7, pp. 275–414.
10. K. F. Herzfeld, 'Theories of Relaxation Times', in *Dispersion and Absorption of Sound by Molecular Processes* (Ed. D. Sette), Academic Press, London, 1963, pp. 272–324.
11. K. Takayanagi, *Progr. Theoret. Phys.* (*Kyoto*), Suppl. **25**, 1 (1963).
12. S. W. Benson, G. C. Berend and J. C. Wu, *J. Chem. Phys.*, **37**, 1386 (1962).
13. S. W. Benson, G. C. Berend and J. C. Wu, *J. Chem. Phys.*, **38**, 25 (1963).
14. S. W. Benson and G. C. Berend, *J. Chem. Phys.*, **39**, 2777 (1963).
15. S. W. Benson and G. C. Berend, *J. Chem. Phys.*, **44**, 470 (1966).
16. J. D. Kelley and M. Wolfsberg, *J. Chem. Phys.*, **44**, 324 (1966).
17. J. G. Parker, *Phys. Fluids*, **2**, 449 (1959).
18. D. Rapp, *J. Chem. Phys.*, **32**, 735 (1960).
19. J. C. Slater and N. H. Frank, *Mechanics*, McGraw-Hill, London, 1947, Chapter 2.
20. R. E. Turner and D. Rapp, *J. Chem. Phys.*, **35**, 1076 (1961).
21. D. Rapp, *J. Chem. Phys.*, **40**, 2813 (1964).
22. D. Rapp and T. E. Sharp, *J. Chem. Phys.*, **38**, 2641 (1963).
23. C. Zener, *Proc. Cambridge Phil. Soc.*, **29**, 136 (1932).
24. F. H. Mies, *J. Chem. Phys.*, **41**, 903 (1964).
25. J. D. Kelley and M. Wolfsberg, *J. Chem. Phys.*, **45**, 3881 (1966).
26. T. E. Sharp and D. Rapp, *J. Chem. Phys.*, **43**, 1233 (1965).
27. D. Rapp and T. E. Sharp, *Symp. Combust.*, *11th*, *Pittsburgh*, *1967*, pp. 77–83.
28. R. J. Cross, *J. Chem. Phys.*, **47**, 3724 (1967).
29. J. M. Jackson and N. F. Mott, *Proc. Roy. Soc.* (*London*), Ser. *A*, **137**, 703 (1932).
30. W. J. Witteman, *J. Chem. Phys.*, **35**, 1 (1961).
31. W. J. Witteman, *Doctorate Thesis*, Technische Hogeschool, Eindhoven, 1963.
32. N. F. Mott and H. S. W. Massey, *The Theory of Atomic Collisions*, 3rd ed., Oxford University Press, 1965.
33. C. E. Treanor, *J. Chem. Phys.*, **43**, 532 (1965) and **44**, 2220 (1966).
34. M. S. Bartlett and J. E. Moyal, *Proc. Cambridge Phil. Soc.*, **45**, 545 (1949).
35. E. H. Kerner, *Can. J. Phys.*, **36**, 371 (1958).
36. R. T. Allen and P. Feuer, *J. Chem. Phys.*, **40**, 2810 (1964).
37. F. Mandl, *Quantum Mechanics*, 2nd ed., Butterworths, London, 1957.
38. F. W. de Wette and Z. I. Slawsky, *Physica*, **20**, 1169 (1954).
39. K. Takayanagi, *Progr. Theoret. Phys.* (Kyoto), **8**, 111 and 497 (1952).
40. K. Takayanagi and T. Kishimoto, *Progr. Theoret. Phys.* (*Kyoto*), **9**, 578 (1953); errata, **10**, 369 (1953).

41. K. Takayanagi, *J. Phys. Soc. Japan*, **14**, 75 (1959).
42. K. Takayanagi, *Sci. Rep. Saitama Univ.*, *Ser. A*, **III, No. 2**, 65 (1959).
43. M. Salkoff and E. Bauer, *J. Chem. Phys.*, **29**, 26 (1958).
44. M. Salkoff and E. Bauer, *J. Chem. Phys.*, **30**, 1614 (1959).
45. R. N. Schwartz and K. F. Herzfeld, *J. Chem. Phys.*, **22**, 767 (1954).
46. S. W. Benson and G. C. Berend, *J. Chem. Phys.*, **44**, 4247 (1966).
47. J. D. Kelley and M. Wolfsberg, *J. Phys. Chem.*, **71**, 2373 (1967).
48. H. C. Longuet-Higgins, *Discussions Faraday Soc.*, **40**, 7 (1965).
49. L. Monchick and E. A. Mason, *J. Chem. Phys.*, **35**, 1676 (1961).
50. J. O. Hirschfelder, C. F. Curtiss and R. B. Bird, *Molecular Theory of Gases and Liquids*, Wiley, London, 1954.
51. P. G. Dickens and A. Ripamonti, *Trans. Faraday Soc.*, **57**, 735 (1961).
52. A. F. Devonshire, *Proc. Roy. Soc. (London)*, *Ser. A*, **158**, 269 (1936).
53. P. G. Dickens and O. Sovers, private communication.
54. D. G. Jones, private communication.
55. J. L. Stretton, *Trans. Faraday Soc.*, **61**, 1053 (1965).
56. D. Schofield, *D. Phil. Thesis*, Oxford University, 1960.
57. R. N. Schwartz, Z. I. Slawsky and K. F. Herzfeld, *J. Chem. Phys.*, **20**, 1591 (1952).
58. F. I. Tanczos, *J. Chem. Phys.*, **25**, 439 (1956).
59. L. Landau and E. Teller, *Z. Phys. Sowjetunion*, **10**, 34 (1936).
60. E. B. Wilson, J. C. Decius and P. C. Cross, *Molecular Vibrations*, McGraw-Hill, London, 1955.
61. K. F. Herzfeld, *Discussions Faraday Soc.*, **33**, 86 (1962).
62. J. L. Stretton, *D. Phil. Thesis*, Oxford University, 1964.
63. E. Bauer and F. W. Cummings, *J. Chem. Phys.*, **36**, 618 (1962).
64. F. H. Mies, *J. Chem. Phys.*, **40**, 523 (1964).
65. F. H. Mies, *J. Chem. Phys.*, **42**, 2709 (1965).
66. M. Krauss and F. H. Mies, *J. Chem. Phys.*, **42**, 2703 (1965).
67. A. K. Jameson, *Dissertation Abstr.*, **24**, 3555 (1964).
68. D. Rapp and P. Englander-Golden, *J. Chem. Phys.*, **40**, 573, (1964); erratum, **40**, 3120 (1964).
   D. Rapp, *J. Chem. Phys.*, **43**, 316 (1965).
69. R. Marriott, *Proc. Phys. Soc. (London)*, **83**, 159 (1964).
70. R. Marriott, *Proc. Phys. Soc. (London)*, **84**, 877 (1964).
71. R. Marriott, *Proc. Phys. Soc. (London)*, **86**, 1041 (1965).
72. R. Marriott, *Proc. Phys. Soc. (London)*, **88**, 83 (1966).
73. R. Marriott, *Proc. Phys. Soc. (London)*, **88**, 617 (1966).
74. D. Secrest and R. B. Johnson, *J. Chem. Phys.*, **45**, 4556 (1966).
75. E. B. Alterman and D. J. Wilson, *J. Chem. Phys.*, **42**, 1957 (1965).
76. K. Takayanagi, *Sci. Rep. Saitama Univ.*, *Ser. A*, **4**, 87 (1963).
77. I. Korobkin and Z. I. Slawsky, *J. Chem. Phys.*, **37**, 226 (1962).
78. e.g. D. W. Placzek, B. S. Rabinovitch, G. Z. Whitten and E. Tschuikow-Roux, *J. Chem. Phys.*, **43**, 4071 (1965).
79. E. E. Nikitin, *Opt. Spectr. (USSR) (English Transl.)*, **9**, 8 (1960).
80. K. L. Wray, *J. Chem. Phys.*, **36**, 2597 (1962).
81. B. Widom, *Discussions Faraday Soc.*, **33**, 37 (1962).

82. H. Shin, *Can. J. Chem.*, **42**, 2351 (1964).
83. H. K. Shin, *J. Chem. Phys.*, **41**, 2864 (1964).
84. H. K. Shin, *J. Chem. Phys.*, **42**, 59 (1964).
85. B. H. Mahan, *J. Chem. Phys.*, **46**, 98 (1967).
86. J. Meixner, *Acustica*, **2**, 101 (1952).
87. R. O. Davies and J. Lamb, *Quart. Rev. (London)*, **11**, 134 (1957).
88. J. Lamb, 'Thermal Relaxation in Liquids', in *Physical Acoustics* (Ed. W. P. Mason), Volume II, Part A, Academic Press, London, 1965, Chapter 4, pp. 203–280.
89. H.-J. Bauer, 'Phenomenological Theory of Multiple Relaxation Process', in *Dispersion and Absorption of Sound by Molecular Processes* (Ed. D. Sette), Academic Press, London, 1963, pp. 325–337.
90. H.-J. Bauer, 'Theory of Relaxation Phenomena in Gases', in *Physical Acoustics* (Ed. W. P. Mason), Volume 2, Part A, Academic Press, London, 1965, Chapter 2, pp. 47–131.
91. E. W. Montroll and K. E. Shuler, *Advan. Chem. Phys.*, **1**, 361 (1958).
92. K. E. Shuler and G. H. Weiss, *J. Chem. Phys.*, **45**, 1105 (1966).
93. A. B. Callear and I. W. M. Smith, *Trans. Faraday Soc.*, **59**, 1735 (1963).
94. W. Roth, *J. Chem. Phys.*, **34**, 2204 (1961).
95. K. F. Herzfeld, *Discussions Faraday Soc.*, **33**, 22 (1962).
96. J. D. Lambert, D. G. Parks-Smith and J. L. Stretton, *Proc. Roy. Soc. (London)*, *Ser. A*, **282**, 380 (1964).
97. J. Meixner, *Kolloid-Z.*, **134**, 3 (1953).
98. D. G. Jones, *D. Phil. Thesis*, Oxford University, 1967.
99. H.-J. Bauer, private communication.
100. H.-J. Bauer and H. Roesler, 'Relaxation of the Vibrational Degrees of Freedom in Binary Mixtures of Diatomic Gases' in *Molecular Relaxation Processes* (Chemical Society Special Publication No. 20) Chemical Society and Academic Press, London, 1966, pp. 245–252.
101. (a) R. J. Rubin and K. E. Shuler, *J. Chem. Phys.*, **25**, 59 (1956); (b) C. C. Rankin and J. C. Light, *J. Chem. Phys.*, **46**, 1305 (1967).
102. W. J. Witteman, *J. Chem. Phys.*, **37**, 655 (1962).
103. J. S. Rowlinson, *Mol. Phys.*, **4**, 317 (1961).
104. S. H. Bauer, *Ann. Rev. Phys. Chem.*, **16**, 245 (1965).
105. R. W. Higgs and R. H. Torborg, *J. Acoust. Soc. Am.*, **42**, 1038 (1967).
106. J. H. Kiefer and R. W. Lutz, *J. Chem. Phys.*, **44**, 658, 668 (1966).
107. J. B. Moreno, *Phys. Fluids*, **9**, 431 (1966).
108. J. H. Kiefer and R. W. Lutz, *J. Chem. Phys.*, **45**, 3888 (1966).
109. F. De Martini and J. Ducuing, *Phys. Rev. Letters*, **17**, 117 (1966).
110. J. G. Parker, *J. Chem. Phys.*, **45**, 3641 (1966).
111. J. G. Parker, *J. Chem. Phys.*, **41**, 1600 (1964).
112. J. B. Calvert and R. C. Amme, *J. Chem. Phys.*, **45**, 4710 (1966).
113. N. A. Generalov, *Dokl. Akad. Nauk SSSR*, **148**, 373 (1963).
114. D. R. White and R. C. Millikan, *J. Chem. Phys.*, **39**, 1803 (1963).
115. R. W. Lutz and J. H. Kiefer, *Phys. Fluids*, **9**, 1638 (1966).
116. F. D. Shields and K. P. Lee, *J. Acoust. Soc. Am.*, **35**, 251 (1963).
117. J. G. Parker and R. H. Swope, *J. Acoust. Soc. Am.*, **37**, 718 (1965).

118. M. C. Henderson, private communication to J. G. Parker (see pp. 1604 and 1605 of Ref. No. 111).
119. H. Roesler, *Acustica*, **17**, 73 (1966).
120. R. Holmes, F. A. Smith and W. Tempest, *Proc. Phys. Soc. (London)*, **81**, 311 (1963).
121. J. G. Parker, *J. Chem. Phys.*, **34**, 1763 (1961).
122. R. Holmes, F. A. Smith and W. Tempest, *Proc. Phys. Soc. (London)*, **83**, 769 (1964).
123. A. G. Gaydon and I. R. Hurle, *Symp. (International) Combust., 8th, Baltimore, 1962*, pp. 309–318.
124. D. L. Matthews, *J. Chem. Phys.*, **34**, 639 (1961).
125. W. J. Hooker and R. C. Millikan, *J. Chem. Phys.*, **38**, 214 (1963).
126. M. G. Ferguson and A. W. Read, *Trans. Faraday Soc.*, **61**, 1559 (1965).
127. L. Doyennette and L. Henry, *J. Phys. Radium*, **27**, 485 (1966).
128. I. R. Hurle, *J. Chem. Phys.*, **41**, 3911 (1964).
129. S. Tsuchiya, *Bull. Chem. Soc. Japan*, **37**, 828 (1964).
130. B. Cary, *Phys. Fluids*, **8**, 28 (1965).
131. S. J. Lukasik and J. E. Young, *J. Chem. Phys.*, **27**, 1149 (1957).
132. M. C. Henderson, *J. Acoust. Soc. Am.*, **34**, 349 (1962).
133. I. R. Hurle and A. L. Russo, *J. Chem. Phys.*, **43**, 4434 (1965); A. L. Russo, *J. Chem. Phys.*, **44**, 1305 (1966).
134. A. R. Blythe, T. L. Cottrell and A. W. Read, *Trans. Faraday Soc.*, **57**, 935 (1961).
135. D. D. Konowalow and J. O. Hirschfelder, *Phys. Fluids*, **4**, 629 (1961).
136. R. C. Millikan and D. R. White, *J. Chem. Phys.*, **39**, 3209 (1963).
137. F. D. Shields and K. P. Lee, *J. Chem. Phys.*, **40**, 737 (1964).
138. D. R. White, *J. Chem. Phys.*, **42**, 447 (1965); erratum, **44**, 430 (1966).
139. D. R. White, *J. Chem. Phys.*, **46**, 2016 (1967).
140. R. C. Millikan, *J. Chem. Phys.*, **40**, 2594 (1964).
141. R. C. Millikan, *J. Chem. Phys.*, **38**, 2855 (1963).
142. D. R. White, *J. Chem. Phys.*, **45**, 1257 (1966).
143. M. G. Ferguson and A. W. Read, *Trans. Faraday Soc.*, **63**, 61 (1967).
144. P. Borrell, 'Measurements of Vibrational Relaxation in Hydrogen Chloride and Bromide by Observation of the Infrared Emission from the Shock-heated Gases', in *Molecular Relaxation Processes* (Chemical Society Special Publication No. 20), Chemical Society and Academic Press, London, 1966, pp. 263–268.
145. C. C. Chow and E. F. Greene, *J. Chem. Phys.*, **43**, 324 (1965).
146. H.-J. Bauer and K.-F. Sahm, *J. Chem. Phys.*, **42**, 3400 (1965).
147. A. B. Callear and G. J. Williams, *Trans. Faraday Soc.*, **62**, 2030 (1966).
148. H. O. Kneser, H.-J. Bauer and H. Kosche, *J. Acoust. Soc. Am.*, **41**, 1029 (1967).
149. F. D. Shields, *J. Acoust. Soc. Am.*, **34**, 271 (1962).
150. F. D. Shields, *J. Acoust. Soc. Am.*, **32**, 180 (1960).
151. E. Sittig, *Acustica*, **10**, 81 (1960).
152. E. F. Smiley and E. H. Winkler, *J. Chem. Phys.*, **22**, 2018 (1954).
153. R. Schulze, *Ann. Physik*, **34**, 41 (1939).
154. A. Eucken and R. Becker, *Z. Physik. Chem.*, **27B**, 235 (1934).

155. H. O. Kneser and H. Roesler, *Acustica*, **9**, 224 (1959) and *Proc. Intern. Comm. Acoustics, 3rd, Stuttgart*, **1**, 515 (1959).
156. H.-J. Bauer and E. Liska, *Proc. Intern. Comm. Acoustics, 4th, Copenhagen*, (1962).
157. T. L. Cottrell, I. M. Macfarlane, A. W. Read and A. H. Young, *Trans. Faraday Soc.*, **62**, 2655 (1966).
158. P. V. Slobodskaia and E. S. Gasilevich, *Opt. Spectr. (USSR) (English Transl.)*, **7**, 58 (1959).
159. M. E. Delany, *Ph.D. Thesis*, University of London, 1958.
160. T. L. Cottrell, I. M. Macfarlane and A. W. Read, *Trans. Faraday Soc.*, **63**, 2093 (1967).
161. P. V. Slobodskaia, *Opt. Spectr. (USSR) (English Transl.)*, **22**, 14, 120 (1967).
162. J. T. Houghton, *Proc. Phys. Soc. (London)*, **91**, 439 (1967).
163. L. O. Hocker, M. A. Kovacs, C. K. Rhodes, G. W. Flynn and A. Javan, *Phys. Rev. Letters*, **17**, 233 (1966).
164. C. B. Moore, R. E. Wood, B.-L. Hu and J. T. Yardley, *J. Chem. Phys.*, **46**, 4222 (1967).
165. C. J. S. M. Simpson, K. B. Bridgman and T. R. D. Chandler, *J. Chem. Phys.*, **49**, 513 (1968).
166. N. H. Johannesen, H. K. Zienkiewicz, P. A. Blythe and J. H. Gerrard, *J. Fluid Mech.*, **13**, 213 (1962).
167. F. D. Shields, *J. Acoust. Soc. Am.*, **29**, 450 (1957) and **31**, 248 (1959).
168. J. T. Yardley and C. B. Moore, *J. Chem. Phys.*, **46**, 4491 (1967).
169. C. J. S. M. Simpson, K. B. Bridgman and T. R. D. Chandler, *J. Chem. Phys.*, **49**, 509 (1968).
170. R. Holmes, H. D. Parbrook and W. Tempest, *Acustica*, **10**, 155 (1960).
171. H.-J. Bauer and E. Liska, *Z. Physik*, **181**, 356 (1964).
172. T. L. Cottrell and M. A. Day, 'The Effect of Noble Gases on Vibrational Relaxation in Carbon Dioxide', in *Molecular Relaxation Processes* (Chemical Society Special Publication No. 20) Chemical Society and Academic Press, London, 1966, pp. 253–256.
173. B. J. Lavercombe, *Nature*, **211**, 63 (1966).
174. J. W. L. Lewis and F. D. Shields, *J. Acoust. Soc. Am.*, **41**, 100 (1967).
175. T. L. Cottrell, I. M. Macfarlane and A. W. Read, *Trans. Faraday Soc.*, **61**, 1632 (1965).
176. H. Roesler and K.-F. Sahm, *J. Acoust. Soc. Am.*, **37**, 386 (1965).
177. K. Yamada and Y. Fujii, *J. Acoust. Soc. Am.*, **39**, 250 (1966).
178. C. S. Tuesday and M. Boudart, *Tech. Note 7, Contract AF 35 (038)-23976*, Princeton University, 1955.
179. M. C. Henderson, A. V. Clark and P. R. Lintz, *J. Acoust. Soc. Am.*, **37**, 457 (1965).
180. D. G. Jones, J. D. Lambert and J. L. Stretton, *Proc. Phys. Soc. (London)*, **86**, 857 (1965).
181. Y. Fujii, R. B. Lindsay and K. Urushihara, *J. Acoust. Soc. Am.*, **35**, 961 (1963).
182. K. Geide, *Acustica*, **13**, 31 (1963).
183. J. D. Lambert and R. Salter, *Proc. Roy. Soc. (London), Ser. A*, **243**, 78 (1957).

184. P. G. Corran, J. D. Lambert, R. Salter and B. Warburton, *Proc. Roy. Soc.* (*London*), *Ser. A*, **244**, 212 (1958).
185. J. C. McCoubrey, R. C. Milward and A. R. Ubbelohde, *Proc. Roy. Soc.* (*London*), *Ser. A*, **264**, 299 (1961).
186. F. D. Shields, *J. Chem. Phys.*, **46**, 1063 (1967).
187. J. D. Lambert, G. A. H. Roberts, J. S. Rowlinson and V. J. Wilkinson, *Proc. Roy. Soc.* (*London*), *Ser. A*, **196**, 113 (1949).
188. D. Pemberton, *Part II Chemistry Thesis*, Oxford, 1959.
189. R. Bass and J. Lamb, *Proc. Roy. Soc.* (*London*), *Ser. A*, **243**, 94 (1957).
190. M. C. Henderson, *Proc. Intern. Comm. Acoustics, 3rd, Stuttgart*, 1959.
191. P. G. Dickens and J. W. .Linnett, *Proc. Roy. Soc.* (*London*), *Ser. A*, **243**, 84 (1957).
192. D. G. Parks-Smith, *D. Phil. Thesis*, Oxford University, 1964.
193. P. G. Dickens, *D. Phil. Thesis*, Oxford University, 1958.
194. P. G. Dickens, *Mol. Phys.*, **2**, 206 (1959).
195. J. G. Strauch, Jr. and J. C. Decius, *J. Chem. Phys.*, **44**, 3319 (1966).
196. T. L. Cottrell and A. J. Matheson, *Trans. Faraday Soc.*, **59**, 824 (1963).
197. T. L. Cottrell, R. C. Dobbie, J. McLain and A. W. Read, *Trans. Faraday Soc.*, **60**, 241 (1964).
198. P. D. Edmonds and J. Lamb, *Proc. Phys. Soc.* (*London*), **72**, 940 (1958).
199. T. L. Cottrell and P. E. Martin, *Trans. Faraday Soc.*, **53**, 1157 (1957).
200. T. L. Cottrell and A. J. Matheson, *Trans. Faraday Soc.*, **58**, 2336 (1962).
201. T. L. Cottrell and M. A. Day, *J. Chem. Phys.*, **43**, 1433 (1965).
202. A. W. Read, private communication.
203. J. C. Gravitt, C. N. Whetstone and R. T. Lagemann, *J. Chem. Phys.*, **44**, 70 (1966).
204. J. G. Parker and R. H. Swope, *J. Chem. Phys.*, **43**, 4427 (1965).
205. W. M. Madigosky, *J. Chem. Phys.*, **39**, 2704 (1963).
206. A. Eucken and S. Aybar, *Z. Physik Chem.*, **46B**, 195 (1940).
207. L. W. Richards and D. H. Sigafoos, *J. Chem. Phys.*, **43**, 492 (1965).
208. T. L. Cottrell and N. Ream, *Trans. Faraday Soc.*, **51**, 1453 (1955).
209. D. G. Jones and J. L. Stretton, unpublished work.
210. J. T. Yardley and C. B. Moore, *J. Chem. Phys.*, **45**, 1066 (1966).
211. T. L. Cottrell, T. F. Hunter and A. W. Read, *Proc. Chem. Soc.*, 1963, 272.
212. D. Sette, A. Busala and J. C. Hubbard, *J. Chem. Phys.*, **23**, 787 (1955).
213. J. L. Hunter and H. D. Dardy, *J. Chem. Phys.*, **42**, 2961 (1965).
214. G. R. Hanes, R. Turner and J. E. Piercy, *J. Acoust. Soc. Am.*, **38**, 1057 (1965).
215. N. A. Clark, C. E. Moeller, J. A. Bucaro and E. F. Carome, *J. Chem. Phys.*, **44**, 2528 (1966).
216. N. J. Meyer, *J. Chem. Phys.*, **33**, 487 (1960).
217. P. G. Dickens and D. Schofield, *J. Chem. Phys.*, **35**, 374 (1961).
218. J. L. Hunter and H. D. Dardy, *J. Chem. Phys.*, **44**, 3637 (1966).
219. T. D. Rossing and S. Legvold, *J. Chem. Phys.*, **23**, 1118 (1955).
220. H. Hinsch, *Acustica*, **11**, 230 (1961).
221. H. Hinsch, *Acustica*, **11**, 426 (1961).
222. E. U. Haebel, *Acustica*, **15**, 426 (1965).
223. J. R. Olsen and S. Legvold, *J. Chem. Phys.*, **39**, 3209 (1963).

224. E. B. Miller, R. R. Boade and S. Legvold, *J. Chem. Phys.*, **42**, 2982 (1965).
225. R. R. Boade, *J. Chem. Phys.*, **42**, 2788 (1965).
226. R. Holmes and W. Tempest, *Proc. Phys. Soc. (London)*, **78**, 1502 (1961).
227. G. H. Hudson, J. C. McCoubrey and A. R. Ubbelohde, *Proc. Roy. Soc. (London)*, *Ser. A*, **264**, 289 (1961).
228. R. Holmes, G. R. Jones and N. Pusat, *J. Chem. Phys.*, **41**, 2512 (1964).
229. J. D. Lambert and R. Salter, *Proc. Roy. Soc. (London)*, *Ser. A*, **253**, 277 (1959).
230. L. M. Valley and S. Legvold, *J. Chem. Phys.*, **36**, 481 (1962).
231. J. L. Wood, *Quart. Rev. (London)*, **17**, 362 (1963).
232. R. Holmes, G. R. Jones and R. Lawrence, *Trans. Faraday Soc.*, **62**, 46 (1966).
233. J. D. Lambert, A. J. Edwards, D. Pemberton and J. L. Stretton, *Discussions Faraday Soc.*, **33**, 61 (1962).
234. S. A. Losev and A. I. Osipov, *Soviet Phys. Usp. (English Transl.)*, **4**, 525 (1962).
235. J. D. Lambert, *Quart. Rev. (London)*, **21**, 67 (1967).
236. H.-J. Bauer, see discussion subsequent to ref. 100.
237. H.-J. Bauer, *Acustica*, **17**, 90 (1966).
238. M. C. Henderson and K. F. Herzfeld, *J. Acoust. Soc. Am.*, **37**, 986 (1965).
239. R. G. Harlow and M. E. Nolan, *J. Acoust. Soc. Am.*, **42**, 899 (1967).
240. J. B. Calvert, *J. Acoust. Soc. Am.*, **37**, 386 (1965).
241. C. M. Harris, *J. Acoust. Soc. Am.*, **35**, 11 (1963).
242. U. E. Schnaus, *J. Acoust. Soc. Am.*, **37**, 1 (1965).
243. D. R. White, *J. Chem. Phys.*, **42**, 2028 (1965).
244. T. L. Cottrell, private communication.
245. J. W. L. Lewis and K. P. Lee, *J. Acoust. Soc. Am.*, **38**, 813 (1965).
246. R. W. Higgs and R. H. Toborg, *J. Acoust. Soc. Am.*, **40**, 1269 (1966).
247. W. J. Witteman, *Phys. Rev. Letters*, **18**, 125 (1965).
     W. J. Witteman and H. W. Werner, *Phys. Rev. Letters*, **A26**, 454 (1968).
248. B. Widom and S. H. Bauer, *J. Chem. Phys.*, **21**, 1670 (1953).
249. W. H. Pielemeir, H. L. Saxton and D. Telfair, *J. Chem. Phys.*, **8**, 106 (1940).
250. A. Eucken and E. Nümann, *Z. Physik Chem. (Leip.)*, **36B**, 163 (1937).
251. A. B. Callear, *Appl. Opt.*, Suppl. **No. 2**, 145 (1965).
252. A. B. Callear, *Discussions Faraday Soc.*, **33**, 28 (1962).
253. N. Basco, A. B. Callear and R. G. Norrish, *Proc. Roy. Soc. (London)*, *Ser. A*, **260**, 459 (1961).
254. R. C. Millikan, *J. Chem. Phys.*, **43**, 1439 (1965).
254a. R. C. Millikan and D. R. White, *J. Chem. Phys.*, **39**, 98 (1963).
255. J. D. Lambert, D. G. Parks-Smith and J. L. Stretton, unpublished work.
256. M. B. McElroy, J. L'Ecuyer and J. W. Chamberlain, *Astrophys. J.*, **141**, 1523 (1965).
257. D. G. Jones, J. D. Lambert and J. L. Stretton, *J. Chem. Phys.*, **43**, 4541 (1965).
258. R. C. Amme and S. Legvold, *J. Chem. Phys.*, **26**, 514 (1957).
259. T. Seshagiri Rao and E. Srinivasachari, *Nature*, **206**, 926 (1965).
260. R. R. Boade and S. Legvold, *J. Chem. Phys.*, **42**, 569 (1965).
261. J. B. Calvert and R. C. Amme, *J. Chem. Phys.*, **33**, 1260 (1960).
262. M. L. Wang, R. R. Boade and S. Legvold, *J. Chem. Phys.*, **41**, 3879 (1964).
263. R. C. Millikan and L. A. Osberg, *J. Chem. Phys.*, **41**, 2196 (1964).
264. C. B. Moore, *J. Chem. Phys.*, **43**, 2979 (1965).

265. T. G. Winter, *J. Chem. Phys.*, **38**, 2761 (1963).
266. J. P. Toennies, H. G. Bennewitz and K. H. Kramer, *Discussions Faraday Soc.*, **33**, 96 (1962).
267. H. Pauly and J. P. Toennies, *Advan. At. Mol. Phys.*, **1**, 195 (1965).
268. J. I. Steinfeld, *J. Chem. Phys.*, **46**, 4550 (1967).
269. B. J. Berne, J. Jortner and R. Gordon, *J. Chem. Phys.*, **47**, 1600 (1967).
270. P. Borrell, *Advan. Mol. Relaxation Processes*, **1**, 69 (1967).
271. B. Hartmann and Z. I. Slawsky, *J. Chem. Phys.*, **47**, 2491 (1967).
272. I. Oppenheim, K. E. Shuler and G. H. Weiss, *Advan. Mol. Relaxation Processes*, **1**, 13 (1967).
273. J. W. Rich and R. G. Rehm, *Symp. Combust.*, *11th*, Pittsburgh, *1967*, pp. 37–48.
274. J. H. Kiefer, *J. Chem. Phys.*, **48**, 2332 (1968).
275. J. P. Appleton, *J. Chem. Phys.*, **47**, 3231 (1967).
276. D. Weaner, J. F. Roach and W. R. Smith, *J. Chem. Phys.*, **47**, 3096 (1967).
277. K. F. Herzfeld, *J. Chem. Phys.*, **47**, 743 (1967).
278. H.-D. Hess and H. Hinsch, *Acustica*, **19**, 344 (1967/68).
279. A. F. Grigor and W. A. Steele, *J. Chem. Phys.*, **48**, 1032, 1038 (1968).
280. G. T. Hageseth, *J. Acoust. Soc. Am.*, **42**, 844 (1967).
281. K. A. Strong and R. M. Brugger, *J. Chem. Phys.*, **47**, 421 (1967).
282. S. Weiss and G. E. Leroi, *J. Chem. Phys.*, **48**, 962 (1968).
283. R. Holmes and M. A. Stott, *J. Sci. Instr.*, **44**, 136 (1967).
284. T. B. Hodkinson and A. M. North, *J. Chem. Soc. (A)*, **1968**, 885.
285. F. D. Shields and J. A. Burks, *J. Acoust. Soc. Am.*, **43**, 510 (1968).
286. J. T. Yardley and C. B. Moore, *J. Chem. Phys.*, **48**, 14 (1968).
287. R. L. Taylor, M. Camac and R. M. Feinberg, *Symp. Combust.*, *11th*, Pittsburgh, *1967*, pp. 49–65.
288. S. J. Colgan and B. P. Levitt, *Trans. Faraday Soc.*, **63**, 2898 (1967).
289. D. R. White, *J. Chem. Phys.*, **48**, 525 (1968).
290. H. K. Shin, *J. Chem. Phys.*, **47**, 3302 (1967).
291. E. Thiele and J. Weare, *J. Chem. Phys.*, **48**, 2324 (1968).
292. C. E. Treanor, J. N. Rich and R. G. Rehm, *J. Chem. Phys.*, **48**, 1789 (1968).

3

The Measurement of Vibrational Relaxation Rates by Shock Techniques

Peter Borrell

3.1 INTRODUCTION

3.1.1 Scope of the Chapter

In the past twenty years the study of the acquisition, loss and exchange of vibrational energy by and between molecules has attracted much attention. The reasons for the attention are: (i) The processes are fairly amenable to study, taking place often in a time scale of 0.1 μs to 1 ms under reasonable experimental conditions. (ii) The theoretical aspects of the problem are also amenable to approximate treatment and perhaps more important (and certainly more unusual) the results from theoretical studies in this area are not only understandable but are also often readily applicable to the results of experiments with which, in many cases, they are found to agree tolerably well. The ready interplay here between theory and experiment provides an encouraging stimulus to workers on both sides. (iii) There is a ready conceptual connection between the acquisition of vibrational energy and the dissociation of the molecules; clearly one can hope to approach the more difficult theoretical problems of dissociation through an understanding of vibrational-energy transfer. In this context, although the conceptual connection is clear, there is a lack of convincing experimental demonstrations between vibrational-energy content and dissociation rate (but see for example Bauer and his coworkers[1,2,3,4,5], and Rabinowitch and his coworkers[6,7,8]). (iv) Last but not least, the initial stimulus for the work was the necessity to measure the rates of vibrational-energy acquisition for the air gases so that some reliable numbers were available for use in solutions of the technological problems of supersonic and hypersonic flight.

In these studies of vibrational relaxation, the shock tube and related techniques have played an important part. This chapter is devoted to describing how vibrational-energy exchange rates may be measured by shock techniques. This is done first by a simple description of a conventional shock tube (Section 3.2.1), by a brief comparative account of other methods of measurement (Section 3.2), by giving a more detailed description of the theory (Section 3.3) and application of the shock-tube (Section 3.4) and finally by reviewing some of the results obtained from shock-tube experiments (Section 3.5 and 3.6).

3.1.2 Some terminology and conventions: Napier Time, Vibrational States and Levels, Relaxation

A linear polyatomic molecule with n atoms has $3n - 5$ vibrational degrees of freedom, a non-linear molecule has $3n - 6$. The vibrational energy contained by a molecule is quantized and each vibration can have an energy $(v + \frac{1}{2})hv$ where v is a quantum number, h is Planck's constant and v is the vibrational frequency. The formula, which is in fact that for an harmonic oscillator, is only an approximate description for the lowest vibrational states of a molecule. For higher vibrational states the energy spacing between states becomes progressively smaller than hv, an effect attributed to anharmonicity of the molecular oscillator. For diatomic molecules with only one vibrational degree of freedom the terms *state* and *level* are used interchangeably to describe each quantum state. The state $v = 0$ is sometimes referred to as the ground vibrational state, while that for $v = 1$ is known as the first vibrational state and so forth. In polyatomic molecules, depending upon the symmetry, some vibrations are degenerate; that is their motion can be resolved into contributions from two or more vibrational states which have equal energies. In this case the terms *state* and *level* are technically not interchangeable since there can be several energy *states* at one energy *level*, but in experiments where there is no method of distinguishing between such states, the terms can still be used indiscriminately without ambiguity.

The population of an energy state, that is the number of molecules with the particular number of vibrational quanta corresponding to this state, is determined by the familiar Boltzmann distribution law. For a single gas at equilibrium the vibrational degrees of freedom are in equilibrium with the translational, rotational and electronic degrees of freedom.

Vibrational relaxation is observed when the populations of the vibrational states are perturbed from equilibrium and describes the return or *relaxation* to equilibrium with the remaining degrees of freedom. Perturbation may be achieved either by changing the population of an individual vibrational level by irradiation of the system with light, or by changing the effective temperature of the other degrees of freedom. Methods which use irradiation to change the population of individual levels are infrared fluorescence (Section 3.2.6), ultraviolet fluorescence (Section 3.2.7), Raman scattering (Section 3.2.8), the spectrophone (Section 3.2.4) and flash photolysis (Section 3.2.5). In ultrasonic (Section 3.2.3) and shock methods (Section 3.2.1), equilibrium is disturbed by changing the translational temperature.

The way in which the relaxation is followed depends upon the characteristics of the molecule itself and the method of perturbation. For heteronuclear diatomic molecules and for polyatomic molecules in which a vibration results in changes in the dipole moment, the population of individual states may be estimated by infrared spectroscopy. Raman spectroscopy (Section 3.2.8) has also been used, as has ultraviolet spectroscopy.

It is the estimation of the rates of vibrational relaxation which is the subject of the present chapter. For simple processes only, such as the relaxation of the population of the upper state of a two-level oscillator or the relaxation of the energy for an harmonic oscillator (Sections 3.5.2 and 3.5.3) immersed in heat baths the temperature of which has been changed initially but then remains constant, the relaxation may be characterized by a quantity τ known as the *relaxation time* defined by the equation.

$$\frac{dx(t)}{dt} = \frac{x(\infty) - x(t)}{\tau} \qquad (3.1)$$

Here $x(t)$ is the value of the physical quantity followed at time t, $x(\infty)$ is the equilibrium value. In many cases relaxation processes may be approximately described by Equation (3.1) and characterized by a relaxation time. However for other situations, such as the relaxation of the upper levels of an harmonic oscillator or the relaxation of a polyatomic molecule where there is transfer of energy between the vibrations, the relaxation cannot necessarily be described by Equation (3.1) or characterized by a single relaxation time.

The term *lifetime* is often encountered in discussion of excited states. From the nature of the processes by which deactivation of the excited states may occur, it is an average quantity and may be roughly defined as the time for the excess concentration of the excited species, in the absence of any further excited species being formed, to fall to $1/e$ of its initial value. In studies of vibrational relaxation, if no radiative process occurs, as for example with N_2, then the lifetime is related to the relaxation time. For molecules which are infrared active there is also a *radiative lifetime*, τ_R, the reciprocal of the Einstein radiative transition probability, which is the lifetime in the absence of any collisions. For most molecules at atmospheric pressure the relaxation time τ is less than τ_R, but CO is a notable exception[9,10,11] at room temperature. A knowledge of τ_R is often of help in determining relaxation times[9,10,11,12], (Sections 3.2.4 and 3.2.6). Similar considerations apply to the determination of relaxation times for excited electronic states where the radiative lifetimes for electronic transitions and also the possibility of the collisional quenching of the electronic state must be taken into account (Sections 3.2.5 and 3.2.7).

In gas-phase systems, relaxation, in which energy is transferred between degrees of freedom, occurs on collision between two molecules by a bimolecular process. Thus τ is pressure dependent. Relaxation times are therefore quoted for standard conditions of pressure or density. Henderson[13] has made an eloquent plea that the standard quantity should be known as the *Napier Time* in honour of John Napier, the 16th century Scots mathematician who invented naperian logarithms and the decimal point. The term *Napier Time* will be used in this chapter.

In a pure gas, bimolecular collisions occur between molecules of the same kind and therefore the Napier time for CO say, is characteristic of energy transfer between CO molecules which takes place for the first vibrational level by reactions such as:

$$CO(v = 0) + CO \underset{k_{10}}{\overset{k_{01}}{\rightleftharpoons}} CO(v = 1) + CO; \qquad \tau(CO(CO)) \qquad (3.2)$$

Here vibrational energy is being transferred to and from the first level of one molecule by collision with another molecule in an undefined state.

In gas mixtures collisions between molecules of different species take place and the efficiency is different for each. It is therefore necessary to specify which relaxation time is meant. For a mixture of N_2 and O_2 there are four characteristic Napier times: two of these correspond to reactions in which the vibrational levels of O_2 are populated by collision with either O_2 or N_2:

$$O_2(v = 0) + O_2 = O_2(v = 1) + O_2; \qquad \tau(O_2(O_2)) \qquad (3.3)$$

$$O_2(v = 0) + N_2 = O_2(v = 1) + N_2; \qquad \tau(O_2(N_2)) \qquad (3.4)$$

The other two characterize reactions in which N_2 acquires vibrational energy:

$$N_2(v = 0) + N_2 = N_2(v = 1) + N_2; \qquad \tau(N_2(N_2)) \qquad (3.5)$$

$$N_2(v = 0) + O_2 = N_2(v = 1) + O_2; \qquad \tau(O_2(N_2)) \qquad (3.6)$$

The Napier times $\tau(O_2(O_2))$ and $\tau(N_2(N_2))$ can be measured for the pure gases independently of the mixture; $\tau(O_2(N_2))$ and $\tau(N_2(O_2))$ are Napier times for the hypothetical situation in which, for $\tau(O_2(N_2))$ say, the relaxation of O_2 occurs in an infinitely dilute mixture of oxygen in nitrogen and energy is transferred to oxygen only by collision with nitrogen. These Napier times cannot be measured directly but can only be calculated by extrapolating the results from mixtures (Section 3.5.4). In general:

$$\tau(A(B)) \neq \tau(B(A)) \qquad (3.7)$$

For a reaction such as (3.2) the Napier time τ is related to the rate constants k_{01} and k_{10} for forward and backward reactions. The actual relation depends upon the model used (see Sections 3.5.2 and 3.5.3); for the Landau–Teller model:

$$\tau = (k_{10} - k_{01})^{-1} \qquad (3.8)$$

and for the two-state model:

$$\tau = (k_{10} + k_{01})^{-1} \qquad (3.9)$$

At low temperatures both may be approximated to

$$\tau = (k_{10})^{-1} \qquad (3.10)$$

so that the Napier time may be identified with the reciprocal of the rate constant for deactivation of the lowest vibrational state. In turn k_{10} expressed in units of collisions s^{-1} is related to the total number of collisions per second undergone by the average molecule, Z, through P^{1-0} the probability of transfer of vibrational energy per collision:

$$k_{10} = ZP^{1-0} \qquad (3.11)$$

Z may either be calculated from collision theory or estimated directly from viscosity measurements. Its reciprocal τ_c is the average time between collisions. The reciprocal of P^{1-0} is Z^{1-0} the number of collisions an average excited molecule undergoes before deactivation.

The quantities P^{1-0} and Z^{1-0} are conveniently known as the *Napier probability* and *Napier number* respectively[13].

A further term occasionally met with is the *cross-section*, σ, for deactivation: it is calculated from:

$$\sigma = P^{1-0}Z/(Nn(8\pi RT/\mu)^{1/2}) \tag{3.12}$$

where N and n are the number of molecules per cubic centimetre of the quenched (or deactivated) and quenching species and μ is the reduced mass of the colliding pair.

It is always assumed that the vibrational energy is transferred by bimolecular reactions (as in Equations 3.2 to 3.6) although this may not necessarily be true for transfer between modes of a polyatomic molecule. So, as has been mentioned, the measured relaxation time depends upon the density of the gas and it is necessary to specify it under standard conditions.

The majority of workers reduce their results to one atmosphere pressure so that the Napier time, τ, has the units: *atmosphere seconds.*

It is unfortunate that unit pressure rather than unit density is used since there is a clear relation (Equations 3.11 and 3.12) between τ and the density of the gas which, at unit pressure, varies with the temperature. Zienkiewicz, Johannesen and their coworkers[14,15,16] have recognized this in their work where the relaxation frequency, the reciprocal of the relaxation time, is recorded at unit density in amagat^{-1} seconds^{-1}.

Even using unit density there is, from the temperature dependence of Z, an automatic $T^{-1/2}$ dependence of τ on temperature quite apart from any variation with temperature of P^{1-0}. The best quantity to calculate would then seem to be P^{1-0} itself but it of course requires accurate values of Z as a function of temperature; these are not readily available for many gases or mixtures.

In this chapter we shall follow the majority and quote all Napier times in atmosphere seconds.

3.2 METHODS FOR STUDYING VIBRATIONAL RELAXATION

In this section a preliminary account of the shock method is presented together, for comparison, with a brief description of other methods employed to study vibrational-energy transfer.

3.2.1 The Simple Shock Tube

A simplified experimental arrangement is illustrated diagrammatically in Figure 3.1. A suitable shock tube would be cylindrical, 5 cm in diameter, and six metres long. The tube is divided by a thin diaphragm of metal or plastic into two sections: the driver or high-pressure section, one metre long, and the test section, five metres long. Tubes up to 60 cm in diameter have been employed[17] and the test and driver sections can be considerably lengthened. The tube diameter is one factor determining the signal obtained from the test gas since time-resolved observations are made across the tube. The length of both test and driver sections are determined by the running time required (Sections 3.3.9 and 3.3.10).

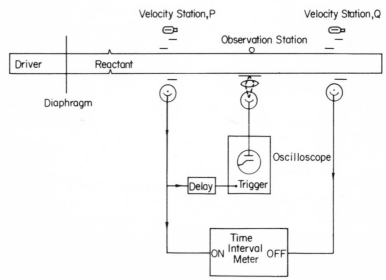

Figure 3.1 A simplified experimental arrangement for a shock tube. The stations P and Q represent light screens for detecting the passage of the shock wave, and by measuring the time difference between arrival at P and Q the velocity can be measured. The oscilloscope is triggered at a preset time to show the electrical output from the detector as the shock passes 0

To operate the tube the test gas is admitted to the test section to say 10 torr pressure. The driver gas, usually a light gas such as H_2 or He, is run into the driver section until the diaphragm bursts. The pressure at which this occurs depends upon the diaphragm material and thickness:

for 0·125 mm Al, the diaphragm would burst at about 6 atm differential pressure. For the conditions: H_2 driver gas (6 atm), HCl test gas (10 torr), a shock wave travelling at about 1·8 mm μs^{-1} (1.8 km s^{-1}) would be propagated into the HCl. The conditions behind the shock front in the shock heated HCl corresponding to this velocity would be: pressure $\sim 0·5$ atm and temperature $\sim 2000°K$.

The formation of a shock wave by bursting the diaphragm may be understood if the driver gas is regarded as a simple piston moving by infinitesimal increments into the test gas, propagating pressure pulses as it progresses. The first pressure pulse moves into the test gas at the speed of sound in the gas, heating and compressing the gas slightly. The second pulse also moves at the speed of sound but now into slightly heated gas. So for the third which moves into hotter gas still and so on. Since the sound speed of a gas increases with the temperature each pressure pulse is moving faster than its predecessor. Thus they all coalesce forming a single sharp front which is the shock wave.

The use of the shock tube in studies of vibrational-energy acquisition is evident: the rise in temperature at the shock front takes place for the rotational and translational degrees of freedom within 10^{-8} s. The molecule then obtains its vibrational energy by exchange with these other degrees of freedom and provided the time for this is longer than 10^{-8} s there will be a relaxation zone behind the shock front where a study of a suitable physical characteristic of the gas will yield information on the relaxation.

The actual conditions of the experiment, temperature, pressure and density, are determined by the velocity of the shock wave and can be calculated from it (Sections 3.3.5 and 3.3.6); it is necessary therefore to measure the velocity accurately. In Figure 3.1 the velocity is calculated from the time taken for the shock wave to pass from P to Q. In normal experiments several measurements would be made to check the constancy of the shock speed at different points in the tube. The calculations are performed assuming a one-dimensional flow within the tube; away from the diaphragm in a tube free from obstructions and protrusions this is a reasonable assumption. The adjustment of the vibrational degrees of freedom does however cause appreciable changes in the translational temperature and density of the gas which have to be accounted for in measurements of rate constants and relaxation times.

Observation of the relaxation is made by studying the emission or absorption of light by the gas or observing the density changes occurring.

Here there are the conflicting requirements of producing enough signal and having the longitudinal slits of the optical system sufficiently small to provide an adequate time resolution. The actual time resolution achievable varies from study to study; a figure of 5×10^{-7} s would be good resolution.

In brief then, the shock tube provides a method for studying relaxation temperatures from about 500 to 5000°K. The lower temperature is usually determined by the extent to which the upper vibrational states can be populated and observed. It provides in addition a method for looking at high vibrational states if they can be observed. The difficulties are the changing conditions of temperature and density as the relaxation proceeds and the relatively poor time resolution which matters more as the temperature is raised and Napier times shorten.

3.2.2   Ancillary Shock Methods: Expansions*

In a simple shock experiment the test gas is at low temperature and the passage of the shock raises the translational temperature so that the excitation of the vibrational degrees of freedom is followed. Another method is to make use of the properties of expansions and, having previously excited the vibrational modes, cool the gas translationally, and so follow the de-excitation of vibration which occurs. The difficulty here is that expansion waves are diffuse and not sharp like shock waves and thus the precise time origin which the passage of a shock provides is lacking for an expansion. Nevertheless various workers, particularly Hurle, Russo[18,19,20,21,22,23] (see also [24,25]) and their coworkers and Holbeche and Woodley[26,27] have used expansions to study vibrational de-excitation and moreover produced puzzling results which pinpoint our lack of understanding of the mechanism of these processes (Sections 3.5.8 and 3.6.4).

In the method used by Hurle, molecules in upper vibrational states are produced at the end of a normal shock tube by the reflected shock wave. In this part of the tube the shock is reflected back into the already heated gas, further raising its temperature (Section 3.3.10). The twice heated gas is normally stationary at the end of the tube, remaining so until the rarefaction or contact surface arrives (Section 3.3.1). The hot gas is expanded through a nozzle and the temperature measured at points downstream within the nozzle. Knowing the shape of the nozzle it is possible to

* See addenda, p. 264.

predict the way in which the temperature should vary with distance downstream for given relaxation times, and by comparison of prediction with experiment the rate of relaxation may be estimated.

In Holbeche's method[26,27], a second diaphragm is placed in the test section. The test gas is introduced into the section between the two diaphragms and the portion beyond the second diaphragm is evacuated. The shock is run in the usual way to excite the gas but the second diaphragm is then ruptured expanding the heated gas into the evacuated section and creating an unsteady expansion. As with the previous method the temperature can be predicted as a function of distance and fitting of the experimental temperatures to the theoretical curves gives the relaxation rates.

3.2.3 Ultrasonic Methods

The theory and practice of ultrasonic methods for studying vibrational relaxation are well described in Chapters 1 and 2 and by Cottrell and McCoubrey[28] and by Herzfeld and Litovitz[29]. The effect of the passage of a train of sound waves through a gas may be imagined as a nearly reversible, adiabatic compression and decompression of the gas at a particular frequency. At low frequencies all the degrees of freedom, translational, rotational and vibrational are able to keep in step, absorbing and losing energy in phase with the applied frequency. If, as the frequency is raised, it becomes impossible for a degree of freedom, a vibration say with a slow relaxation rate, to keep up it falls out of phase; energy is then lost from the sound wave and dissipated in the gas. At a still higher frequency there is insufficient time within the period of compression and decompression for energy to be gained or lost by the vibration: the vibration is then inactive and absorption of energy decreases once more. Now, however, because the inactive vibration no longer contributes to the heat capacity of the gas, the sound wave travels through the gas with a higher velocity.

So, measurement of either the velocity or the absorption of sound in a gas as a function of frequency can yield information on relaxation processes, and Napier times for vibrational, rotational and electronic relaxation have been measured.

One important condition for it to be possible to study the relaxation of a particular degree of freedom by ultrasonic methods is that the degree of freedom must contribute appreciably to the heat capacity at the temperature of measurement; in other words for a vibration, there must be an appreciable fraction of the molecules in the first vibrational level. Molecules with large energy spacings between their vibrational levels such as H_2 or HCl are excluded from study near room temperature. Another point is that in polyatomic molecules with many vibrations, it is only the lowest vibrational states which are generally studied; but this is scarcely an objection since for the few molecules which have been examined for transfer of energy between vibrations, the process occurs rapidly compared with gain or loss of energy from the weakest vibrational mode. It is then the rate constants for the weakest modes which are of primary importance for the transfer of vibrational energy to and from the molecule.

The majority of ultrasonic studies, because of the nature of the apparatus, are performed in the temperature range 270 to 550°K. This limits the method in comparison with the shock tube since it is clear from the studies made that the temperature dependence is of as great an interest as the absolute magnitude of the Napier time when comparing the results with those from theoretical treatments. On the other hand with ultrasonics it is possible to measure relaxation processes occurring in shorter times than can be measured with a shock tube.

3.2.4 The Spectrophone

In this device the test gas, which must absorb infrared radiation, is excited by a beam of infrared light of a suitable wavelength, chopped at known frequency. The energy is absorbed in one of the vibrational modes of the molecule and is lost from this either by relaxation, where it appears as translation energy in the gas, or by fluorescence. If we ignore losses by fluorescence, the chopped beam produces a series of pressure pulses in the gas which can be 'heard' by a suitable microphone. Due to the time taken for the relaxation there is a phase difference between the chopped beam and the sound output, and from this the rate of relaxation may be estimated[30]. There are various experimental problems both in the instrument itself due to unexpected acoustic resonances and in the phase detection but these have generally been overcome and the instrument has yielded useful results.

For molecules in which the relaxation time is comparable with the radiative lifetime, Read and Ferguson[10,12] have developed a simpler technique in which only the amplitude of the spectrophone signal is observed. For these molecules much of the energy is lost by fluorescence but introduction of foreign gases which increase the relaxation rate increases the spectrophone signal until all the energy is dissipated in the gas. By comparing the signals with and without foreign gas and knowing the radiative lifetime it is possible to estimate the Napier time.

Like ultrasonic apparatus the spectrophone can only be used over a limited temperature range and only infrared-active molecules may be studied. The frequency range available is smaller too.

However with the spectrophone not only can simple relaxation of a vibrational mode be studied, but transfer between modes can also be examined as it is possible to irradiate the system in separate experiments with different wavelengths so exciting various infrared-active vibrational modes[31]. Comparison of the rates of relaxation provides information on transfer between the modes. See also the addenda, p. 264.

3.2.5 Flash Photolysis

Vibrational excitation is achieved in this method by exposing the test gas to an intense flash of visible and ultraviolet light for a period of a few microseconds[32,33]. Absorption of the light populates electronically excited states in the absorbing molecule which may decompose, fluoresce or be deactivated. For suitable molecules, vibrational relaxation in both excited and ground electronic states can be studied by absorption or emission spectroscopy, either observing the system with a photomultiplier at a suitable wavelength or using the techniques of kinetic spectroscopy. Callear and his coworkers[34,35,36] have measured relaxation times for several excited states of NO in this way.

Flash photolysis can also be used to study vibrational relaxation in cases where the products from a chemical reaction initiated by the flash are vibrationally excited. Thus vibrationally excited products have been observed by Norrish and his coworkers[32] in a number of reactions. The difficulty here is that the initial distribution of vibrationally excited states is not known and the vibrational rates cannot be calculated. Bugrim and his coworkers have suggested a method for treatment of results in such cases[37].

Flash photolysis is not a general method for studying relaxation since whether vibrational states can be observed depends upon the chemical and spectroscopic properties of the individual systems. The temperature range is also limited, since to control the temperature of the experiment it is necessary to use the isothermal method in which the reactant is mixed with an excess of diluent inert gas; then most experiments are performed near room temperature.

3.2.6 Infrared Fluorescence

Millikan[11] used this method first for carbon monoxide but it should be applicable to any infrared-active gas where fluorescence can be observed, i.e. usually where the radiative lifetime is comparable to the relaxation time. In Millikan's experiment a stream of CO was continuously irradiated from a source of intense infrared radiation. The fluorescence from CO was observed as a function of pressure of added gases and by means of a Stern–Volmer plot, knowing the radiative lifetime, it was possible to calculate the rate of relaxation via CO-added gas collisions. The collisional deactivation of CO by itself could only be roughly estimated because the Napier time is, unusually, much larger than the radiative lifetime[10,11].

The technique has been extended by Yardley and Moore[38] to study energy transfer between vibrational modes in CH_4. One mode (v_3) was excited by a chopped He/Ne laser beam ($3.39\,\mu$) and the fluorescence from v_4 was observed at $8\,\mu$. The phase difference between input and output provided a measure of the transfer efficiency. Similarly Hocker and his coworkers[39] have observed the relaxation of the (001) state of CO_2 by stimulating the (100) → (001) transition with a Q-switched laser.

This method is clearly limited to infrared-active gases and as with the other methods the temperature range will be confined to near room temperature.

3.2.7 Ultraviolet Fluorescence

It is similarly possible to study vibrational relaxation in excited electronic states by observation of the fluorescence from individual vibrational levels. Klemperer and his coworkers[40,41] have demonstrated this elegantly for I_2 ($^2\Pi$ state) which was excited by irradiation with the sodium D lines and with the mercury green line. These lines excite particular levels of the upper state and, by studying the fluorescence from other levels as a function of the pressure of foreign gases, it was possible to determine cross-sections for rotational- and vibrational-energy transfer.

Here again studies have been confined to low temperatures and this method is clearly only applicable to molecules with discrete ultraviolet spectra.

3.2.8 A Raman Method

It is possible to produce an overpopulation of the first vibrational level by Raman scattering with a laser beam. Using a Q-switched laser De

Martini and Ducuing[42] irradiated hydrogen at 26 atm pressure. The vibrational population was followed by observing the anti-stokes Raman scattering with a C.W. laser and, for the intensities obtained, this could be studied as a function of time after the initial excitation by the Q-switched laser so yielding a relaxation time. This is clearly a general method but requires a high pressure of test gas to obtain a sufficient signal. The time resolution depends upon the sharpness of the initial pulse of radiation and should be less than a microsecond.

3.2.9   Summary

The principal advantage of the shock tube over the other methods which have been described is the large temperature range available for study. It is also, like the ultrasonic techniques, a general method for studying relaxation if one is content to study density changes only, which limits the results usually to the lowest vibrational modes and levels. To study higher vibrational levels and particular modes of vibration in polyatomic molecules then, as for several of the methods mentioned, the choice of molecules for study is limited by the physical characteristics, usually the spectroscopic properties, of the molecules themselves. In comparison with the ultrasonic and some fluorescence studies, the time resolution achievable with a shock tube is not as good which is unfortunate since relaxation times shorten at high temperatures.

3.3   ELEMENTARY SHOCK-TUBE THEORY

An account is given in this section of the wave patterns occurring in a simple shock tube and the elementary relationships from which the conditions of shock-tube experiments are determined. The content is clearly derived from previous authors[43,44,45,46,47,48,49] who give fuller derivations of some equations and a variety of other conditions which can be met with in shock-tube experiments.

3.3.1   The Wave Pattern in a Simple Shock Tube

The processes occurring after the rupture of the diaphragm in a shock tube such as that described in Section 3.2.1, may be illustrated by means of an x–t diagram, Figure 3.2; x is the distance along the shock tube and t the time elapsed after ideal bursting.

After the diaphragm has burst the shock wave (S) is propagated into

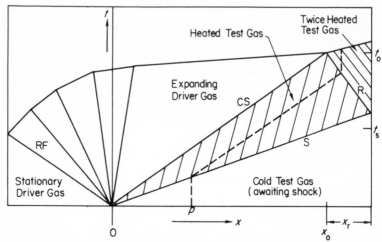

Figure 3.2 *x–t* diagram for a shock tube showing the position of the waves in the tube at progressive times after the diaphragm bursts. S is the incident shock, R the reflected shock, RF the rarefaction fan and CS the contact surface

the test gas with a velocity $(\mathrm{d}t/\mathrm{d}x)^{-1}$, and continues uniformly until it is reflected from the end plate. The reflected shock (R) has a lower velocity and moves into the heated gas, heating it a second time. Also propagated simultaneously from the diaphragm are the contact surface (C) and the rarefaction fan (RF). The contact surface, the junction between the shock-compressed test gas and the driver gas, moves with a velocity slower than that of the initial or incident shock but, except for very weak shocks, at a supersonic speed. Unlike the shock wave (Section 3.2.1) the system of waves forming the rarefaction fan cannot coalesce but spreads out; initially the head moves back into the driver at the sound speed of that gas. On reflection at the end plate, it continues to move at the sound speed but now it travels into gas which is already moving down the tube. The rarefaction head accelerates as it goes through the fan and its final velocity at the sound speed of the rapidly moving gas is faster than that of the incident shock wave (Section 3.3.4).

Subsequent events depend upon the length of tube. For that shown in the diagram the termination of the sequence begins with the simultaneous interaction of the contact surface, rarefaction and the reflected shock. In a longer tube the rarefaction would overtake the contact surface and meet the reflected shock. In a much longer tube the rarefaction would overtake the incident shock wave itself. All these interactions lead finally to the quenching of the high temperature and the cessation of movement in the tube, but the actual processes occurring after each individual interaction are complicated: the collision of two waves or a wave and the contact surface can lead to the formation of further waves, and whether these occur depends upon their speeds and the physical properties of the gases. The region beyond the interaction has therefore been left blank.

The paths of a particle in the test sections is shown by the point P. Gas at point P is accelerated by the shock and stopped by the reflected shock. The effect of the shocked test gas moving down the tube is to shorten the time available to study a given process, by reducing the time resolution of the measurement (Section 3.3.7).

The maximum observation time is given by the distance t_0 and depends upon the lengths of the two sections and the properties of the gases used (Sections 3.3.9 and 3.3.10).

The conditions obtaining across the shock wave can be calculated readily (Sections 3.3.5 and 3.3.6); those across the other waves are more complicated to derive but below approximate expressions are given to compute a maximum observation time (Section 3.3.9).

3.3.2 Coordinate Systems and Nomenclature

The natural coordinate system in use is that shown in Figure 3.3(a) where the shock wave is moving with velocity W_1 into stationary test gas.

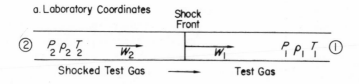

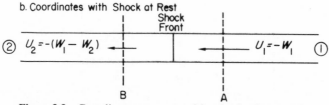

Figure 3.3 Coordinate systems used for the shock equations

The particle velocity (W_2) is the speed of the gas flow behind the shock front.

It is easier to formulate the equations in terms of shock-fixed co-ordinates, that is regarding the shock front as stationary. Then the flows are described by velocities u_1 and u_2 as shown in Figure 3.3(b).

The usual designation of the regions in a shock tube is shown in Figure 3.2. The regions are:

Rest test gas: 1

Shocked test gas: 2

Driver gas in motion: 3

Driver gas stationary: 4

The symbols p, ρ and T refer to pressure, density and absolute temperature respectively.

3.3.3   Conservation Equations: The Rankine–Hugoniot Relations

The basic equations relating the physical conditions before and behind the shock front are obtained by assuming that the overall flow is one-dimensional, that there is no interaction with the walls of the tube and thus that mass, momentum and energy are conserved in the gas after passage through the front.

The front itself does not enter into the calculation and the flow of these quantities is considered across two planes (A, B, Figure 3.3) on either side of the front. This has the advantage that for a relaxing gas it is possible to define two regions separated by the relaxation zone known as the frozen and equilibrium regions. The *frozen* gas is the shocked gas before vibrational relaxation has occurred (i.e. the vibrations are 'frozen'). The equilibrium region is where relaxation is complete (Figure 3.8). By performing separate calculations in which the plane behind the front (Figure 3.3b) is placed first before the relaxation zone and second behind the zone, it is possible to calculate the conditions in each region.

The relation for conservation of mass is reached by considering the mass crossing a unit area per second equal to ρu or

$$\rho_1 u_1 = \rho_2 u_2 = m \qquad \text{(dimensions: ml}^{-2}\text{t}^{-1}\text{)} \qquad (3.13)$$

For a shock wave $u_1 > u_2$ and $\rho_2 > \rho_1$ so that the shock is compressive.

The momentum per second crossing a unit plane is mu ($= \rho u^2$) and the change (decrease) in momentum is thus $\rho_1 u_1{}^2 - \rho_2 u_2{}^2$. For conservation the decrease of momentum is equal to an increase in pressure $p_2 - p_1$

or

$$p_1 + \rho_1 u_1{}^2 = p_2 + \rho_2 u_2{}^2 \qquad (\text{ml}^{-1}\text{t}^{-2}) \tag{3.14}$$

The kinetic energy per second crossing the unit plane is $\frac{1}{2}mv^2$ ($= \frac{1}{2}\rho u^3$) and the decrease in kinetic energy is $\frac{1}{2}(\rho_1 u_1{}^3 - \rho_2 u_2{}^3)$. For conservation this is equal to the increase in enthalpy of the gas, $(h_2 - h_1)$ where h is the specific enthalpy, the enthalpy per gram of substance ($h = H/M_w$). Thus

$$\tfrac{1}{2}u_1{}^2 + h_1 = \tfrac{1}{2}u_2{}^2 + h_2 \qquad (\text{l}^2\text{t}^{-2}) \tag{3.15}$$

Since $u_1 > u_2$, $h_2 > h_1$ and the shock heats the gas. Equations (3.13), (3.14), (3.15) are sometimes known as the Rankine–Hugoniot relations.

There are eight unknowns in these three equations: the enthalpy is related to the temperature by $h = f(T)$ and thus by specifying the initial conditions p_1, ρ_1, T_1, five are left. The equation of state for the gas relates three of these. Here the equation for a perfect gas will be assumed:

$$p = \rho r T \tag{3.16}$$

where r is a constant for a given gas $= (R/M_w)$. Three of the remaining variables may be specified in terms of one other, which is usually the shock velocity. Therefore by measuring the initial conditions p_1, ρ_1, T_1 and the shock velocity the remaining quantities may be determined.

The difference between the conditions obtaining in a frozen and a relaxed gas can be traced to h_2. In the frozen gas $h_2 - h_1$ is taken to be $c_p(T_2 - T_1)$. c_p is the gramme specific heat at constant pressure; it contains only contributions from the active degrees of freedom and is constant with temperature. The enthalpy of the relaxed gas contains contributions from the vibration and is not a simple function of temperature (Section 3.3.6).

It is possible then to solve the equations for an ideal gas, or to obtain the frozen conditions (Section 3.3.5); but for the equilibrium condition where vibrational relaxation is complete, $h = f(T)$ must be known for the individual gas and the equations are then solved by graphical or iterative methods (Section 3.3.6).

3.3.4 The Hugoniot Relation

Some further insight into the nature of the shock transition may be obtained by examining the Hugoniot for the gas.

Equation (3.14) may be rearranged and combined with Equation (3.13) to give

$$p_2 - p_1 = m(u_1 - u_2) \tag{3.17}$$

which indicates again that for $u_1 > u_2$ the shock is compressive. Using Equation (3.13) again the mass of gas flowing through the front is given by:

$$m^2 = (p_2 - p_1)/(1/\rho_1 - 1/\rho_2) \tag{3.18}$$

The enthalpy difference across the front is obtained from Equation (3.15)

$$h_2 - h_1 = \tfrac{1}{2}(u_1^2 - u_2^2) \tag{3.19}$$

which by using Equation (3.13) and substituting Equation (3.18) yields

$$h_2 - h_1 = \tfrac{1}{2}(p_2 - p_1)(1/\rho_1 + 1/\rho_2) \tag{3.20}$$

The difference in internal energy across the front may be obtained by using the relations

$$H = U + PV \tag{3.21}$$

or

$$h = e + p/\rho \tag{3.22}$$

where e, the specific internal energy, is U/M_w. From Equations (3.20) and (3.22) the difference is

$$e_2 - e_1 = \tfrac{1}{2}(p_2 + p_1)(1/\rho_1 - 1/\rho_2) \tag{3.23}$$

Equation (3.23) is known as the Hugoniot relation. For an ideal gas in which $e = c_v T$, and (3.16) applies, using the relation $\gamma = C_p/C_v$, Equation (3.23) may be rearranged to give:

$$p_2/p_1 = \left[\left(\frac{\gamma+1}{\gamma-1} \right) \frac{1}{\rho_1} - \frac{1}{\rho_2} \right] \Big/ \left[\left(\frac{\gamma+1}{\gamma-1} \right) \frac{1}{\rho_2} - \frac{1}{\rho_1} \right] \tag{3.24}$$

For given values of p_1, ρ_1 and γ, p_2 may be plotted as a function of $1/\rho_2$ yielding a curve known as the Hugoniot (or sometimes the Hugoniot adiabatic). A Hugoniot for 'frozen' HCl is shown in Figure 3.4 together with the familiar adiabatic curve for an ideal gas:

$$p_2/p_1 = (\rho_2/\rho_1)^\gamma \tag{3.25}$$

The Hugoniot differs from the adiabatic in that in shock compression there is an increase in entropy whereas the adiabatic is an isentropic curve and relates points which may be reached from each other by reversible

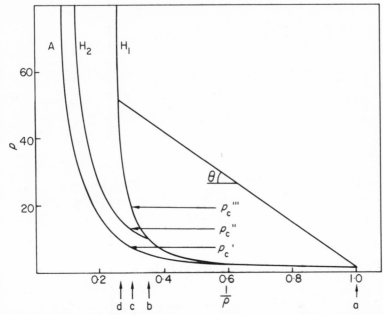

Figure 3.4 Comparison of Hugoniot and reversible adiabatic. A is the adiabatic and H_1 the Hugoniot (3.24). For the adiabatic, compression from a to b and then to c yields the same final pressure as compression from a to c (p'_c). For the Hugoniot compression from a yields p'''_c while compression from c occurs along the Hugoniot H_2 and yields p''_c

adiabatic processes. The Hugoniot on the other hand is only the locus of points which may be reached from $(p_1, 1/\rho_1)$ by shock waves of different velocities. The compression does not take place along the Hugoniot as the (hypothetical) reversible adiabatic compression would along the adiabatic. The difference may be illustrated by considering the compression of the gas from $1/\rho_a$ to $1/\rho_c$ in two separate experiments: in the first it is compressed directly; in the second in two stages by compression from $1/\rho_a$ to $1/\rho_b$ and then $1/\rho_b$ to $1/\rho_c$. For the adiabatic compression the pressure reached is p'_c which is the same for both experiments. Shock compression gives rise to two different final pressures p''_c and p'''_c since for compression from the point $(p_b, 1/\rho_b)$ another Hugoniot shown on the diagram exists on which $(p''_c, 1/\rho_c)$ is to be found.

For very weak shocks the adiabatic and Hugoniot coincide and have the same initial slope $(-\gamma p_1 \rho_1)$ and so a very weak shock is isentropic. This may also be seen from Equation (3.23) since for a weak shock where state 2 is close to state 1 the equation reduces to:

$$de = -p \, d(1/\rho) \tag{3.26}$$

Comparison with the thermodynamic equation for the first law:

$$de = T \, ds - p \, d(1/\rho) \tag{3.27}$$

shows that the limiting process is isentropic.

The shock velocity u_1 is given by m/ρ_1 (Equation 3.13) and substitution for m with (3.18) gives

$$u_1 = (1/\rho_1)[(p_2 - p_1)/(1/\rho_1 - 1/\rho_2)]^{1/2} \tag{3.28}$$

The velocity is therefore given by the slope of the chord linking the initial and final conditions on the Hugoniot for Figure 3.4 since

$$\tan \theta = (p_d - p_a)(1/\rho_a - 1/\rho_d) \tag{3.29}$$

$$u_1 = |\tan \theta|^{1/2}/\rho_a \tag{3.30}$$

These relations can be turned round so that the final conditions behind the shock may be determined for the initial conditions and shock speed, if the Hugoniot has been calculated.

For the limiting condition of a very weak isentropic shock Equation (3.28) reduces to

$$u_1 = -(1/\rho_1)[\partial p/\partial(1/\rho)]_s^{1/2} \tag{3.31}$$

or

$$u_1 = \left(\frac{\partial p}{\partial \rho}\right)_s^{1/2} = a_1 \tag{3.32}$$

which is the sound speed of the gas, a_1. Thus the limiting condition of a very weak shock wave is a sound wave, and the sound speed is directly related to the tangent to both Hugoniot and adiabatic at $p_1, 1/\rho_1$.

By inspection of Figure 3.4 it can be seen that the slope of the chord is always greater than that of the tangent so that the shock speed is always greater than the sound speed of the gas:

$$u_1 > a_1 \tag{3.33}$$

Points on the Hugoniot beyond $1/\rho_1$ ($> 1/\rho_1$) cannot represent any physical process since they correspond to decreases in entropy.

The entropy increase for a shock wave may be evaluated from the first law (3.27) since

$$T\Delta s = (e_2 - e_1) + \int_{1/\rho_1}^{1/\rho_2} p\, d(1/\rho) \tag{3.34}$$

The term $e_2 - e_1$ is given by Equation (3.23) and is the area under the chord between (p_1, ρ_1) and (p_2, ρ_2). The second term which is negative is the area under the Hugoniot. Therefore the entropy increase in the shock is the area enclosed by the Hugoniot and chord, divided by a suitably averaged temperature. Again, points $(> 1/\rho_1)$ give Δs to be negative. A more useful formula for the entropy increase is given by Equation (3.31).

The sound speed in the compressed gas, a_2, may be obtained from the tangent to the Hugoniot at p_2, $1/\rho_2$ as for Equation (3.28).

$$u_2 = m/\rho_2 = \frac{1}{\rho_2}\left[(p_2 - p_1)\Big/\left(\frac{1}{\rho_1} - \frac{1}{\rho_2}\right)\right]^{1/2} \tag{3.35}$$

In the limit $\rho_1 \to \rho_2$, $p_1 \to p_2$ then

$$u_2 = -(1/\rho_2)[\partial p/\partial(1/p)_s]^{1/2} = a_2 \tag{3.36}$$

the slope at $p_2 \cdot 1/\rho_2$ is greater than that of the chord and so

$$a_2 > u_1 \tag{3.37}$$

The velocity of the shock wave is always less than the sound speed of the hot gas and so any wave which follows the shock at the velocity of sound of the compressed gas ultimately overtakes the shock. This fact has already been mentioned for the rarefaction wave (Section 3.3.1).

3.3.5   Solution of the Conservation Equations for Ideal Gases

For gases which are ideal in the sense that they obey the equation of state Equation (3.16) and have a simple relation between e and c_v such as $e = c_v T$, the conservation Equations (3.13), (3.14), (3.15) may be solved explicitly in terms of a single variable if $p_1 \rho_1$ and T_1 are specified. The most convenient variable to choose is the pressure ratio p_2/p_1. The density ratio (ρ_2/ρ_1) is readily found by rearrangement of Equation (3.24), which was derived for an ideal gas:

$$\frac{\rho_2}{\rho_1} = \left[\left(\frac{\gamma+1}{\gamma-1}\right)\frac{p_2}{p_1} + 1\right]\Big/\left[\left(\frac{\gamma+1}{\gamma-1}\right) + \frac{p_2}{p_1}\right] \tag{3.38}$$

The ratio of temperatures (T_2/T_1) may now be obtained using Equation (3.16)

$$T_2/T_1 = (p_2/p_1)(\rho_1/\rho_2) \tag{3.39}$$

or

$$\frac{T_2}{T_1} = \left[\frac{p_2}{p_1}\right]\left[\left(\frac{\gamma+1}{\gamma-1}\right)+\frac{p_2}{p_1}\right]\Big/\left[\left(\frac{\gamma+1}{\gamma-1}\right)\frac{p_2}{p_1}+1\right] \tag{3.40}$$

The shock velocity u_1 is given by Equation (3.28). The density terms can be eliminated with Equation (3.38) to yield:

$$u_1^2 = (p_1/\rho_1)[(\gamma-1)/2]\{[(\gamma+1/\gamma-1)p_2/p_1]+1\} \tag{3.41}$$

Equation (3.41) may be simplified if u_1 is replaced by the dimensionless Mach number M, the ratio of the shock velocity to the sound speed in the test gas:

$$M = u_1/a_1 \tag{3.42}$$

where Equations (3.32 and 3.26)

$$a_1 = (\partial p/\partial \rho)_s^{1/2} = (\gamma rT)^{1/2} = (\gamma p/\rho)^{1/2} \tag{3.43}$$

The Mach number of the shock wave is then:

$$M^2 = \left[\frac{\gamma-1}{2\gamma}\right]\left[\left(\frac{\gamma+1}{\gamma-1}\right)\frac{p_2}{p_1}+1\right] \tag{3.44}$$

or

$$= \left[\left(\frac{\gamma+1}{\gamma-1}\right)\frac{p_2}{p_1}+1\right]\Big/\left[\left(\frac{\gamma+1}{\gamma-1}\right)+1\right] \tag{3.45}$$

With the aid of Equations (3.38), (3.40), (3.43) and (3.45) the various quantities may be found knowing the initial conditions p_1, ρ_1 and T_1 and either p_2, ρ_2 or M.

In Figure 3.5 the ratios (p_2/p_1), (ρ_2/ρ_1) and (T_2/T_1) are plotted as functions of Mach number for an ideal monatomic gas and in Figure 3.6, the ratios (ρ_2/ρ_1) and (T_2/T_1) are shown as a function of (p_2/p_1) together with similar plots for an adiabatic compression obtained from Equations (3.25) and (3.16).

The most striking feature of the plots is that although for a given compression a higher temperature is achieved with a shock wave than in a

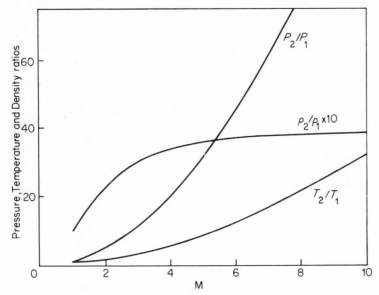

Figure 3.5   Shock ratios p_2/p_1, T_2/T_1 and ρ_2/ρ_1 as a function of Mach number for an ideal gas with $\gamma = 1\cdot67$

reversible adiabatic compression, the density ratio is much less and in fact for strong shock tends to a limit.

For strong shocks $[(p_2/p_1) \gg 1]$ the equations may be approximated to:

$$(3.38) \quad (\rho_2/\rho_1) = (\gamma+1)/(\gamma-1) \tag{3.46}$$

$$(3.40) \quad (T_2/T_1) = [(\gamma-1)/(\gamma+1)](p_2/p_1) \tag{3.47}$$

$$(3.44) \quad M^2 = [(\gamma+1)/2\gamma](p_2/p_1) \tag{3.48}$$

The smaller density ratio in the shocked gas than in the adiabatically compressed gas can be seen as a direct consequence of the real processes occurring in the shock rather than the hypothetical process of reversible compression.   The entropy change for the change $(p_1, \rho_1) \rightarrow (p_2, \rho_2)$ is given by

$$s_2 - s_1 = [r/(\gamma-1)] \ln[(p_2/p_1)(\rho_1/\rho_2)^\gamma] \tag{3.49}$$

For reversible adiabatic compression use of Equation (3.25) shows that $\Delta s = 0$. For a real process, in which there is no interaction outside the

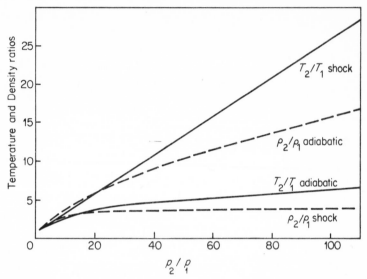

Figure 3.6 A comparison of the ratios T_2/T_1, ρ_2/ρ_1 from shock (full line) and adiabatic (dashed line) compression

system, $\Delta s > 0$, and thus from Equation (3.49) for a given (p_2/p_1),

$$(\rho_2/\rho_1)_{\text{shock}} < (\rho_2/\rho_1)_{\text{adiabatic}} \qquad (3.50)$$

and

$$(T_2/T_1)_{\text{shock}} > (T_2/T_1)_{\text{adiabatic}} \qquad (3.51)$$

The entropy change occurring across the front is obtained from Equations (3.38 and 3.49):

$$s = \left[\frac{r}{(\gamma-1)}\right] \ln \left[\frac{p_2}{p_1}\right] \left[\frac{\dfrac{\gamma+1}{\gamma-1}+\dfrac{p_2}{p_1}}{\dfrac{\gamma+1}{\gamma-1}\dfrac{p_2}{p_1}+1}\right]^{\gamma} \qquad (3.52)$$

Values of p_2/p_1, ρ_2/ρ_1 and T_2/T_1 for an ideal gas with $\gamma = 1.4$ and 1.67 have been tabulated as a function of Mach number by Schapiro[50] and Zucrow[51].

3.3.6 Solution of the Conservation Equations for Gases with Active Vibrations

For weak shocks in some diatomic and polyatomic gases, the equations in the previous section may be used to calculate the conditions behind the shock front because the vibrational modes are inactive at lower temperatures. In addition they may be used to calculate the conditions in the 'frozen' shocked gas, that is when the translational and rotational degrees of freedom are assumed to have relaxed perhaps together with the weaker vibrations in a polyatomic molecule. In these cases

$$h = [c_p \text{ (translation)} + c_p \text{ (rotation)} + c_p \text{ (weaker vibration)}] \, T \qquad (3.53)$$

and the procedure is as before.

If the temperature rise in the shock encompasses a region in which a vibration becomes active then

$$h = f(T) \qquad (3.54)$$

and it is not possible to solve Equations (3.13), (3.14) and (3.15) explicitly and the solution must be found by some iterative method.

Equations (3.20) and (3.28) can be combined to give

$$h_2 - h_1 = \frac{u_1^2}{2} \left[1 - \left(\frac{\rho_1}{\rho_2} \right)^2 \right] \qquad (3.55)$$

and Equations (3.13) and (3.18) to give:

$$\frac{p_2}{p_1} = 1 + \frac{u_1^2}{r T_1} \left(1 - \frac{\rho_1}{\rho_2} \right) \qquad (3.56)$$

If u_1 is expressed in $\mathrm{Km\,s^{-1}}$ ($\mathrm{mm\,\mu s^{-1}}$) and H in $\mathrm{cal\,mole^{-1}}$ then these become:

$$H_2 - H_1 = 119 \cdot 503 u_1^2 (M_w) \left[1 - \left(\frac{\rho_1}{\rho_2} \right)^2 \right] \qquad (3.57)$$

$$\frac{p_2}{p_1} = 1 + \frac{120 \cdot 274 (M_w) u_1^2}{T_1} \left[1 - \left(\frac{\rho_1}{\rho_2} \right) \right] \qquad (3.58)$$

To these can be added

$$\frac{p_2}{p_1} = \frac{\rho_2}{\rho_1} \frac{T_2}{T_1} \qquad (3.59)$$

A variety of methods have been suggested for the solution of these equations and these are fully dealt with by Bradley[43], Gaydon and Hurle[44], Greene and Toennies[46] and Glass[45]. Most of these require tables of values for

$$h = f(T) \qquad (3.60)$$

which are readily available[52,53,54].

Millikan[55,56] has suggested an iterative method, suitable for use with a computer for solving these equations. First the frozen conditions, $p_2(f)$, $\rho_2(f)$ and $T_2(f)$ are found for a given velocity and these are used to find ΔH, the enthalpy difference across the shock front from Equation (3.57). Another enthalpy difference $\Delta H''$ for the temperature change is computed using Equation (3.60). The difference $E = \Delta H'' - \Delta H'$ is the error in the calculation and a rough new temperature T' is determined from $T_2 = T(f)$ $-(E/C_p)$ where C_p is the computed heat capacity at the temperature. T_2 is then used to calculate ρ'_2 using $p_2(f)$ Equation (3.59) and then p'_2 is calculated from Equation (3.58). This completes the first iteration, and a new one is begun to recalculate E. A programme to effect this calculation has been published[55,56].

Campbell and Klimas[57] have recently suggested an analytical method for solving the equations.

The results of such a calculation for HCl are shown in Figure 3.7. The density and pressure are larger in the relaxed gas than the frozen gas, while the temperature is less. These lead to transitions throughout the relaxation zone such as those shown in Figure 3.8. Within the zone, the details of temperature, pressure and density clearly depend upon the rate and mechanism of relaxation.

3.3.7 Compression of the Observation Time

When a process occurring behind the shock wave is measured by making observations at a particular point in the tube, the time elapsed after the passage of the shock in the laboratory is shorter than the *particle time*, that is the time that has elapsed since the gas being observed passed through the shock wave. Figure 3.9 illustrates the effect. The particle time is the time for the gas being observed to reach the observer; during this time the shock has progressed the distance $w_1 t_p$ and the gas $w_2 t_p$. The laboratory time is the time since the shock passed the observer when it has progressed $w_1 t_1$

$$w_1 t_p = w_2 t_p + w_1 t_1 \qquad (3.61)$$

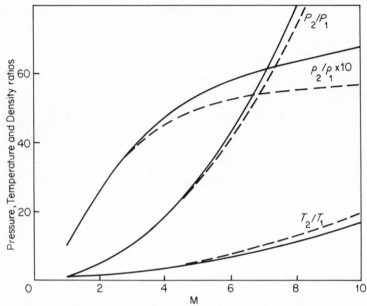

Figure 3.7　Ratios p_2/p_1, T_2/T_1, ρ_2/ρ_1 for HCl. The dashed lines are for the frozen conditions ($\gamma = 1\cdot40$) and the full line for equilibrium conditions. The initial temperature $T_1 = 300°K$. The calculations were for pure HCl with no dissociation

or

$$\frac{t_p}{t_1} = \frac{w_1}{w_1 - w_2} \tag{3.62}$$

For shock fixed coordinates (Section 3.3.2) and Equation (3.13),

$$\frac{t_p}{t_1} = \frac{u_1}{u_2} = \frac{\rho_2}{\rho_1} \tag{3.63}$$

If the time resolution of the observation is about one μs then the processes occurring must take place in longer than $\sim 6\ \mu$s for a diatomic gas. Mirels[58], and Fox, McLaren and Hobson[59] have shown that the compression may be greater than this factor in short shock tubes used at low pressures due to interaction of the contact surface with the boundary layer.

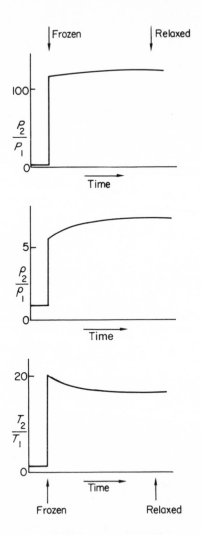

Figure 3.8 The variation of p, ρ and T through the relaxation zone from frozen to relaxed conditions. The ordinates are to scale for a Mach 10 shock in HCl

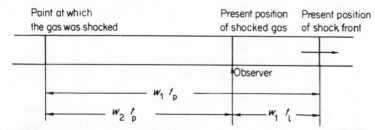

Figure 3.9   Diagram to show the compression of the observation time for measurements made in a fixed laboratory. Due to the motion of the gas behind the front the gas observed in the laboratory time has been shocked for a longer period

3.3.8   Shock Experiments in Real Gases

The equations derived have been for the incident shock in ideal gases which were able to relax. To test them experimentally requires independent measurements of the temperature, density, pressure and shock speed. Density measurements are the easiest to make (Sections 3.4.3, 3.4.4) but temperature measurements can be made by line reversal (Section 3.4.5) and other spectroscopic methods; it is possible to make reasonable estimates of the pressure with suitable transducers. To the accuracy of the experiments performed so far, the incident shock equations are found to be true and in most studies now, where independent checks are perhaps not possible, they are assumed to hold. Deviations can however be expected in a variety of circumstances: first if some chemical reaction such as dissociation occurs. This is not technically a deviation since the treatment can readily be modified to include chemical reaction[43,44,46] but clearly an unsuspected reaction in a relaxation study can lead to the calculation of erroneous reaction conditions as well as affecting the relaxation time.

A uniform velocity was assumed in deriving the equations. Shock waves are found to undergo some deceleration due to interaction with the walls of the tube which results in a modification of the conditions behind the shock.

The equations were derived for a plane shock wave. Bursting of the diaphragm initially produces a spherical front although this is found to become rapidly planar as the shock progresses down the tube. An acceleration has also been observed as the shock moves away from the diaphragm. It is therefore necessary to observe well down the tube away from the diaphragm. Observations are commonly made at distances from 50 to 150 tube diameters downstream from the diaphragm[45].

A further effect is due to the boundary layer which lies between the heated shocked gas and the cold wall of the tube. For low-pressure operation $(p_1 < 1$ torr) in small tubes $(d < 5$ cm) it has been observed that the boundary layer grows after the shock has passed and in some cases closure has been seen in which the whole tube cross-section has been filled with cooler gas. In any study therefore it is necessary to make test measurements to guard against these deviations. These effects have been discussed by Mirels[58] and Fox, McLaren and Hobson[59].

3.3.9   Miscellaneous Equations for Calculation of Shock-tube Parameters

It is possible to derive equations for the other conditions in the tube (section 3.3.1) such as the velocities of the contact surface and the head and tail of the rarefaction fan. Although these can be quite rigorous, they do assume ideal behaviour of the diaphragm and contact surface and unlike the equations for the incident shock are only poor approximations. For this reason and for reasons of space no derivation will be shown here but several results are quoted which are helpful in the design of shock tubes. Derivations may be found in references 43, 44, 45 and 46.

The pressure ratio across the diaphragm p_4/p_1 may be found in terms of the shock Mach number:

$$\frac{p_4}{p_1} = \left[\frac{2\gamma_1 M_1{}^2 - (\gamma_1 - 1)}{(\gamma + 1)}\right]\left[1 - \left(\frac{\gamma_4 - 1}{\gamma_1 + 1}\right)\left(\frac{a_1}{a_4}\right)\left(M_1 - \frac{1}{M_1}\right)\right]^{[(-2\gamma_4)/(\gamma_4 - 1)]}$$

(3.64)

Here γ_4 is the heat capacity of the driver gas and a_4 its sound speed. Due to lack of ideality in the diaphragm burst, the actual (p_4/p_1) required will be higher than that given by Equation (3.64) to achieve a desired shock velocity.

The maximum time for observation depends upon the shock-tube length and the properties of the gas used. Gaydon and Hurle[44] have derived equations to calculate these.

The observation time t_0 (Figure 3.2) at distance x_0 is determined by the difference in arrival times of shock, t_s and contact surface t_c. The quantities t_c, x_0 and t_0 $(= t_c - t_s)$ are given by:

$$t_c = \frac{2x_d}{a_4}\left[1 - \left(\frac{\gamma_4 - 1}{\gamma_1 + 1}\right)\left(\frac{a_1}{a_4}\right)\left(1 - \frac{1}{M^2}\right)\right]^{[(\gamma_4 + 1)/(2\gamma_4 - 1)]}$$

(3.65)

$$x_0 = t_c \left[\frac{2a_1(M^2-1)}{(\gamma+1)M} \right] \tag{3.66}$$

$$t_0 = t_c \left[1 - \frac{2(M_1{}^2-1)}{(\gamma+1)M^2} \right] \tag{3.67}$$

x_d is the length of the driver section. To use these: Equation (3.67) gives t_c for the desired observation time in terms of Mach number; the minimum driver and test lengths can then be calculated from Equations (3.65) and (3.66).

These relations derived for ideal tube behaviour can only be approximate and are subject to the deviations mentioned in Section 3.3.8. In particular the contact surface is not a discontinuity, and mixing of the driver and test gases has been observed which shortens the observation time.

Boundary layer closure must also shorten this time. The combined effects can result in up to 50% reduction of the time calculated.

3.3.10   The Reflected Shock

Equations for the reflected shock qualities have not been given, although they too may readily be derived, since the majority of relaxation studies are made in the incident shock region. The zone behind the reflected shock is doubly affected by deviations although many satisfactory chemical studies have been made using reflected shocks.

In designing a tube, if the observation station is too close to the end plate, then the observation time t_0 will be shortened. The minimum distance from the end of the tube to the observation station x_R (Figure 3.2) can be calculated (Gaydon and Hurle[44]) from:

$$x_R = \frac{(t_c + x_0/W_R)}{(1/W_s + 1/W_R)} \tag{3.68}$$

where W_s and W_R are the velocities of the incident and reflected shocks. The reflected shock velocity W_R is given by

$$\frac{W_R}{W_s} = \frac{\left[2 + \left(\frac{2\gamma}{\gamma-1} \right) \left(\frac{p_1}{p_2} \right) \right]}{\left(\frac{\gamma+1}{\gamma-1} \right) - \left(\frac{p_1}{p_2} \right)} \tag{3.69}$$

3.4   EXPERIMENTAL METHODS IN SHOCK TUBES

The major difference between various shock studies of relaxation is to be found in the detection system used. In this section the briefest account of shock-tube construction and measurement is given together with a fuller discussion of detection techniques.

3.4.1   Construction

The elementary details of the construction and operation of a simple shock tube were given in Section 3.2.1. Shock tubes differ in details such as diameter, dictated frequently by the optical path length across the tube necessary for observation or the desire to be free from wall effects at low pressures; length, decided by the running time desired; and construction material, chosen usually for chemical inertness to the reactant gas. There are also special methods for producing strong shocks using two diaphragms, changes in the area of the tube, combustion and electromagnetic drivers. These variations are well covered by the authors given in references 43, 44, 45 and 46.

3.4.2   Velocity Measurements

The key measurement which provides the conditions of the experiment is the velocity. What is required to record the passing of the shock is a device which responds rapidly, preferably within 0·1 μs, to the change in pressure, temperature or density at the shock front. Pressure transducers, thin-film platinum temperature strips and a variety of schlieren methods have been used to provide an electrical signal as a shock passes. For strong shocks electrical ionization probes have been used; for gases with the appropriate spectroscopic characteristics, such as bromine, the change in light absorption at the front gives a pulse from a photomultiplier, and in experiments where light is emitted observation of the emission provides the necessary pulse. For luminous gases, photographic techniques using drum or framing cameras are interesting but not general ways for measuring velocity (see references in Section 3.4.1).

Two devices feeding a time-interval meter allow the velocity to be estimated, but since a check should always be made on attenuation, it is necessary to have several velocity stations feeding several time-interval meters or a common input of an oscilloscope. The disadvantage of using an oscilloscope for velocity measurements is the poor accuracy compared with a time-interval meter, but the accuracy may be improved by the use of raster techniques or spiral traces (see[60]) which increase the length of the trace per sweep.

3.4.3   Techniques for Relaxation Measurements

As for velocity measurements, any technique which responds to the changes in density or temperature in the relaxation zone (Figure 3.8) may be used to follow the relaxation. The change in pressure is too small to be useful. In addition any absorption or emission of light, the intensity of which depends upon populations of individual vibrational levels, will also yield information on the Napier times. The principal techniques which have been used are interferometry, schlieren and electron-beam methods which respond to changes in density; the line-reversal technique which responds to changes in temperature and ultraviolet and infrared absorption and fluorescence with which populations of individual vibrational levels may be followed. These are described in the subsequent sections.

3.4.4   Interferometry

Interferometry was one of the earliest methods used to study density changes in the relaxation zone[16,61,62,63]. The most common arrangement

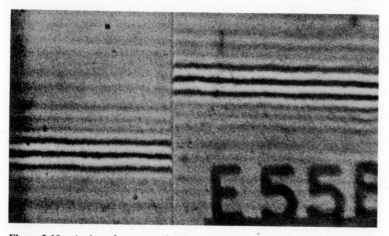

Figure 3.10   An interferogram of a shock into N_2. The shock wave is passing from right to left across the figure and vertical shift in the horizontal interference fringes gives a measure of the density change across the front. There is no relaxation in this figure since the temperature of the shocked gas is low. This figure was kindly supplied by Dr H. K. Zienkiewicz

used is a Mach-Zender interferometer[64]. Here a beam of monochromatic light from a spark source or flash lamp is split, one half passing through the shock-tube windows and the other through a compensator. The beams are recombined to produce interference fringes upon a photographic plate. Density changes within the shock tube cause a fringe shift. The light source which usually has a duration of 0·1 μs is flashed as the shock passes the windows so that a photograph of the density profile across the front is obtained. Figure 3.10 is an example. For the N_2 record the transition is sudden and there is no further change in the shocked gas at this temperature, for there is no appreciable population of the upper vibrational levels. For CO_2 the front is once again sharp as the translational and rotational degrees of freedom are excited, but the vibration relaxes more slowly taking place in the curving region behind the front. Analysis of the curvature yields the relaxation time.

There is an obvious difficulty of estimating the fringe shift at the shock front but this can be circumvented either by making separate experiments with a white light source, or by arranging to produce offset fringes where the total change may more readily be measured.

The method follows the changes in density and, since the contribution to the density change by a vibrational level is determined by the change in population of the level, it is usually the relaxation of the lowest vibrational levels and the weaker modes which are observed. However Zienkiewicz and Johannesen[16] have refined the technique so that they were able to observe the relaxation of all three vibrational modes of CO_2.

A new interferometric technique illustrated in Figure 3.12 is the laser interferometer developed by Besse and Kelley[65] and McChesney and Bristow[66]. Here the light beam is split, one half being passed across the shock tube and back, and then combined with the other half to produce fringes on a field, part of which is observed by a photomultiplier. A change in density in the tube causes the fringes to shift across the field and as they pass the photomultiplier aperture an electrical output is produced which is displayed on an oscilloscope. The fringe shift can readily be counted as pulses in the trace produced.

3.4.5 Schlieren and other Densitometry Techniques; The Theory of Density Changes

The schlieren technique is an optical method in which light from a source is passed through the shock tube and imaged on a photographic plate. In front of the plate is a knife edge positioned so that no light reaches the

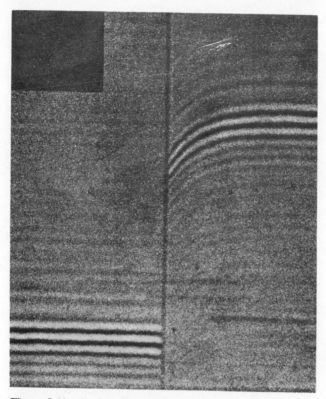

Figure 3.11   An interferogram of a shock in CO_2. As for Figure 3.10 the front is shown by the vertical shift in the horizontal fringes. The relaxation zone from the frozen to the equilibrium conditions is shown by the curved fringes. Analysis of the curvature gives the relaxation time. This figure was kindly supplied by Dr H. K. Zienkiewicz

photographic plate. A density change in the shock tube will deflect a little light over the knife edge on to the photographic plate. In this way photographs of the density changes may be made.

It is possible to arrange that a photomultiplier receives the light and thus obtain an electrical output, the amplitude of which is a function of the density change. Kiefer and Lutz[67,68,69] have studied the relaxation of H_2 and D_2 by this technique using a laser light source, and Witteman[70], Daen

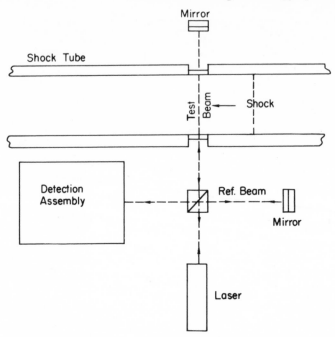

Figure 3.12 A diagram of a laser interferometer which provides a photomultiplier output. The reference and test beams are combined and produce interference fringes. The movement of the fringes can be estimated from the output. Reproduced by permission of the author and the Editor of *J. Chem. Phys.*

and de Boer[71] used it to examine CO_2. Moreno[72] used a broad-beam technique to measure D_2.

The treatment of density profiles in the relaxation zone to yield the relaxation time relies on Equation (3.2) where $x(t)$ is the total vibrational energy at time t (Section 3.5.2). There are nearly as many methods for reducing the data as authors who have studied density methods, but Blythe[73] has reviewed these and compared them with the 'exact' method developed by Johannesen[74]. The reader is referred to the summary*.

3.4.6   Line- and Band-reversal Techniques

The line-reversal technique was first devised by Gaydon, Hurle and their coworkers[44,75] to measure the temperature of shock-heated gases. The principle of the method is the same as that of the optical pyrometer: the

* See addenda, p. 264.

light from a continuous source with a characteristic temperature T_s is passed through the hot gas to which has been added a material with a suitable line emission spectrum, Na for example. The light is then viewed with a photomultiplier through an interference filter which passes only the wavelength of a chosen emission line (for Na say 5889 Å or 5895 Å). If the shocked gas has a temperature T_g the same as that of the source $T_g = T_s$, then the light will pass through as in the absence of any gas. If $T_g < T_s$ then some light from the source will be absorbed by the cooler gas and the signal from the photomultiplier attenuated. If $T_g > T_s$ then the additional emission from the hotter gas will enhance the output signal. Thus in a series of experiments the temperature of the shocked gas may be estimated by comparing the output for different source temperatures, and the variation of temperature may be followed as a function of time by displaying the output on an oscilloscope, which is triggered as the shock approaches the observation station.

The emitters most frequently used have been sodium and chromium which have been added as NaI and $Cr(CO)_6$. The NaI is added as a fine smoke in the gas.

An experimental arrangement used by Gaydon and Hurle is shown in Figure 3.13 where the double-beam method which they developed is shown. One beam operates as described, the reference temperature being that of the lamp. A neutral filter is placed in the other beam so that source temperature is effectively reduced. By comparison of the outputs the temperature may be determined more accurately.

The useful temperature range of the method is 2000°K to 6000°K, the lower limit being set by the emission available from the additive, and the upper temperature, the light sources currently available. The bottom limit could be lowered by using emitters in the infrared region of the spectrum.

The advantage of this method for relaxation studies is that the measured temperature follows the effective vibrational temperature of the shocked gas, and not the translational temperature. Thus for N_2 and CO it was found that there was a lag in the attainment of the equilibrium temperature which could be correlated with the vibrational relaxation of the gas, and thus the method could be used directly to study relaxation times. The results obtained generally agree well with those from other methods[76].

The mechanism which leads to the equality of vibrational temperature with the electronic temperature of the emitting species is not clear. The emission for Na is due to the transition $(3^2P \rightarrow 3^2S)$ and as Hurle[76] points out the energy differences for this transition and N_2 ($v = 7$ or $8 \rightarrow v = 0$)

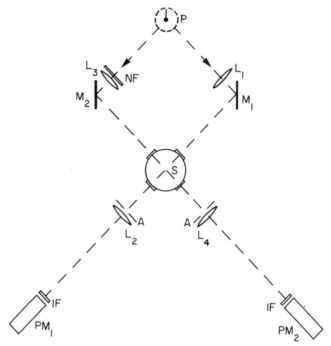

Figure 3.13 The experimental arrangement used for the sodium-
line reversal technique (after Gaydon and Hurle). Light from the
source (P) is passed through the shock tube (S) by two paths to the
two photomultipliers (PM). In the first path the same lamp provides
the reference temperature; in the second the reference temperature
is effectively reduced by interposition of the neutral filter (NF)

are nearly equal. The energy-exchange reaction:

$$N_2(v = 7) + Na(3^2S) \rightleftharpoons N_2(v = 0) + Na(3^2P) \qquad (3.70)$$

should thus be fast. The effectiveness of nitrogen in quenching the
fluorescence of Na[77,78,79] has also been explained by this exchange reaction in which the electronic energy of the sodium is transferred to the
vibration of N_2. However if the reaction were simply this then the emission
would be a function of the population of the seventh vibrational level of N_2
and the relaxation would not be expected to have the simple form (Section
3.1.2) which is actually observed.

Callear[80], generalizing on the results of the theory of Dickens, Linnett and Sovers[81] which suggested that in the transfer of electronic to vibrational energy only single quanta of vibrational energy may be gained or lost, has suggested a mechanism in which energy is transferred between the $v = 0$ and $v = 1$ states of N_2 and the sodium excited state. Generally:

$$N_2(v = 0, 1) + Na(^2S) = N_2(v = 0, 1) + Na(^2P) \tag{3.71}$$

With the aid of an elementary analysis Callear was able to show that the observed temperature should lie in between the translational and vibrational temperatures of the shock-heated gas. The idea does not however explain why it is specifically the vibrational temperature that is followed by the emission: in the suggested reactions a large amount of translational energy must be converted to electronic energy and this would seem to be the determining factor rather than the vibrational-energy content. Nor does the mechanism explain the higher quenching cross-section of N_2 for $Na(3^2P)$ than Ar for example: it seems insufficient to say that N_2 is chemically more reactive.

The use of sodium salts has been criticized because of the inhomogeneous system created[82] but the amounts used in current experiments[76] are much smaller than in the earlier work. Similarly difficulties can be imagined due to the finite time for the decomposition of $Cr(CO)_6$ but results from both methods agree with one another and with other methods. It has recently been shown that the temperatures in neon, heated in a reflected shock using $Cr(CO)_6$ line reversal, agree well with those determined by an ultrasonic method[83].

A 'band reversal' technique has been used by Russo[23] to follow relaxation in CO. Here light from a continuous source is passed through a filter which isolates the region corresponding to the fundamental, $\Delta v = 1$, band of CO in the infrared region. The light is then passed through the shock-heated CO and so the vibrational temperature is measured directly.

3.4.7 Infrared Fluorescence

For molecules with infrared-active vibrations the populations of individual levels may be followed by observing the infrared emission from them. The first work in this direction was by Windsor, Davidson and Taylor[84,85] who measured the relaxation time of CO by observing the emission from the overtone ($v = 2 \rightarrow v = 0$) band with a PbS detector. Since then infrared fluorescence has been used to study fundamental and overtone emission from several molecules[86,87,88].

The technique is straightforward: the emission from the shock tube is passed through a filter or monochromator to isolate the wavelength region of interest and on to a detector, which is usually of the photoconductive or photovoltaic type. PbS detectors have too long a time constant for general use and generally InSb or Ge/Au detectors are used which have time constants ($< 1 \mu s$) comparable with that of the optical system used and are suitable for wavelengths below 6μ. With Ge/Cu detectors emission up to 30μ may be studied. The output is displayed with an oscilloscope triggered to operate as the shock approaches the observation station.

The fundamental band emitted contains contributions from all transitions ($v = i + 1 \rightarrow v = i$); because of the Boltzmann equilibrium the populations n_i are $n_i > n_{i+1} > n_{i+2}$ and so on, and what is mainly observed at lower temperatures, if the whole band is received, is the emission $v = 1 \rightarrow v = 0$: it is then the relaxation of the first level which is followed (but see Section 3.5.6 and Decius[89]). It would be possible to observe upper levels if the contributions from say $v = 2 \rightarrow v = 1$ could be resolved as a band but, since the anharmonicity is not great, the bands overlap and only individual rotational lines of the desired transition may be isolated. The loss in signal in moving from observation of the whole band to an individual line is usually too great and makes the observation of upper bands in the fundamental impracticable. The upper levels may be observed however by measuring the emission from overtones such as $v = i + 2 \rightarrow v = i$. Although completely forbidden for an harmonic oscillator, overtone emission may be observed for molecules but the intensity is low in comparison with the fundamental[90]. Observations have been made on the $2 \rightarrow 0$ transitions of CO[84,85,87] and HCl[91]. The $3 \rightarrow 0$ transitions for HCl have been observed[86] at $1\cdot2 \mu$ using a photomultiplier with an S1 photocathode, but the signals were very weak due to low populations and transition probabilities which both fall off for higher levels.

Similarly for polyatomic molecules the relaxation of individual vibrational modes may be observed and particular modes have been observed in CO_2[15,91,92,93] and N_2O[14,91].

Apart from the usual difficulties of signal-to-noise ratio, one particular problem in emission is the possibility of pre-emission phenomena. Just before the shock arrives, the gas at the observation window is subject to irradiation from the column of heated gas behind the oncoming shock waves. Clearly it is possible for the fluorescence of the irradiated gas to be observed[94]. In addition reflections in the tube may give rise to spurious emissions before the arrival of the shock and therefore it is helpful in analysis of the results to record the arrival of the shock as well as the emission which occurs, and then any pre-emission effects can be allowed for.

3.4.8 Ultraviolet Fluorescence and Absorption

Observation of the ultraviolet emission from individual vibrational levels of excited electronic states provides a method for measuring relaxation in suitable gases. Roth[95] has observed the $(^2\Sigma \rightarrow ^2\Pi)$ transition of NO and measured the relaxation times for the first and second levels of $NO(^2\Sigma)$ between 5000 and 10,000°K. Earlier he had observed a relaxation for the radical CN. Although this is almost the only way of measuring the relaxation time of free radicals, there is a difficulty in disentangling the relaxation from the chemical processes forming the radicals.

Levitt and Sheen[96,97] have introduced SO_2 ($<10\%$) into N_2 and by observing the emission from SO_2 have measured the relaxation time for N_2/SO_2 mixtures. SO_2 is only five times more efficient than N_2 itself in exchanging vibrational energy with N_2, so that at low concentrations of SO_2 emission acts effectively as an indicator for the nitrogen relaxation.

It is also possible to study the relaxation of individual vibrational levels of the electronic ground states by following the change in the ultraviolet absorption spectra. Robben[98,99] has used this technique for NO. Even for molecules without a discrete spectrum it is possible to associate particular regions with absorption from individual vibrational levels. In this way Richards and Sigafoos[100] attributed the change in the ultraviolet absorption of CH_4 at 1470 Å with temperature to changes in vibrational populations and measured the relaxation in the shock-heated gas.

Chow and Greene[101] in a careful study measured the relaxation time of several levels of HI by observing the change in absorption at various wavelengths in the continuous spectrum. Their potential diagram for the molecule is shown in Figure 3.14 where the transitions for the first and second levels are also indicated. The wavelengths were chosen so that the maximum in the vibrational wave function of say the first level corresponded with a node in that for the second level, so that the contribution to the absorption by the second level was minimized. In this way it was possible to study the relaxation of the first four vibrational levels.

Appleton and Steinberg in a similar study observed the relaxation of the thirteenth vibrational level in N_2 by monitoring the change in absorption in the vacuum ultraviolet region of the spectrum[102].

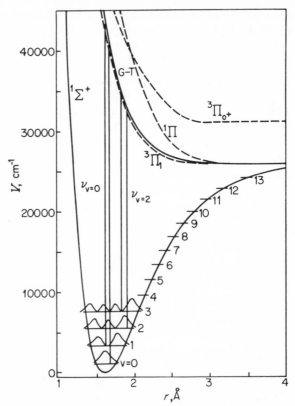

Figure 3.14 The potential-energy diagram used by Chow and Greene to analyse the absorption spectrum of HI. By choosing appropriate wavelengths, indicated by the vertical lines, the absorption characteristic of each of the first four vibrational levels could be studied as a function of time. Reproduced by permission of the author and the Editor of *J. Chem. Phys.*

3.5 THE ADEQUACY OF MODELS USED IN THE INTERPRETATION OF NAPIER TIMES

3.5.1 Introduction

Experimental results have usually been analysed by applying Equation (3.1):

$$dx(t)/dt = [x(\infty) - x(t)]/\tau \qquad (3.72)$$

where $x(t)$ is the value of a physical quantity at time t after the perturbation, $x(\infty)$ is its eventual value at equilibrium and τ is the relaxation time.

Equation (3.72) may be obtained from irreversible thermodynamics or from kinetic models two of which, the two-state vibrator and the harmonic oscillator, are now considered (Sections 3.5.2 and 3.5.3).

The dependence of τ upon temperature and its absolute value may be calculated theoretically and the background to such calculations is discussed briefly in Section 3.5.4. Empirical suggestions are dealt with in Section 3.5.5. An account of the adequacy of these models is given in Sections 3.5.6–3.5.10.

3.5.2   Two-state Model

In the two-state model the molecule has two vibrational levels, the ground and first, having populations n_0 and n_1. Energy is exchanged on collision and the rate of excitation in the pure gas is then:

$$\frac{dn_1}{dt} = k'_{01}n_0n - k'_{10}n_1n \tag{3.73}$$

Here $n = n_1 + n_0$ since collision may take place with either species and is assumed to be equally effective with both. The bimolecular rate constants k'_{01}, k'_{10} may be replaced by concentration-dependent quantities k_{01} and k_{10} equal to $k'_{01}n$ and $k'_{10}n$ respectively, so that Equation (3.73) becomes:

$$\frac{dn_1}{dt} = k_{01}n - (k_{01} + k_{10})n_1 \tag{3.74}$$

At equilibrium, $n_1 = n_1(\infty)$, and Equation (3.74) may be rearranged to have the same form as Equation (3.73):

$$\frac{dn_1}{dt} = (k_{01} + k_{10})[n_1(\infty) - n_1] \tag{3.75}$$

The relaxation time is related to the rate constants:

$$\tau = (k_{01} + k_{10})^{-1} \tag{3.76}$$

and is concentration dependent. At equilibrium the ratio of rate constants is given by the ratio of the equilibrium populations, $n_1(\infty)$ and $n_0(\infty)$, which in turn is given by the Boltzmann relation:

$$\frac{k_{01}}{k_{10}} = \frac{n_1(\infty)}{n_0(\infty)} = \exp\left(\frac{-hv}{kT}\right) \tag{3.77}$$

where v is the vibrational frequency,
then

$$\tau^{-1} = k_{10}[1 + \exp(-hv/kT)] \tag{3.78}$$

or for

$$T \ll hv/k, \qquad \tau = k_{10}^{-1}. \tag{3.79}$$

The reciprocal relaxation time, like k_{10}, is dependent upon density.

The rate constants are temperature dependent and so the equations are only valid for two-state oscillators relaxing in a constant-temperature heat bath. For many diatomic molecules the two-state model is not an unreasonable approximation for the relaxation of the first vibrational level at lower temperatures well below (hv/k). At these temperatures $n_1 \ll n_0$ and in a shock tube the approximation of the constant-temperature heat bath is reasonable too.

3.5.3   Landau–Teller Model for an Harmonic Oscillator

A more realistic model for a molecular vibrator is the harmonic oscillator. For this Landau and Teller[103] derived the relaxation equation for the total vibrational energy.

Here it is assumed that transitions can occur only between neighbouring vibrational levels. Infrared spectroscopic transitions for an harmonic oscillator are governed by a similar selection rule; this is obeyed approximately by real molecules[90] for which non-nearest neighbour transitions, $(\Delta v > 1)$ have a lower probability than $\Delta v = 1$ transitions. The suggestion can be supported theoretically but is a direct consequence of using first-order perturbation theory to treat the problem.

A further condition is that the collisional transition probabilities and hence the rate constants bear the simple relation to one another:

$$k_{10}:k_{21}:k_{32}:k_{43} \text{ etc} = k_{01}:k_{12}:k_{23}:k_{34} \text{ etc} = 1:2:3:4 \text{ etc} \tag{3.80}$$

and as for the two-state case (equation 3.77):

$$k_{01}/k_{10} = \exp(-hv/kT) \tag{3.81}$$

The nearest neighbour condition is given by:

$$k_{ij} = 0, \qquad i-j \neq 1. \tag{3.82}$$

For these conditions the equation for the change of population of the ith level is:

$$\frac{dn_i}{dt} = k_{i-1,i}n_{i-1} + k_{i+1,i}n_{i+1} - (k_{i,i-1} + k_{i,i+1})n_i \qquad (3.83)$$

and for the lowest state:

$$\frac{dn_0}{dt} = k_{10}n_1 - k_{01}n_0 \qquad (3.84)$$

Since a molecule in the ith level has i vibrational quanta the change in the number of quanta in level i is given by $i\,dn_i/dt$ and the change in the total number of quanta is obtained by summing Equations (3.83) and (3.84) over all the levels, which having separated two of the terms gives:

$$\sum_{i=0}^{\infty} i\frac{dn_i}{dt} = \sum_{1}^{\infty}(i-1)k_{i-1,i}n_{i-1} + \sum_{1}^{\infty}k_{i-1,i}n_{i-1} + \sum_{0}^{\infty}(i+1)k_{i+1,i}n_{i+1}$$

$$- \sum_{0}^{\infty}k_{i+1,i}n_{i+1} - \sum_{0}^{\infty}ik_{i,i-1}n_i - \sum_{1}^{\infty}ik_{i,i+1}n_i \qquad (3.85)$$

Terms 1, 3, 5 and 6 cancel giving the change in the number of vibrational quanta

$$\sum_{i=0}^{\infty} i\frac{dn_i}{dt} = \sum_{1}^{\infty}k_{i-1,i}n_{i-1} - \sum_{0}^{\infty}k_{i+1,i}n_{i+1} \qquad (3.86)$$

or

$$= \sum_{0}^{\infty}k_{i,i+1}n_i - \sum_{1}^{\infty}k_{i,i-1}n_i \qquad (3.87)$$

With equation (3.80)

$$\sum_{i=0}^{\infty} i\frac{dn_i}{dt} = \sum_{0}^{\infty}(i+1)k_{01}n_i - \sum_{1}^{\infty}ik_{10}n_i$$

so that with the total number of vibrational quanta ($V_q = \Sigma in_i$) and number of molecules ($n = \Sigma n_i$)

$$\frac{dV_q}{dt} = k_{01}V_q + k_{01}n - k_{10}V_q \qquad (3.88)$$

At equilibrium

$$\frac{dV_q}{dt} = 0; \qquad V_q(\infty) = k_{01}n/(k_{10}-k_{01}) \tag{3.89}$$

and thus

$$\frac{dV_q}{dt} = (k_{10}-k_{01})[V_q(\infty)-V_q] \tag{3.90}$$

Multiplication of V_q in Equation (3.90) by $h\nu$ gives a similar equation for the total vibrational energy $E_{vib} = V_q h\nu$. Either equation is of the same form as Equation (3.72); the relaxation time is given by:

$$\tau = (k_{10}-k_{01})^{-1} \tag{3.91}$$

which can be contrasted with Equation (3.76); as before

$$\tau^{-1} = k_{10}[1 - \exp(-h\nu/kT)] \tag{3.92}$$

and for

$$T \ll h\nu/k, \qquad \tau = k_{10}^{-1} \tag{3.93}$$

As for the two-state gas the relaxation time depends upon concentration. By substitution, Equation (3.90) can also be applied to the vibrational temperature since for a small temperature change

$$E'_{vib} - E''_{vib} = C_{vib}(T'_{vib} - T''_{vib}) \tag{3.94}$$

where C_{vib} is the vibrational heat capacity. As C_{vib} is a function of T_{vib} itself, Equation (3.90) for the vibrational temperature can only be valid for small excursions of the vibrational temperature. The concept of a vibrational temperature requires that a Boltzmann equilibrium is maintained between the vibrational levels throughout relaxation; this has been shown to be true for the relaxation of a Landau–Teller oscillator by Herman and Rubin[104]. As for the two-state gas, Equation (3.90) is valid only for oscillators relaxing in a heat bath the temperature of which is constant after the initial perturbation.

3.5.4 Mixtures

The relaxation of a gas in a mixture may be seen most easily for a two-state gas (A) mixed with an inert gas (B); [CO(Ar)] would be such a mixture.

The analogous equation to (3.73) is:

$$\frac{dn_1{}^A}{dt} = k_{01}^{AA'} n_A n_A(0) + k_{01}^{AB'} n_B n_A(0) - k_{10}^{AA'} n_A n_A(1) - k_{10}^{AB'} n_B n_A(1) \qquad (3.95)$$

Where $k_{01}^{AA'}$ refers to collision between molecules of type A and $k_{01}^{AB'}$ refers to collisions of molecules of A with molecules of B and so on. Proceeding as before the relaxation time of the mixture is:

$$\tau^{-1} = (k_{01}^{AA} + k_{10}^{AA})n_A + (k_{01}^{AB} + k_{10}^{AB})n_B \qquad (3.96)$$

or

$$\tau^{-1} = n_A/\tau_{AA} + n_B/\tau_{AB} \qquad (3.97)$$

where τ_{AB} is the hypothetical relaxation time for gas A which relaxes only by collision with B. The equation may be rearranged using the mole fraction $X_A = (n_A/n_A + n_B)$

$$[(n_A + n_B)\tau]^{-1} = X_A/\tau_{AA} + X_B/\tau_{AB} \qquad (3.98)$$

which for unit concentration gives

$$\frac{1}{\tau_{mix}} = \frac{X_A}{\tau_{AA}} + \frac{X_B}{\tau_{AB}} \qquad (3.99)$$

A similar expression for the vibrational energy of a Landau–Teller harmonic oscillator may also be derived.

Equation (3.99) allows τ_{AB} to be determined from the relaxation time for the mixture and the relaxation time for pure A(τ_{AA}).

For a mixture of gases which may both relax ($O_2 - N_2$ for example) an equivalent expression for B may be derived:

$$\frac{1}{\tau_B} = \frac{X_B}{\tau_{BB}} + \frac{X_B}{\tau_{BA}} \qquad (3.100)$$

Notice that generally $\tau_{BA} \neq \tau_{AB}$.

However, in such a mixture other processes may occur as well as simple collision excitation. For example vibration–vibration energy transfer may occur:

$$A(v = n) + B(v = m) = A(v = n+1) + B(v = m-1) \qquad (3.101)$$

In this case the equations will not apply and only by careful measurement of the relaxation time over a range of concentrations and comparison with the relaxation times for the pure compounds will it be possible for some idea of the mechanism to be deduced.

The cases mentioned of separate collisional excitation of the two molecules and collisional excitation of one followed by vibrational-energy transfer are formally analogous to parallel and series excitation in polyatomic molecules. Here, where a molecule has several modes of vibration, the vibrations may be excited separately by collisional processes (parallel excitation) or one mode may be excited and then energy transferred from this to another mode (series excitation). Mathematical treatments of these mechanisms have been given by Cottrell and McCoubrey[28] and Herzfeld and Litovitz[29].

3.5.5   Absolute Calculations of Napier Times

Chapter 2 deals fully with the theory of vibrational relaxation and there have been reviews by Herzfeld[105] and Takayanagi[106,107]. Here only the principal result which is used in the subsequent sections is mentioned.

The first approaches to the problem were made by Zener[108] and Landau and Teller[103,109]. The latter authors started from the assumption that the probability for collisional de-excitation of the first vibrational level, P_{10}, would be given by

$$P_{10} = C\,e^{-X} \qquad (3.102)$$

C is a factor to allow for the geometry of the collision. X is given by

$$X = t_c/t_0 \qquad (3.103)$$

where t_0 is the natural period of the vibration and t_c is the duration of the collision. t_0 itself is given by x/v where x is the distance travelled by the colliding partners in the repulsive force field and v is the velocity of collision. It can be seen immediately that the highest probabilities of transfer will be for low frequency vibrations and collisions with high velocities. The intermolecular potential should also fall sharply. Landau and Teller then went on to calculate classically the dynamics of the collision and after averaging over all velocities, finally showed the rate constant for de-excitation k_{10} to be proportional to

$$k_{10} \sim \exp[-(\text{constant})T^{-1/3}] \qquad (3.104)$$

Later workers have elaborated these calculations and also treated the whole problem quantum mechanically, but in most approaches the $T^{-1/3}$

dependence has resulted and in general the Napier time τ can be expressed as

$$\log \tau = A + BT^{-1/3} \tag{3.105}$$

where A and B are constants which have only slight temperature dependences.

The most widely applied approach is that of Swartz, Slawsky and Herzfeld familiarly known as the SSH theory and described by Herzfeld and Litovitz[29]. This is a three-dimensional quantum-mechanical treatment which yields a prediction for the relaxation time in terms of the mass of the colliding molecules, the temperature, the vibration frequency and parameters for an exponential repulsive potential between the collision partners. Since parameters are usually available for the Lennard-Jones intermolecular potential, various methods have been advocated[28,29] for fitting the two potentials together so that Lennard-Jones parameters can be used. When this is done absolute values for the Napier times may readily be calculated.

The approaches referred to in the previous paragraphs concerned relaxation by exchange of energy between translation and vibration. Due to discrepancies between theory and experiment for a number of hydrides, Cottrell and his coworkers[110,111,112] suggested that relaxation could occur by exchange between rotation and vibration, particularly where the velocity of rotation of the peripheral atoms is high due to the low moment of inertia. Moore[113] has adopted this suggestion and made a semiclassical calculation which was applied to a large number of molecules with some success (Section 3.5.9).

The only other widely applied calculations are those of Benson and Berend[114] who solved the classical equations of motion for individual collisions between molecules with Morse and Lennard-Jones interaction potentials and averaged the results over a large number of initial conditions. They have applied their method to a variety of molecules and have also considered the effects of rotation–vibration exchange as well as translation–vibration[115].

3.5.6 Empirical Suggestions

Millikan and White[116] were able to fit a large number of experimental Napier times to the formula:

$$\log_{10} \tau(\text{atm s}) = (5 \cdot 00 \times 10^{-4}) \mu^{1/2} \theta^{4/3} (T^{-1/3} - 0 \cdot 015 \, \mu^{1/4}) - 8 \cdot 00 \tag{3.106}$$

where μ is the reduced mass of the collision partners $(m_1 m_2/(m_1 + m_2)$ for masses m_1 and m_2) and θ the characteristic temperature of the vibration (hv/k). Equation (3.106) was produced by fitting and by qualitative consideration of the theoretical approaches mentioned previously. The formula was fitted for both pure compounds and some mixtures. Phinney[117] has fitted results for N_2, O_2 and Cl_2 to a common plot with $T^{-1/3}$ temperature dependence. It seems unlikely that this will fit over the very wide range of measurements now available or that many of the substances will fit since the plot has a single slope.

Boade[118] has suggested a plot similar to but more complicated than that of Millikan and White for use with halomethanes.

Lambert and Salter[119] have also suggested a correlation for molecules at 300°K. This has mainly been applied to results from ultrasonic experiments and was considered in Chapter 2.

3.5.7   The Applicability of the Two-state and Landau–Teller Models

In a shock-tube experiment vibrational energy is obtained at the expense of translational energy and the temperature changes throughout the relaxation (Figure 3.8). Strictly then, the equation for relaxation (Equation 3.72) obtained for either the two-state Equation (3.75) or Landau–Teller model Equation (3.90) should not be applicable, since both were derived for a system relaxing in a constant-temperature heat bath. Nevertheless most workers treat their results using Equation (3.72) and specify them either for the initial frozen temperature or for the average of the frozen and relaxed temperature.

Johannesen[74] has developed a method for estimating Napier times which takes advantage of the change in temperature during relaxation to obtain whole series of Napier times from a single record. Blythe[73] has compared this with other approximate methods.

Zienkiewicz and Johannesen[16] treated their interferometric results from both oxygen and carbon dioxide so that the variation of Napier time during the whole relaxation is shown. Their results for oxygen are shown in Figure 3.15; all the segments should overlap to form a continuous line if the relaxation time does not depend upon the magnitude of the excursion from equilibrium. There is an obvious lack of continuity between the segments which is even more striking with their results from carbon dioxide[15]. The authors question the validity of the Landau–Teller approach. The experiments have however been repeated by Lutz and Kiefer[120] whose results

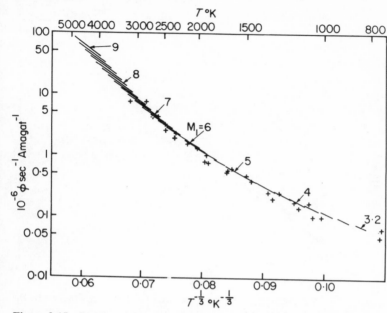

Figure 3.15 Results of Zienkiewicz and Johannesen for O_2. Each result is treated to yield the relaxation frequency ($= 1/\tau$) throughout the relaxation. The segments would normally be expected to form a continuous line, see Figure 3.16. Reproduced by permission of the author and the Editor, *J. Fluid Mech.*

for oxygen obtained with a laser schlieren technique are shown in Figure 3.16. Here the segments do line up and the authors claim a better sensitivity in their measurements; on the other hand the scatter of their results is noticeably greater than that obtained by Zienkiewicz and Johannesen*.

In spectroscopic experiments which observe the population of the first vibrational level, the Napier time might be expected to be related to the rate constants by Equation (3.76) for the two-state model:

$$\tau = (k_{10} + k_{01})^{-1} \tag{3.76}$$

In other experiments where total properties of the system such as vibrational temperature or density are observed, the Landau–Teller model Equation (3.91) is more appropriate:

$$\tau = (k_{10} - k_{01})^{-1} \tag{3.91}$$

* But see addenda, p. 264.

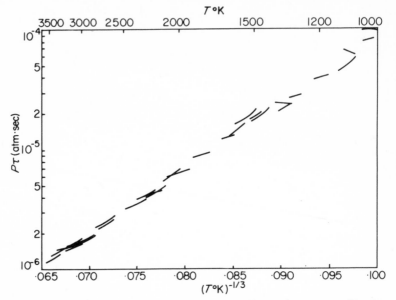

Figure 3.16   Results of Lutz and Kiefer for O_2. The results are treated by the method of Johannesen (Figure 3.15). The authors claim that the segments form a continuous line in contrast to those of Zienkiewicz and Johannesen (Figure 3.15). Reproduced by permission of the author and the Editor, *Phys. Fluids*

At low temperatures both reduce to

$$\tau = k_{10}^{-1} \qquad\qquad (3.79, 3.93)$$

but at higher temperatures, relaxation times obtained by the different methods should deviate from one another. At high temperatures however the two-state model provides a poorer description of the first vibrational level due to the appreciable population of the upper levels, and in general the scatter of most experimental results is too great to attribute any differences between models. Decius[89] has pointed out that observations of complete vibration–rotation bands in the infrared to which, for the fundamental say, there are contributions from all $\Delta v = 1$ transitions, the relaxation time is related to the rate constants by Equations (3.91) and not Equation (3.76) as might be expected.

3.5.8   Higher Vibrational Levels

Vibrational levels higher than the first are assumed in the Landau–Teller model to be populated by a step-wise mechanism since only $\Delta v = 1$ transi-

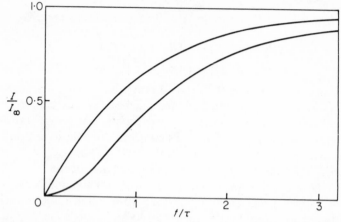

Figure 3.17 Ideal relaxation curves for the first (upper curve) and second (lower curve) vibrational levels of a three-state system. The curve for the first level $[I/I_\infty = 1 - \exp(t/\tau)]$ is that normally observed. The lower curve for the second level $\{I/I_\infty = [1 - \exp(t/\tau)]^2\}$ is obtained by either assuming a Landau–Teller relation between the rate constants for the successive steps or by assuming that the second level is populated by a resonance exchange process (see text)

tions are allowed. In addition the rate constants for the individual steps have simple-arithmetic relations with one another (Equation 3.80). The evidence available with which to examine this assumption is rather sparse. Callear and Smith[35] in a flash-photolysis experiment were able to interpret their observations of the vibrational levels of NO $(A^2\Sigma^+$ state) if $k_{32}:k_{21}:k_{10}$ were in roughly the correct ratio. Fitzsimmons and Bair[121] observed the relaxation of vibrationally excited O_2 from the flash photolysis of ozone. Their results for the thirteenth to nineteenth levels were consistent with a single-step quantum-jump model. Hooker and Millikan[87] observing the infrared emission from the second level of shock-heated CO could fit the observed relaxation curves to this model (Figure 3.17). Windsor, Davidson and Taylor[84,85] had earlier assumed this for CO but were not able to observe the infrared emission from the first level simultaneously. The proposed mechanism for the population of the second state is:

$$2CO(v = 0) \underset{k_{10}}{\overset{k_{01}}{\rightleftharpoons}} CO(v = 1) + CO(v = 0) \qquad (3.107)$$

$$CO(v = 1) + CO(v = 0) \underset{k_{21}}{\overset{k_{12}}{\rightleftharpoons}} CO(v = 2) + CO(v = 0) \quad (3.108)$$

where $k_{01}:k_{12} = 1:2$ and so on. However an alternative mechanism also fits

$$2CO(v = 0) = CO(v = 1) + CO(v = 0) \qquad (3.109)$$

$$2CO(v = 1) = CO(v = 2) + CO(v = 0) \qquad (3.110)$$

The second reaction here (Equation 3.110) is a *resonance-exchange* reaction, in which a single quantum of vibrational energy is exchanged between the molecules with no interchange of energy between vibration and translation. Resonance-exchange reactions are thought to be much more rapid than translation–vibration transfer.

Brown and Klemperer[40] and Steinfeld and Klemperer[41] studied the quenching of the fluorescence of I_2, excited with sodium D or mercury lines, by added gases. They found that $\Delta v = 2$ transitions contributed appreciably ($\sim 20\%$) to the measured cross-sections for vibrational-energy transfer from the fifteenth and twenty-fifth vibrational levels of I_2.

In investigations of the relaxation of the second, third and fourth levels of shock-heated HI (Section 3.4.7) Chow and Greene[101] observed that the levels relaxed at the same rate with a simple form of relaxation (Figure 3.17), normally characteristic of the first level, and not in accordance with the Landau–Teller model. Borrell and Gutteridge[86] found similar behaviour for the third level of HCl. These observations can be explained by direct, $\Delta v > 1$, transitions to the upper levels with rate constant k_{20} equal to k_{10}. Chow and Greene preferred the mechanism

$$HI(v = 0) + M \xrightarrow{k_1} HI(v = 1) + M \xrightarrow{k_2} HI(v = 2) + M \xrightarrow{k_3} etc. \qquad (3.111)$$

where $k_1 \gg k_2, k_3$ etc. but the nature of the steps was undefined. Appleton and Steinberg[102] were able to observe the absorption of the thirteenth vibrational level of N_2 in the vacuum ultraviolet (Section 3.4.8) and follow the relaxation. The relaxation time was in rough agreement with other measurements (Section 3.6.4). Here again with the Landau–Teller model a simple form of relaxation would not be expected although an apparently larger relaxation time would result if a complex curve was analysed as a simple exponential decay.

The results obtained with expansions (Sections 3.2.2 and 3.6.4) in which N_2 is shock heated, allowed to relax and then expanded, and where the de-excitation is followed, seem to yield enormously fast relaxation times in that the observed vibrational temperatures throughout the expansion are much lower than expected*. Hurle[122] has suggested that the effect

* But see addenda, p. 264.

could be explained by $\Delta v > 2$ transitions among the upper levels. However Rich and Rehm[123,124] in a theoretical study have shown that rapid
cooling in expansions could result for anharmonic oscillators under
extreme conditions even if only $\Delta v = 1$ transitions are allowed. (See also
Rankin and Light[125].)

At the moment then, whether the Landau–Teller condition is applicable
to any or all molecules is not clear. Certainly the excellent agreement over
a wide temperature range between results for CO say from infrared emission where the population of the first level is followed by interferometry,
which measures the density changes and is thus indicative of the total
vibrational-energy content, and line reversal which gives the vibrational
temperature, has been correctly taken as an indication that the Landau–
Teller model is adequate. On the other hand, there might be other relations
between rate constants which also lead to the result that measurements on
individual levels and on total energies yield the same Napier time.

3.5.9   Temperature Dependence

Most workers plot the temperature dependence of their results according
to Equation (3.105) and that this is generally justified is demonstrated for
N_2 in Figure 3.18 where the results for very high temperatures match up
with that at room temperature. The large number of compounds and
mixtures fitted by the Millikan and White correlation (Equation 3.106) is a
further illustration that the $\log \tau/T^{-1/3}$ dependence is adequate. Some
early workers attempted to fit their results to a $\log \tau/T^{-1}$ dependence with
some success but they were unable to correlate the 'Activation Energy'
obtained with the energy separation of the vibrational levels[126]. Widom
has pointed out in connection with several theoretical approaches that he
has tried, that Blackman's early results for oxygen[127] could be fitted to
$\log \tau/-T^{1/3}$ and $\log \tau/-T$ plots as well as Equation (3.105), and one is left
to wonder how large a range of powers of T would fit many results within
the experimental error. Benson and Berend's[114] calculations do not yield
an explicit temperature dependence for τ but they were able to fit the
results for the molecules which they have treated to their predicted curves.
They point out that $\log \tau/T^{-1/3}$ plots should be S shaped. The agreement
with experiments is illustrated by Figure 3.19 where the points off the curve
are all the results of experiments for which the water content is high. The
expected temperature at which extreme curvature should take place

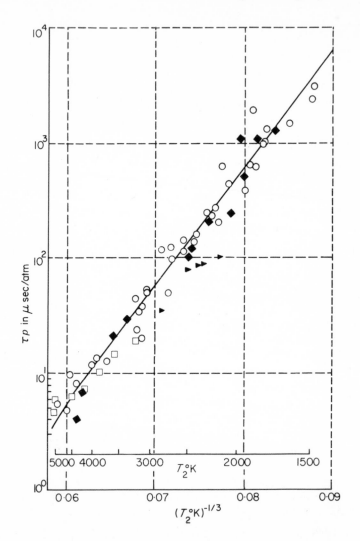

Figure 3.18   Results for N_2 (after Hurle) showing the good agreement between different methods. $\square$: Interferometry, Blackman; $\bigcirc$: interferometry and sensitized infrared emission with CO, Millikan and White. $\blacklozenge$, $\blacktriangleright$ Hurle, sodium-line reversal. Reproduced by permission of the author and the Editor, *J. Chem. Phys.*

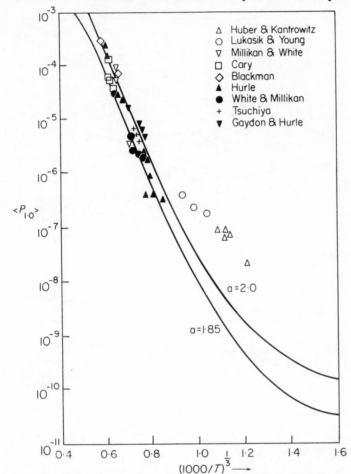

Figure 3.19 Results and theoretical predictions for N_2 (after Benson and Berend). The predicted results should be between the two curves. The results at lower temperatures ($T^{-1/3} > 0.9$) are thought to be high due to water impurity in the test gas. Reproduced by permission of the author and the Editor, *J. Chem. Phys.*

varies from molecule to molecule but is frequently outside the range of the measurements made.

Not unreasonably, success is claimed for the theoretical approaches, which predict a linear $\log \tau / T^{-1/3}$ dependence. With the SSH approach particularly, not only the temperature dependence but also the absolute

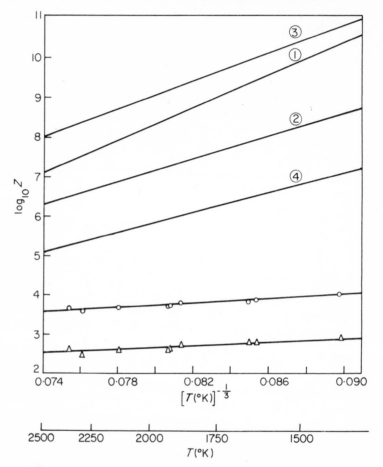

Figure 3.20 Results of Chow and Greene for HI(HI), triangle, and HI(Ar), circles, together with the predictions from the Millikan–White correlation [1:HI(HI); 2:HI(Ar)] and the SSH approach [3:HI(HI); 4:HI(Ar)]. In this graph $\log_{10} Z$ is plotted rather than $\log_{10} p\tau$ (Section 3.1.2). Reproduced by permission of the author and the Editor, *J. Chem. Phys.*

value of the relaxation time of many molecules and mixtures may be predicted within an order of magnitude.

Definite exceptions are NO, the hydrogen halides and some other hydrides. Figure 3.20 shows the departure from theory for HI(HI) and HI(Ar) collisions. The Napier time for NO can be predicted if a special potential function is used which allows for the incipient dimerization or the formation of an electronically excited state (Section 3.6.8). With the hydrogen halides (Section 3.6.9) there is the possibility mentioned before of relaxation occurring by rotation–vibration rather than translation–vibration energy exchange which has been dealt with by Moore's theory. The success of Moore's theory[113] is shown in Figure 3.21 where molecules for which the average peripheral rotational velocity of the hydrogen atoms is greater than the average translational velocity have been fitted. The hydrogen halides themselves do not fit well since the rotational spacing is large and therefore requires that the problem should be tackled by a full quantum-mechanical treatment. Calculations by Benson and Berend[115] also support Moore's approach.

An interesting observation by Millikan and Osburg[128] with a bearing on the possible importance of rotation–vibration transfer is that at room

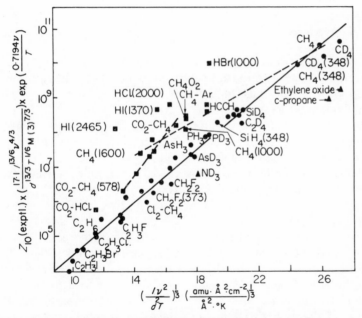

Figure 3.21 A comparison of Moore's theory for rotation–vibration energy transfer with experiment. An explanation of the axes and the references for the compounds will be found in the original paper. Reproduced by permission of the author and the Editor, *J. Chem. Phys.*

temperature the Napier time for CO (*para*-hydrogen) is less than that for CO (*ortho*-hydrogen).

The Millikan and White correlation will only fit molecules which undergo translation–vibration energy exchange: Benson and Berend[114] have criticized the choice of results from which the formula was derived. It does seem to weight the vibrational frequency too heavily[91] since HBr is predicted to relax more slowly than nitrogen which seems unlikely on chemical grounds, even if there were no possibility of rotation–vibration exchange interfering.

In conclusion, Equation (3.105) provides an adequate description of the temperature dependence of the Napier time for many molecules and mixtures and the constants can be effectively predicted with SSH theory. One should be cautious when dealing with hydrides, molecules under extreme conditions and molecules where there is a possibility of non-adiabatic collisions occurring which lead to the formation of electronically excited states.

3.5.10   Mixtures

Relaxation of a gas in the presence of foreign gases is more easily studied under the constant-temperature conditions obtainable with say ultrasonic experiments rather than in a shock tube. Experiments with added monatomic inert gases have been interpreted with the aid of Equation (3.99) and show that this equation is adequate. The Napier times for relaxation of diatomic molecules with inert gases decrease in the order He < Ne < Ar < Kr (Sections 3.6.5 and 3.6.7).

This simple mass effect also occurs with added D_2 which at temperatures below 1600°K has the same effect as He in the relaxation of O_2 and CO. For CO_2, on the other hand, D_2 is more effective than He in removing vibrational energy[129].

Above 1600°K the Napier time for $CO(D_2)$ falls below the extrapolated value expected from the lower-temperature measurements. White[130] has pointed out that only above 1600°K are the vibrational levels of D_2 other than the ground state appreciably populated. It is hard to understand qualitatively what the effect of this would be: clearly vibration–vibration transfer of energy from D_2 to CO would be facilitated, but the reverse reaction leading to de-excitation of CO which normally makes the largest contribution to the rate constant should be operative at the lower temperatures too. White's interesting observation deserves further study.

When diatomic and polyatomic molecules are added to a relaxing system there is not only the possibility of simple collisional exchange but vibration–vibration transfer may occur[131]. For N_2 and CO, for example, N_2 may relax by:

$$N_2(v = 0) + N_2 = N_2(v = 1) + N_2 \tag{3.112}$$

$$N_2(v = 0) + CO = N_2(v = 1) + CO \tag{3.113}$$

$$N_2(v = 0) + CO(v = 1) = N_2(v = 1) + CO(v = 0) \tag{3.114}$$

Equations (3.112) and (3.113) are the normal collisional processes; Equation (3.114) is an exchange reaction which in this case is very favoured because of the close coincidence between the energy separations of the vibrational levels in CO and N_2. Millikan and White[88] demonstrated the efficiency of the transfer reaction and used it to follow the relaxation of N_2 by observing the change in infrared emission from CO.

The effect of such vibration–vibration transfer reactions on the measured Napier times for mixtures is to give a non-linear dependence of τ on the mole fraction: Equation (3.99) is not obeyed. Quadratic dependences have been observed for $O_2(H_2O)$ and O_2 (hydrocarbon) mixtures: see Section 3.6.5. It is necessary therefore when dealing with mixtures to make a thorough check of the relation between τ and mole fraction before using Equation (3.99).

3.6 RESULTS FOR INDIVIDUAL MOLECULES

3.6.1 Table of Napier Times

Table 3.1 gives a Napier time for molecules which have been studied in shock tubes. Values for other systems are given elsewhere in the volume (Chapter 2) and by Read[132] and Borrell[133]. The values in the table are generally interpolated from the author's experimental results using the expression:

$$\log_{10} \tau = A + BT^{-1/3} \tag{3.115}$$

The values of A and B are given in the following sections where the experimental results are discussed.

Table 3.1

Napier times for individual molecules

Napier times in μs

$T/°K$	300	1000	1500	2000	2500	5000	References
Molecule (Collision partners)							
$H_2(H_2)$	1060	—	2·4	1·1	0·6	—	42, 68
$H_2(Ar)$	—	—	9·8	4·5	2·45	—	68
$D_2(D_2)$	—	—	4·4	1·3	0·8	—	67, 72
$D_2(Ar)$	—	—	21·2	7·8	3·8	—	67, 72
$D_2(Kr)$	—	—	72·1	26·6	—	—	72
$N_2(N_2)$	—	—	4680	854	143	5·5	76, 88, 102
$N_2(H_2)$	—	—	2·75	—	—	—	134
$O_2(O_2)$	42000	86·1	18·0	8·0	3·1	0·5	120, 135, 136, 137
$O_2(H_2)$	2·2	0·4	0·2	—	—	—	63
$O_2(D_2)$	—	2·4	1·0	—	—	—	63
$O_2(He)$	42·5	2·2	—	—	—	—	138
$O_2(Ar)$	—	1000	126	34·7	14·5	—	135, 138
$O_2(CH_4)$	0·97	0·1	—	—	—	—	139
$O_2(C_2H_4)$	—	0·1	—	—	—	—	139
$O_2(C_2H_2)$	—	0·3	—	—	—	—	139
$CO(CO)$	[2·4 × 10⁶]	2000	257	74·1	29·5	2·3	70, 84, 87 140, 141 142, 143
$CO(H_2)$	70·8	2·6	1·1	0·7	—	—	87, 130
$CO(D_2)$	—	30·2	—	—	—	—	130
$CO(He)$	2290	30·2	10·0	—	—	—	143
$CO(Ne)$	—	—	—	55	24	—	143
$CO(Ar)$	—	—	—	347	123	—	143
$CO(Kr)$	—	—	—	589	429	12·0	143
$NO(^2\Pi)(NO)$	0·4	0·1	0·07	0·05	0·04	0·005	95, 144, 145, 146
$NO(^2\Pi)(Ar)$	—	—	—	5·9	2·8	0·33	95
$NO(^2\Sigma)(NO)$	—	—	—	—	—	28	95
$HCl(HCl)$	>10^4	0·71	0·32	0·19	—	—	86, 91, A2–4
$DCl(DCl)$	—	1·59	0·74	0·31	—	—	A2
$HBr(HBr)$	>1·5 × 10^3	—	16	—	—	—	91
$HI(HI)$	—	—	0·4	0·51	—	—	101
$HI(Ar)$	—	—	2·25	1·8	—	—	101
$Cl_2(Cl_2)$	4·3	0·14	0·07	—	—	—	62, 132
$CN(Xe)$	—	—	—	—	—	112	147
$N_2O(N_2O)$	1·25	0·2	—	—	—	—	14
$CO_2(CO_2)$	4·7	0·77	0·48	0·36	0·29	0·17	148
$CH_4(CH_4)$	1·8	0·25	0·2	—	—	—	100
$CH_4(Ar)$	—	0·9	0·4	—	—	—	100

3.6.2   Hydrogen

(See Figure 3.22). The earliest measurement was by Gaydon and Hurle[140] who, using the sodium-line reversal technique, obtained a Napier time of 0·8 μs atm at 2900°K. White[63], using interferometry to study the relaxation of O_2 in the presence of H_2 and D_2, obtained an upper limit of 2 μs atm

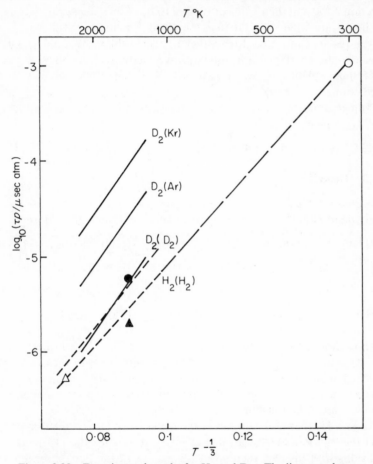

Figure 3.22   Experimental results for H_2 and D_2. The lines are drawn from the equations in the text. The line for $H_2(H_2)$ has been extrapolated to embrace the room temperature result of De Martini and Ducuing. The filled points are estimates of White. The unshaded triangle is an early result of Gaydon and Hurle

at 1400°K, by a laser schlieren technique. Kiefer and Lutz[68] observed the relaxation of H_2/Ar mixtures and found the Napier time given by:

$$H_2(H_2): \log_{10} \tau(\text{atm s}) = 43{\cdot}4T^{-1/3} - 9{\cdot}409 \qquad (1100-2700°\text{K}) \qquad (3.116)$$

De Martini and Ducuing[42] were able to measure the Napier time at room temperature by populating the first vibrational level by Raman scattering (Section 3.2.8) and their value of $1060(\pm 100)\,\mu\text{s atm}$ is encouragingly close to the extrapolated value ($1040\,\mu\text{s atm}$) from Equation (3.116).

The results cannot be fitted either by the Millikan–White correlation which predicts a Napier time too large by a factor of about forty, or by the SSH theory which is nearly two orders of magnitude too low in its prediction.

Napier times for $H_2(\text{Ar})$ were found to be four times as long as for $H_2(H_2)$. The equation is[68]:

$$H_2(\text{Ar}): \log_{10} \tau(\text{atm s}) = 44{\cdot}0T^{-1/3} - 9{\cdot}41 \qquad (1100-2700°\text{K}) \qquad (3.117)$$

3.6.3 Deuterium

(See Figure 3.22). The Napier times for deuterium have been estimated by studying mixtures with the rare gases. The equations obtained were:

$$D_2(D_2)^{67}: \log \tau(\text{atm s}) = 48{\cdot}1T^{-1/3} - 9{\cdot}57 \qquad (1100-3000°\text{K}) \qquad (3.118)$$

$$D_2(D_2)^{72}: \log \tau(\text{atm s}) = 54{\cdot}3T^{-1/3} - 10{\cdot}102 \qquad (1200-2300°\text{K}) \qquad (3.119)$$

$$D_2(\text{Ar})^{72}: \log \tau(\text{atm s}) = 54{.}3T^{-1/3} - 9{\cdot}420 \qquad (1200-3000°\text{K}) \qquad (3.120)$$

$$D_2(\text{Kr})^{72}: \log \tau(\text{atm s}) = 54{\cdot}3T^{-1/3} - 8{\cdot}886 \qquad (1200-2300°\text{K}) \qquad (3.121)$$

Slightly shorter times were found by Kiefer and Lutz[67] for $D_2(\text{Ar})$.

Kiefer and Lutz found a discrepancy in their density measurements which they attributed to a faster relaxation of *para*-D_2 than *ortho*-D_2. Further analysis of the data[69] revealed an error so that the possibility of double relaxation was eliminated.

As with H_2 the results do not fit the Millikan–White correlation or the SSH theory. Parker[149] was able to fit the results for the D_2 (inert gas) collisions with his own theory but the results for the pure gases did not agree. Since the theory included the possibility of rotation–vibration transfer, he put forward the rather curious suggestion that the ability to exchange rotational energy between the collision partners seemed to impede the exchange of vibrational energy.

3.6.4 Nitrogen

(See Figures 3.18 and 3.19). There have been extensive studies of the relaxation of N_2. Blackman's early work[61] using interferometry still stands at the highest temperature (5000°K), but the small water contents of the nitrogen used gave Napier times which were too low at the lower temperature (3500°K). The best results now are those of Hurle[76,141] who used sodium-line reversal and Millikan and White[88] who used interferometry and a sensitized-infrared-emission technique. In this method a small amount of CO (1%) was added to the nitrogen and the infrared emission from CO was observed. The observed relaxation exactly paralleled that observed by interferometry. It was concluded that there was ready exchange between CO and N_2:

$$N_2(v = 1) + CO(v = 0) = N_2(v = 0) + CO(v = 1) \qquad (3.122)$$

CO having a long Napier time itself does not, at a 1% concentration, appreciably affect the relaxation of N_2. The Napier time fits the equation:

$$N_2(N_2): \log_{10} \tau(\text{atm s}) = 102T^{-1/3} - 11 \cdot 24 \qquad (1500–6000°K) \quad (3.123)$$

Appleton and Steinberg have extended the range beyond 6000°K[102] by observing the vacuum ultraviolet absorption at a wavelength which corresponds to transitions from the thirteenth vibrational level. Their results agree with the extrapolation of Equation (3.122) although the Napier times could be up to a factor of three longer. At low temperatures the Napier time has been measured with the impact tube[150] and by a resonance method[151]. The values obtained are shorter than those predicted by extrapolating Equation (3.122) and the discrepancy is generally attributed to the presence of impurities in the N_2 used[132].

Nitrogen has been the gas primarily used in the expansion experiments where anomalously short relaxation times have been found. In these experiments the Napier time is not estimated directly: the observed vibrational temperatures in the nozzle are fitted with a model for the expansion in which the Napier time is used as a parameter. It has been found by Hurle, Russo and Hall[18,21,23,122] that the results could only be fitted with a Napier time which was one hundredth of the values given by Equation (3.122). A similar result was obtained in unsteady expansion experiments by Holbeche and Woodley[26,27] who also used line reversal to measure the vibrational temperature. Sebacher[25] has also observed the effect in experiments where the temperature was measured by observing the fluorescence of N_2 excited by an electron beam. Both Hurle[19] and Russo[22] have checked the conditions of their experiments very carefully and have eliminated the possibilities that either free electrons or impurities are interfering in the experiments. Guenoche and Chareyre[152] have criticized the use of the line-reversal technique in these experiments but their criticism was effectively answered by Hurle[141].

The explanation for this effect is not yet clear* : Hurle, as has been mentioned (Section 3.5.7), has suggested that $\Delta v > 1$ transitions in the higher levels may be responsible. Rich and Rehm[123,124] have suggested that the effect could occur for an anharmonic oscillator model ($\Delta v = 1$ transitions only) under extreme conditions (see also Rankin and Light[125]).

Both the Millikan and White correlation and the SSH theory successfully predict the Napier time for N_2.

The Napier time for $N_2(H_2)$ collisions is much lower than that for $N_2(N_2)$ (Table 3.1). White[134] has studied N_2/H_2 mixtures in the range 1150–1600°K and found that the equation fitted:

$$N_2(H_2): \log_{10} \tau(\text{atm s}) = 35 T^{-1/3} - 8.62 \qquad (3.124)$$

The points are scattered however and any extrapolation of Equation (3.124) may give erroneous results.

The ready exchange of vibrational energy between N_2 and CO has already been mentioned[88] although this gives no measure of τ for $N_2(CO)$.

Water is extremely effective in the removal of vibrational energy from N_2. Estimates of the relative collision efficiency of $H_2O:N_2$ vary from ~ 10–100 at 2500°K[149] to 1100 at 560°K[150]. At higher temperatures the effect of impurities such as O_2, H_2O and C_2H_2 on N_2 relaxation has been examined up to concentrations of 0·1%. They had little effect[22].

Levitt and Sheen[96,97] measured the Napier time for N_2 by adding SO_2 (1%) and observing the ultraviolet emission. Their results were in reasonable agreement with those of other workers and they concluded that SO_2 was about one fifth as effective as N_2 itself in exciting N_2; $5\tau[N_2(SO_2)] = \tau[N_2(N_2)]$.

Modica and La Graff[153] have studied the dissociation of C_2H_4 diluted with N_2. From the analysis of their results they suggested that the probability of removal of vibrational energy from nitrogen by CF_2 and C_2F_4 was nearly one thousand times greater than by nitrogen itself.

The effect of N_2 relaxation on the dissociation of N_2[17,154,155] and on the oxidation of ammonia in a nitrogen atmosphere[156] has been examined.

* But see addenda, p. 264.

There seems little evidence for coupling between relaxation and chemical reaction in these cases.

3.6.5 Oxygen

Lutz and Kiefer[120] have determined the Napier time of O_2 over the range 1000–3000°K with their laser schlieren technique and found that their results agreed well with White and Millikan[137] who used interferometry (600–2600°K) and Generalov and Losev[136] (1200–7000°K) who observed the relaxation by following the change in the ultraviolet-absorption spectrum. The results then for the temperature range 600 to 7000°K (Figure 3.23) fit the formula:

$$O_2(O_2): \log_{10} \tau(\text{atm s}) = 54 \cdot 7 T^{-1/3} - 9 \cdot 535 \qquad (3.125)$$

The equation also fits the earlier results of Blackman[61] at high temperatures although his results at lower temperatures ($\sim 1000°K$) are low. It has been mentioned (Section 3.5.7) that Lutz and Kiefer treated the whole

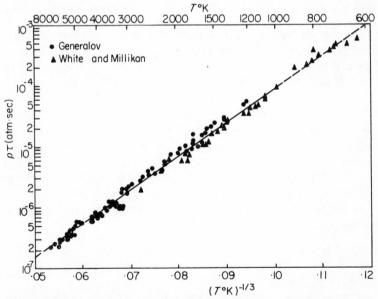

Figure 3.23 Results for O_2 (after Lutz and Kiefer). The solid line shows the least squares line for the Lutz and Kiefer measurements together with extrapolation. Reproduced by permission of the author and the Editor, *J. Chem. Phys.*

of the relaxation in each experiment rather than determining a single point from each run. As they point out, if the other workers had done the same the excellent agreement would be marred. The extrapolation of equation (3.125) to 300°K gives a value twice as large as that obtained by Parker (1.8×10^4 μs at 295°) using an acoustic resonance tube[157] and by Holmes, Smith and Tempest (1.6×10^4 μs at 300°K)[158].

Napier times for pure oxygen agree well with predictions from both SSH theory and the Millikan–White correlation.

Millikan and White have made several studies of the effect of additives on the relaxation of oxygen (Table 3.1). Light gases such as H_2, D_2 and He all reduce the relaxation time considerably: the equations for the Napier times are[63,138,139,159]:

$$O_2(H_2): \log_{10} \tau(\text{atm s}) = 15.62 T^{-1/3} - 8.00 \qquad (400–3000°K) \qquad (3.126)$$

$$O_2(D_2): \log_{10} \tau(\text{atm s}) = 27.3 T^{-1/3} - 8.35 \qquad (400–1000°K) \qquad (3.127)$$

$$O_2(He): \log_{10} \tau(\text{atm s}) = 26.2 T^{-1/3} - 8.28 \qquad (400–1200°K) \qquad (3.128)$$

$$O_2(Ar): \log_{10} \tau(\text{atm s}) = 70 T^{-1/3} - 10.00 \qquad (1000–2700°K) \qquad (3.129)$$

The value for $O_2(He)$ at 300°K extrapolated from equation (3.128) is in good agreement with those obtained by Holmes, Smith and Tempest[158], 45 μs, and by Parker[157], 32 μs. It also fits the results of Shields and Lee[160] at 480°K, as do the extrapolations for $O_2(H_2)$ and $O_2(D_2)$.

It can be seen that He and D_2 have similar effects, which suggests that for molecules with small interaction potentials the mass is the predominant factor in determining the relaxation time. For the combustible gases the interferograms show that the relaxation is complete before chemical reaction occurs, but above 1500°K the apparent Napier time for $O_2(D_2)$ falls below the extrapolation of equation (3.127); the relaxation is probably shortened by the presence of combustion products.

The Napier times for mixtures of oxygen and hydrocarbons have also been measured[139]:

$$O_2(CH_4): \log_{10} \tau(\text{atm s}) = 18 T^{-1/3} - 8.7 \qquad (450–1300°K) \qquad (3.130)$$

$$O_2(C_2H_4): \log_{10} \tau(\text{atm s}) = -6.92 \qquad (450–1300°K) \qquad (3.131)$$

$$O_2(C_2H_2): \log_{10} \tau(\text{atm s}) = 7.5 T^{-1/3} - 7.3 \qquad (450–1300°K) \qquad (3.132)$$

The absence of a temperature dependence for the $O_2(C_2H_4)$ Napier time is notable. The times (Table 3.1) are very short as might be expected with polyatomic molecules. The extrapolated value for $O_2(CH_4)$ agrees with the result of Schnaus[161]. It is possible that the longer time for $O_2(C_2H_2)$ may be due to the lack of small moment of inertia in C_2H_2, which both CH_4 and C_2H_4 possess and which might preclude the possibility of any rotation–vibration exchange (Sections 3.5.5 and 3.5.9). At higher temperatures the relaxation is accelerated by chemical reaction.

The results for the hydrocarbons were obtained with the aid of Equation (3.100) assuming a linear dependence of the reciprocal relaxation time upon mole fraction. Jones, Lambert and Stretton[162] have criticized this procedure: with mixtures of oxygen and water a quadratic dependence has been found[163,164] which is explicable if energy is transferred from vibration to vibration as well as from vibration to translation (Section 3.5.10). As White himself pointed out[139] this mechanism may well be operative with O_2 (hydrocarbon) mixtures and if so, the Equations (3.130), (3.131) and (3.132) are in error. The results of Schnaus[161] for $O_2(CH_4)$ and $O_2(CD_4)$ at room temperature could be fitted to such a quadratic plot.

A general study of the effect on the relaxation of O_2 of a number of possible additives at concentrations equivalent to impurity level has been made by Generalov[165].

3.6.6   Air

Air is a special case of a mixture of nitrogen and oxygen and its relaxation has been observed by White and Millikan using interferometry[166]. Two processes can be seen: the rapid relaxation of O_2 followed by the slow relaxation of N_2. In these experiments the relaxation time of N_2 was faster than that for pure N_2:

$$\tau[O_2(\text{mixture})] < \tau[N_2(\text{mixture})] < \tau[N_2(\text{pure})] \qquad (3.133)$$

and it was concluded the relaxation of N_2 is accelerated by exchange of vibrational energy with O_2. Bauer and Roesler obtained a similar result from ultrasonic measurements at lower temperatures[167].

3.6.7   Carbon Monoxide

The first study of CO in a shock tube was made by Windsor, Davidson and Taylor[84,85] who observed the infrared emission at $2\cdot3\,\mu$ from the reflected shock in the gas. Their times were found to be a little long by later workers. Matthews[142] used interferometry, Gaydon and Hurle[140] the line-reversal technique and Hooker and Millikan[87,143] observed the infrared emission from the fundamental band. The Napier time is given by the equation:

$$CO(CO): \log_{10} \tau(\text{atm s}) = 69 \cdot 5 T^{-1/3} - 9 \cdot 65 \qquad (1000-6000°K) \qquad (3.134)$$

The Napier time at room temperature is difficult to determine since the molecule is unusual in that its radiative lifetime is shorter than its Napier time. In addition the relaxation is considerably accelerated by small traces of impurity. Millikan[11] using his vibrational fluorescence method (Section 3.2.6) and Ferguson and Read[10] with a spectrophone have estimated the order of magnitude of the Napier time. The extrapolated shock value (Table 3.1) is in the correct range. Doyenette and Henry[9] have estimated the value to be 6 μs.

The Napier time for CO is satisfactorily predicted by SSH theory, and fits the Millikan–White correlation. The shock excitation of the second vibrational level observed by Hooker and Millikan[87] is dealt with in Section 3.5.7.

The effect of additives on the relaxation has been extensively studied by Millikan and White[11,130,143]. Figure 3.24 shows some of their results. The equations are:

$$CO(H_2): \log_{10} \tau(\text{atm s}) = 29 \cdot 1 T^{-1/3} - 8 \cdot 49 \qquad (500-2000°K) \qquad (3.135)$$

$$CO(He): \log_{10} \tau(\text{atm s}) = 37 \cdot 8 T^{-1/3} - 8 \cdot 30 \qquad (580-1500°K) \qquad (3.136)$$

$$CO(D_2): \text{as for } CO(He) \qquad (940-1600°K) \qquad (3.137)$$

$$CO(Ne): \log_{10} \tau(\text{atm s}) = 61 \cdot 7 T^{-1/3} - 9 \cdot 16 \qquad (1400-3000°K) \qquad (3.138)$$

$$CO(Ar): \log_{10} \tau(\text{atm s}) = 79 \cdot 1 T^{-1/3} - 9 \cdot 74 \qquad (1700-2700°K) \qquad (3.139)$$

$$CO(Kr): \log_{10} \tau(\text{atm s}) = 81 \cdot 8 T^{-1/3} - 9 \cdot 70 \qquad (2100-7000°K) \qquad (3.140)$$

The equations for $CO(H_2)$ and $CO(He)$ may be extrapolated to 300°K to give values in agreement with those obtained from vibrational fluorescence[11]. The extrapolated value at 300°K for $CO(Ne)$ is much longer than the measured value which is now thought to be in error.

Below 1600°K the Napier time for $CO(D_2)$ is the same as that for $CO(He)$ demonstrating once again, as do the other results for CO (inert gas), the predominant effect of the mass of the collision partners on the relaxation time where the interaction potentials are small. Between 1600°K and 2800°K the relaxation for $CO(D_2)$ is faster than that given by Equation (3.137). White[130] suggested that this may be due to vibration–

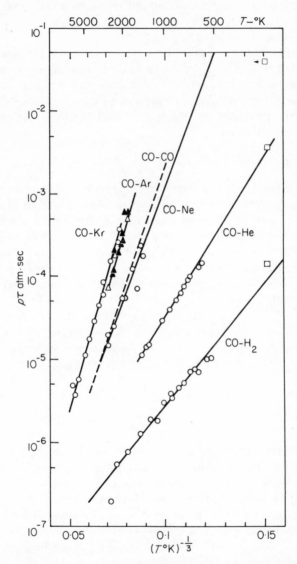

Figure 3.24 Results for CO with various added gases.
Reproduced by permission of the author and the Editor,
J. Chem. Phys.

vibration energy transfer pointing out that only above 1600°K is there any appreciable population in the first and higher vibrational states of D_2. This has been discussed in Section 3.5.10.

The rapid transfer of vibrational energy between CO and N_2 which enabled the Napier times for N_2 to be determined by observation of the relaxation of CO in the mixture[88] has been mentioned in Section 3.6.4. Exchange of vibrational energy between CO and CH_4 has been demonstrated by vibrational fluorescence[168] and between NO and CO by flash photolysis[34].

3.6.8   Nitric Oxide

Both the ground ($^2\Pi$) and first excited ($^2\Sigma$) states of NO have been studied by flash photolysis and shock techniques. Robben studied the relaxation of the ($^2\Pi$) state by observing the change in ultraviolet absorption[98] and infrared emission in the shock-heated gas. Wray[146] observed the change in ultraviolet absorption at 1270 Å and so measured the relaxation time over the range 1000–7000°K. The results were somewhat lower than those of Robben where the measurements overlapped; both could be extrapolated to agree with the room-temperature results obtained by Bauer and his coworkers[145,169] using ultrasonic techniques and Basco, Callear and Norrish [144] who used flash photolysis. As can be seen from Figure 3.25, Wray's results cannot be fitted to a single simple equation. They can usefully be approximated by two equations of the usual form:

$$NO(^2\Pi)(NO): \log_{10} \tau(\text{atm s}) = 12\cdot8 T^{-1/3} - 8\cdot3 \quad (300–2000°K)$$

$$= 56 T^{-1/3} - 11\cdot4 \quad (3000–7000°K) \quad (3.141)$$

These results are considerably lower than those predicted by SSH theory using the Lennard-Jones parameters. Robben[98] was able to obtain a better fit at low temperatures using a potential in which the possibility of dimer formation was taken into account. The SSH theory was developed on the assumption that collisions were adiabatic, that is there is no change of electronic configuration on collision. Nikitin[170] has pointed out that for NO, collisions are likely to be non-adiabatic and he has calculated the rate of such processes. Wray[146] has shown that even when this is taken into account there is still a discrepancy between the predicted and experimental results. The results and theoretical predictions are illustrated in Figure 3.25.

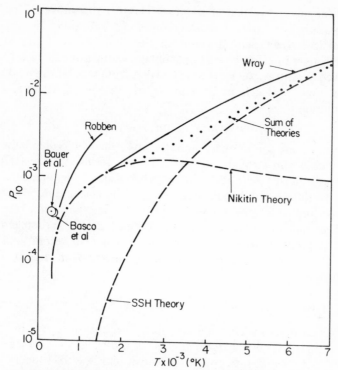

Figure 3.25 Results for NO (after Wray) obtained by various workers compared with theoretical predictions. Reproduced by permission of the author and the Editor, *Phys. Fluids*

Wray's measurements were made with mixtures of NO and Ar and thus the Napier time for NO(Ar) could be estimated:

$$NO(^2\Pi)(Ar): \log_{10} \tau(\text{atm s}) = 61 T^{-1/3} - 10.05 \qquad (2000\text{--}7000°\text{K})$$

$$(3.142)$$

At room temperature[171] the Napier time for NO(Ar) has also been shown to be much longer than that for NO(NO). This is in contrast to relaxation times for rotational and electronic relaxation where NO and Ar were found to be equally effective collision partners. The difference can be attributed to non-adiabatic collisions enhancing the effect of NO–NO collisions as has been mentioned.

Callear has shown that at room temperature there is ready exchange of vibrational energy between NO and CO^{34} and also[36] that hydrides are particularly efficient in removing vibrational energy from $NO(^2\Pi)$.

The Napier time for the $(^2\Sigma)$ state of NO was measured by Roth[95] who observed the ultraviolet emission at 2479 Å from shock-heated NO. The Napier times varied from 28 μs at 5000°K to 5 μs at 10,000°K and fit the equation:

$$NO(^2\Sigma)(NO): \log_{10} \tau(atm \ s) = 60T^{-1/3} - 8 \cdot 05 \qquad (3.143)$$

These times are longer than those for any other molecule at the same temperature and strongly contrast with the results for the $(^2\Pi)$ state which are shorter than other diatomic molecules. The difference could lie in the different electronic state; the Nikitin calculations[170] have yet to be performed for this system.

Callear and Smith[35] have shown that for the $(^2\Sigma)$ state at room temperature the probabilities for de-excitation of the first, second and third vibrational levels are roughly in the ratio predicted by the Landau–Teller theory.

3.6.9   Hydrogen Halides*

The Napier times for HCl and HBr have been measured by observation of the infrared emission from the shock-heated gases. The result for $HCl^{86,91}$ at 2000°K is in rough agreement with the SSH theory, Table 3.1, but a $\log \tau/T^{-1/3}$ plot will have to be curved if the result of Ferguson and Read[12] with the spectrophone is to fit. The result for HBr[91] is much lower than that predicted by SSH theory and both HCl and HBr fall several orders of magnitude below the predictions of the Millikan–White correlation. These discrepancies may be due to the relaxation occurring through a mechanism of rotation–vibration exchange (Section 3.5.9).

Chow and Greene[101] measured the rate constants for the removal of vibrational energy in mixtures of HI with Ar (see Figure 3.20). The Napier times can be approximately fitted by the equations:

$$HI(HI): \log_{10} \tau(atm \ s) = 28 \cdot 4T^{-1/3} - 8 \cdot 86 \qquad (1400-2400°K) \qquad (3.144)$$

$$HI(Ar): \log_{10} \tau(atm \ s) = 17 \cdot 6T^{-1/3} - 7 \cdot 14 \qquad (1400-2400°K) \qquad (3.145)$$

Both Chow and Greene[101] and Borrell and Gutteridge[86] have observed that the first, second and third levels of HCl and HI relax together with

* See addenda, p. 264.

the first. The implications for the Landau–Teller model are discussed in section 3.5.8.

3.6.10 Chlorine

Smiley and Winkler[62] studied the relaxation of chlorine using interferometry: the Napier times can be fitted with the equation:

$$Cl_2(Cl_2): \log_{10} \tau(atm\ s) = 30T^{-1/3} - 9.85 \qquad (500–1500°K) \qquad (3.146)$$

The extrapolated value agrees well with the results at room temperature[132] but the results are shorter than those which Shields[172] obtained at higher temperatures by sound absorption measurements.

3.6.11 Cyanide Radical

The ultraviolet emission from impure Xe was found to be due to the CN radical, and the change with the time was attributed to vibrational relaxation[147]. The Napier time fits the equation:

$$CN(Xe): \log_{10} \tau(atm\ s) = 77T^{-1/3} - 8.47 \qquad (6000–10,000°K) \qquad (3.147)$$

The time (table 3.1) seems very long in comparison with those measured for less reactive molecules.

3.6.12 Nitrous Oxide

N_2O was first studied in a shock tube by Griffith, Brickl and Blackman[173] using interferometry. Their results agreed with those of Euken and Numann[174] who had studied the sound absorption over a wide temperature range. Bhangu also used interferometry[14]; he analysed his results using the method of Johannesen[74] (see Section 3.5.7) in which, instead of obtaining a single point for each experiment, the record yields a continuous series of results for the temperature variation in the experiment. As with CO_2 and O_2 the records did not coalesce to form a single line as they should if Landau–Teller theory is obeyed*. The results, in near agreement with those of Griffith, Brickl and Blackman[173] may be fitted approximately by the equation:

$$N_2O(N_2O): \log_{10} \tau(atm\ s) = 15.6T^{-1/3} - 8.25 \qquad (400–1000°K) \quad (3.148)$$

The extrapolated room temperature result is in rough agreement with the ultrasonic results at room temperature[175,176].

* See addenda, p. 264.

With a polyatomic molecule there is the possibility that individual vibrational modes may have different Napier times. Although at lower temperatures it is the v_1 and v_2 modes of N_2O which make the greatest contribution to the specific heat and observed interferometric changes, Bhangu[14] was able to resolve the contribution of v_3 and reported that all the modes relaxed together. However a result obtained from observation of infrared emission of the shock-heated gas[91] gives a longer value for v_3 at 650°K. At room temperature Cottrell and his coworkers[31] using a spectrophone found that the v_3 vibration had a longer Napier time than the other vibrations.

3.6.13   Carbon Dioxide

CO_2 was first studied by Smiley and Winkler[62] using interferometry. Subsequent interferometric studies have been made by Griffith, Brickl and Blackman[173] and by Johannesen, Zienkiewicz, Blythe and Gerrard[15]. Witteman[70] has studied CO_2 using an integrated schlieren technique. The times obtained by Smiley and Winkler are shorter than those of the other workers and this could well be due to a higher water content of their CO_2. The results of Johannesen and others agree with Griffith and coworkers and both overlap with the ultrasonic results at low temperatures. Those of Witteman may also overlap with the low temperature results but fall increasingly below those of other workers at higher temperatures. The results of Daen and de Boer[71] on the other hand are higher than those of any other workers and do not agree with the extensive results from ultrasonic measurements at room temperature. Neither the results of Witteman nor those of Johannesen fit a straight line log $\tau/T^{-1/3}$ plot since both fall away at the higher temperature recorded. Camac[148] has measured Napier times above 2000°K. It can be seen from Figure 3.26 that these agree roughly with those at lower temperatures although the scatter is larger. The Napier time for Camac's line is expressed by:

$$CO_2(CO_2): \log_{10} \tau(\text{atm s}) = 15{\cdot}9T^{-1/3} - 7{\cdot}70 \qquad (300\text{–}5000°\text{K}) \qquad (3.149)$$

This gives a result which is roughly a factor of two too small at room temperature.

As with N_2O (Section 3.6.12) there is the possibility that the different vibrational modes will not have the same Napier time. In an early measurement Greenspan and Blackman[177] claimed that each mode had a separate Napier time but this has been discounted by Johannesen and

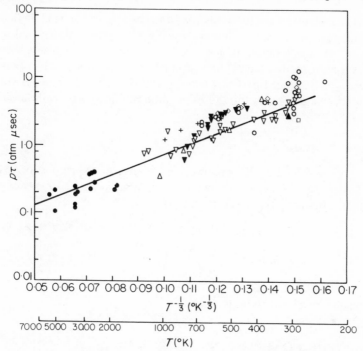

Figure 3.26 Experimental results for CO_2 (after Camac). The fitted points are Camac's results; the remainder are the results of other workers. Reproduced by permission of the author and the Cornell University Press ©

coworkers[15]. Hurle and Gaydon[178] observed a relaxation in CO_2 at high temperatures (2500°K) with the line-reversal method and attributed this to relaxation of the v_3 vibration. The time was several orders of magnitude longer than that predicted for v_1 and v_2 at this temperature and as long as times for 'inert' molecules like CO_2 and N_2. Borrell[91] observed the infrared emission from the v_3 bend of shock-heated CO_2: the Napier time was about twice the time obtained from interferometry for the v_1 and v_2. Johannesen and others[15] from their interferometric work concluded that all three modes must relax with approximately the same Napier time: recently they too have observed the infrared emission and have also obtained a longer value for v_3[179]. Simpson[93] has contradicted this. Above 2000°K all modes are thought to relax together[148]. Cottrell and others[31] working with a spectrophone at room temperature found the same Napier time for v_3 as the other modes but suggested that the errors were such that v_3 could possibly relax more slowly. Slobodskaya[180,181] has found a longer time for v_3 than v_1 at room temperature. Hocker and his coworkers[39] have studied the relaxation of v_3 by observing fluorescence induced by Q-switched laser. Their result for v_3 is shorter than the Napier times for the other modes. Cheo has made a similar study[182].

3.6.14   Methane

Richards and Sigafoos[100] observed the change in absorption of CH_4 at 1470 Å, in shock-heated mixtures of CH_4 and Ar and so measured the Napier times for $CH_4(CH_4)$ and $CH_4(Ar)$:

$$CH_4(CH_4): \log_{10} \tau(\text{atm s}) = 17 \cdot 4T^{-1/3} - 8 \cdot 34 \qquad (740–1600°\text{K}) \qquad (3.150)$$

$$CH_4(Ar): \log_{10} \tau(\text{atm s}) = 26 \cdot 6T^{-1/3} - 8 \cdot 7 \qquad (740–1600°\text{K}) \qquad (3.151)$$

The extrapolated values at 300°K from equation (3.150) agree well with ultrasonic results at room temperature[132].

3.7   SOME CONCLUSIONS

The present situation in the prediction and measurement of Napier times is fairly satisfactory: it is possible to estimate the Napier times of single stable molecules to within an order of magnitude by use of theory except when the molecule contains hydrogen. Similar satisfactory estimates can be made from empirical formulae. Few measurements have yet been made of the Napier times of free radicals and excited states, and those that have point to difficulties in the system where the intermolecular potentials are likely to be unusual and are unmeasurable by normal means.

The anomalous results for hydrogen and hydrogen-containing molecules seems attributable on a qualitative basis to the possibility of energy exchange between rotation and vibration, but a theory of the same applicability as the SSH approach is still lacking.

The adequacy of the Landau–Teller model is questionable: clearly the results of Johannesen and his coworkers[15,16] and Lutz and Kiefer[120] need following up. Urgently required are further results for the rate of population of higher levels, perhaps by methods similar to that of Appleton and Steinberg[102] and Chow and Greene[101].

The anomalous effects in cooling seem well established and if the theoretical conclusions are correct they may be explained if the occurrence of $\Delta v > 1$ transitions and faster transitions than the Landau–Teller theory predicts are allowed for higher levels.

Mixtures seem less well understood, partly because of the difficulty of distinguishing energy exchange by vibration–vibration transfer from the normal vibration–translation process. The differences in Napier time for $CO_2(He)$ and $CO_2(D_2)$ noticed by Cottrell and Day[129] require explanation in the light of other experiments in which the Napier times for mixtures with He and D_2 have been found to be the same.

The solution to problems of mixtures may be better tackled by ultrasonic measurements at room temperature, but clearly the effect of temperature is of much interest in this context. The effects noticed in $CO(D_2)$ mixtures by White[130] require further study.

The exchange of energy in polyatomic molecules is of direct interest in chemical kinetics but because of the shortness of the Napier times at higher temperatures is probably best studied by methods other than shocks.

REFERENCES

1. S. H. Bauer and K. Kuratani, *J. Am. Chem. Soc.*, **87**, 150 (1965).
2. S. H. Bauer, A. Lifschitz and C. Lifschitz, *J. Am. Chem. Soc.*, **87**, 143 (1965).
3. S. H. Bauer and E. Ossa, *J. Chem. Phys.*, **45**, 434 (1966).
4. S. H. Bauer and E. L. Resler, *Science*, **146**, 1045 (1965).
5. S. H. Bauer, W. S. Watt, P. Borrell and D. Lewis, *J. Chem. Phys.*, **45**, 444 (1966).
6. G. H. Kohlmaier and B. S. Rabinovitch, *J. Chem. Phys.*, **38**, 1709 (1963).
7. D. W. Setser, B. S. Rabinovitch and J. W. Simons, *J. Chem. Phys.*, **40**, 1751 (1964).
8. J. W. Simons, B. S. Rabinovitch and D. W. Setser, *J. Chem. Phys.*, **41**, 800 (1964).
9. L. Doyennette and L. Henry, *J. Phys.* (*Paris*), **27**, 485 (1966).
10. M. G. Ferguson and A. W. Read, *Trans. Faraday Soc.*, **61**, 1559 (1965).
11. R. C. Millikan, *J. Chem. Phys.*, **38**, 2855 (1963).
12. A. W. Read and M. G. Ferguson, *Trans. Faraday Soc.*, **63**, 61 (1967).
13. M. C. Henderson, *Phys. Today*, **16** No. 1, 84 (1963).
14. J. K. Bhangu, *J. Fluid Mech.*, **25**, 817 (1966).
15. N. H. Johannesen, H. K. Zienkiewicz, P. A. Blythe and J. H. Gerrard, *J. Fluid Mech.*, **13**, 213 (1962).
16. H. K. Zienkiewicz and N. H. Johannesen, *J. Fluid Mech.*, **17**, 499 (1963).
17. B. Cary, *Phys. Fluids*, **8**, 26 (1965).
18. J. G. Hall and A. L. Russo, *Recent Advances in Aerothermochemistry*, Conference Proceedings, 12, AGARD, Paris, 1967, p. 443.
19. I. R. Hurle, *J. Chem. Phys.*, **41**, 3592 (1964).

20. I. R. Hurle and A. L. Russo, *J. Chem. Phys.*, **43**, 4434 (1965).
21. I. R. Hurle, A. L. Russo and J. G. Hall, *J. Chem. Phys.*, **40**, 2076 (1964).
22. A. L. Russo, *J. Chem. Phys.*, **44**, 1305 (1966).
23. A. L. Russo, *Cornell Aero. Laboratory Report*, AD-1689-A-8, (1967).
24. E. L. Harris and L. M. Albacete, *NASA Rept. AD 601590*, 1964.
25. D. I. Sebacher, *Proc. Heat Transfer and Fluid Mechanics Institute*, Stanford University Press, Stanford, 1966, p. 135.
26. T. A. Holbeche, *Nature*, **203**, 476 (1964).
27. T. A. Holbeche and J. G. Woodley, *Recent Advances in Aerothermochemistry*, Conference Proceedings, 12, AGARD, Paris, 1967, p. 507.
28. T. L. Cottrell and J. C. McCoubrey, *Molecular Energy Transfer in Gases*, Butterworths, London, 1961.
29. K. F. Herzfeld and T. A. Litovitz, *Absorption and Dispersion of Ultrasonic Waves*, Academic Press, New York, 1959.
30. R. Kaiser, *Can. J. Phys.*, **37**, 1499 (1959).
31. T. L. Cottrell, I. M. Macfarlane, A. W. Read and A. H. Young, *Trans. Faraday Soc.*, **62**, 2655 (1966).
32. R. G. W. Norrish, *Proc. 12th Solvay Conference on Chemistry*, Interscience, New York, 1963, p. 91
33. G. Porter, *Technique of Organic Chemistry*, Vol. 8 (part II), 2nd ed., Interscience, New York, 1963, p. 1055.
34. A. B. Callear, *Discussions Faraday Soc.*, **33**, 28 (1962).
35. A. B. Callear and W. M. Smith, *Trans. Faraday Soc.*, **59**, 1735 (1963).
36. A. B. Callear and G. J. Williams, *Trans. Faraday Soc.*, **62**, 2030 (1966).
37. E. D. Bugrim, A. I. Liutyi and V. S. Rossikhin, *Opt. i Spektroskopiya*, **21**, 27 (1966).
38. J. T. Yardley and C. B. Moore, *J. Chem. Phys.*, **45**, 1066 (1966).
39. L. O. Hocker, M. A. Kovacs, C. K. Rhodes, G. W. Flynn and A. Javan, *Phys. Rev. Letters*, **17**, 233 (1966).
40. R. L. Brown and W. Klemperer, *J. Chem. Phys.*, **41**, 3072 (1964).
41. J. I. Steinfeld and W. Klemperer, *J. Chem. Phys.*, **42**, 3475 (1965).
42. F. De Martini and J. Ducuing, *Phys. Rev. Letters*, **17**, 117 (1966).
43. J. N. Bradley, *Shock Waves in Chemistry and Physics*, Methuen, London, 1962.
44. A. G. Gaydon and I. R. Hurle, *The Shock Tube in High Temperature Chemical Physics*, Chapman & Hall, London, 1963.
45. I. I. Glass, *Theory and Performance of Simple Shock Tubes, Vols. I and II*, Institute of Aerophysics, Report No. 12, 1958.
46. E. F. Greene and J. P. Toennies, *Chemical Reactions in Shock Waves*, Arnold Publishers, London, 1964.
47. W. G. Vincenti and C. H. Kruger, *Introduction to Physical Gas Dynamics*, Wiley, New York, 1965.
48. J. K. Wright, *Shock Tubes*, Methuen, London, 1961.
49. I. B. Zeldovitch and A. S. Kompaneets, *Theory of Detonation*, Academic Press, New York, 1960.
50. A. H. Schapiro, *Compressible Fluid Flows, Vols. I and II*, Ronald Press, New York, 1953.
51. M. J. Zucrow, *Aircraft and Missile Propulsion*, Vol. I, Wiley, New York, 1958.

52. B. J. McBride, S. Heimel, J. G. Ehlers and S. Gordon, *N.A.S.A. S.P.-3001*, Washington, 1963.
53. *Landholt Bornstein Tables*, Vol. 2, Part 4, Springer Verlag, Berlin, 1961.
54. *JANAF Thermochemical Tables*, The Dow Chemical Company, Midland Michigan, 1965.
55. R. C. Millikan, *Report 64-RL-3700C*, General Electric Co., Schenectady, New York, 1964.
56. R. C. Millikan, *Report 65-C-106*, General Electric Co., Schenectady, New York, 1965.
57. G. S. Campbell and P. C. Klimas, *AIAA Journal*, **5**, 1713 (1967).
58. H. Mirels, *Phys. Fluids*, **6**, 1201 (1963).
59. J. N. Fox, T. I. McLaren and R. M. Hobson, *Phys. Fluids*, **9**, 2345 (1966).
60. See reference 44, p. 166.
61. V. H. Blackman, *J. Fluid Mech.*, **1**, 61 (1956).
62. E. F. Smiley and E. H. Winkler, *J. Chem. Phys.*, **22**, 2018 (1954).
63. D. R. White, *J. Chem. Phys.*, **42**, 447 (1965).
64. *Physical Measurements in Gas Dynamics and Combustion. Vol. IX of High Speed Aerodynamics*, Oxford University Press, 1955.
65. A. L. Besse and J. G. Kelley, *Rev. Sci. Instr.*, **37**, 1497 (1966).
66. M. McChesney and M. Bristow, private communication.
67. J. H. Kiefer and R. W. Lutz, *J. Chem. Phys.*, **44**, 658 (1966).
68. J. H. Kiefer and R. W. Lutz, *J. Chem. Phys.*, **44**, 668 (1966).
69. J. H. Kiefer and R. W. Lutz, *J. Chem. Phys.*, **45**, 3888 (1966).
70. W. J. Witteman, *J. Chem. Phys.*, **37**, 655 (1962).
71. J. Daen and P. C. T. de Boer, *J. Chem. Phys.*, **36**, 1222 (1962).
72. J. B. Moreno, *Phys. Fluids*, **9**, 431 (1966).
73. P. A. Blythe, *J. Fluid Mech.*, **10**, 33 (1961).
74. N. H. Johannesen, *J. Fluid Mech.*, **10**, 25 (1961).
75. A. G. Gaydon and I. R. Hurle, *Proc. Roy. Soc. (London), Ser. A*, **262**, 38 (1961).
76. I. R. Hurle, *J. Chem. Phys.*, **41**, 3911 (1964).
77. D. R. Jenkins, *Proc. Roy. Soc. (London) Ser. A*, **293**, 493 (1966).
78. R. G. W. Norrish and W. MacF. Smith, *Proc. Roy. Soc. (London), Ser. A*, **176**, 295 (1940).
79. S. Tsuchiya and K. Kuratani, *Combust. Flame*, **8**, 299 (1964).
80. A. B. Callear, *Appl. Opt. Suppl.*, **2**, 145–70 (1965).
81. P. G. Dickens, J. W. Linnett and O. Sovers, *Discussions Faraday Soc.*, **33**, 52 (1962).
82. N. N. Sobolev, A. V. Potapov, V. F. Kitaeva, F. S. Faizullov, V. N. Alyamovskii, E. T. Antropov and I. L. Isaev, *Opt. i Spectroskopiya*, **6**, 284 (1959).
83. E. H. Carnevale, S. Wolnik, G. Larson, C. Carey and G. W. Wares, *Phys. Fluids*, **10**, 1459 (1967).
84. M. H. Windsor, N. Davidson and R. Taylor, *J. Chem. Phys.*, **27**, 315 (1957).
85. M. H. Windsor, N. Davidson and R. Taylor, *7th Symp. Combustion*, Butterworths, London, 1959, p. 80.
86. P. Borrell and R. Gutteridge, *Recent Advances in Aerothermochemistry*, Conference Proceedings 12, AGARD, Paris, 1967, p. 131.
87. W. J. Hooker and R. C. Millikan, *J. Chem. Phys.*, **38**, 214 (1963).

88. R. C. Millikan and D. R. White, *J. Chem. Phys.*, **39**, 98 (1963).
89. J. C. Decius, *J. Chem. Phys.*, **32**, 1262 (1960).
90. G. Herzberg, *Spectra of Diatomic Molecules*, Van Nostrand, New York, U.S.A., 1950.
91. P. Borrell, *Chem. Soc. (London) Spec. Publ.* **20**, 1966, p. 263.
92. J. C. Breeze and C. C. Ferriso, *J. Chem. Phys.*, **39**, 2619 (1963).
93. C. J. S. M. Simpson, private communication.
94. M. Lapp, *Phys. Fluids*, **7**, 1233 (1964).
95. W. Roth, *J. Chem. Phys.*, **34**, 999 and 2204 (1961).
96. B. P. Levitt, private communication.
97. B. P. Levitt and D. B. Sheen, *Chem. Soc. (London) Spec. Publ.* **20**, 1966, p. 269.
98. F. Robben, *J. Chem. Phys.*, **31**, 420 (1959).
99. F. Robben, P. R. Monson and J. J. Allport, *J. Chem. Phys.*, **33**, 630 (1960).
100. L. W. Richards and D. H. Sigafoos, *J. Chem. Phys.*, **43**, 492 (1965).
101. C. C. Chow and E. F. Greene, *J. Chem. Phys.*, **43**, 324 (1965).
102. J. P. Appleton and M. Steinberg, *J. Chem. Phys.*, **46**, 1521 (1967).
103. L. D. Landau and E. Teller, *Z. Phys.*, *Sowjetunion*, **10**, 34 (1936).
104. R. Herman and R. J. Rubin, *Phys. Fluids*, **2**, 547 (1959).
105. K. F. Herzfeld, *Proc. Internat. School of Physics*, 'Enrico Fermi', 272, 1963.
106. K. Takayanagi, *Progr. Theoret. Phys. (Kyoto)*, Suppl. No. 25, 1 (1963).
107. K. Takayanagi, *Advan. At. Mol. Phys.*, **1**, 149 (1965).
108. C. Zener, *Phys. Rev.*, **37**, 556 (1931).
109. H. A. Bethe and E. Teller, *Aberdeen Proving Ground Rep. X115.*
110. T. L. Cottrell, R. C. Dobbie, J. McLain and A. W. Read, *Trans. Faraday Soc.*, **60**, 241 (1964).
111. T. L. Cottrell and A. J. Matheson, *Trans. Faraday Soc.*, **58**, 2336 (1962).
112. T. L. Cottrell and A. J. Matheson, *Trans. Faraday Soc.*, **59**, 824 (1963).
113. C. B. Moore, *J. Chem. Phys.*, **43**, 2979 (1965).
114. S. W. Benson and G. C. Berend, *J. Chem. Phys.*, **44**, 470 (1966).
115. S. W. Benson and G. C. Berend, *J. Chem. Phys.*, **44**, 4247 (1966).
116. R. C. Millikan and D. R. White, *J. Chem. Phys.*, **39**, 3209 (1963).
117. R. Phinney, *AIAA Journal*, **2**, 240 (1964).
118. R. R. Boade, *J. Chem. Phys.*, **42**, 2788 (1965).
119. J. D. Lambert and R. Salter, *Proc. Roy. Soc. (London)*, Ser. A, **253**, 277 (1959).
120. R. W. Lutz and J. H. Kiefer, *Phys. Fluids*, **9**, 1638 (1966).
121. R. V. Fitzsimmons and E. J. Bair, *J. Chem. Phys.*, **40**, 451 (1964).
122. I. R. Hurle, *Chem. Soc. (London), Spec. Publ.* **20**, 1966, p. 277.
123. J. W. Rich and R. G. Rehm, *Cornell Aero. Laboratory Report*, AF-2022-A-1, (1966).
124. J. W. Rich and R. G. Rehm, *Cornell Aero. Laboratory Report*, AF-2022-A-2, (1967).
125. C. C. Rankin and J. C. Light, *J. Chem. Phys.*, **46**, 1305 (1967).
126. J. W. Arnold, J. C. McCoubrey and A. R. Ubbelohde. *Trans. Faraday Soc.*, **53**, 738 (1957).
127. B. Widom, *J. Chem. Phys.*, **27**, 940 (1957).
128. R. C. Millikan and L. A. Osburg, *J. Chem. Phys.*, **41**, 2196 (1964).
129. T. L. Cottrell and M. A. Day, *Chem. Soc. (London), Spec. Publ.* **20**, 1966, p. 253.

130. D. R. White, *J. Chem. Phys.*, **45**, 1257 (1966).
131. J. D. Lambert, *Quart. Rev. (London)*, **21**, 67 (1967).
132. A. W. Read, *Progr. in Reaction Kinetics*, **3**, 203 (1965).
133. P. Borrell, *Advances in Molecular Relaxation Studies*, **1**, 69 (1967).
134. D. R. White, *J. Chem. Phys.*, **46**, 2016 (1967).
135. M. Camac, *J. Chem. Phys.*, **34**, 448 (1961).
136. N. A. Generalov and S. A. Losev, *J. Quant. Spectr. Radiative Transfer*, **6**, 101 (1966).
137. D. R. White and R. C. Millikan, *J. Chem. Phys.*, **39**, 1803 (1963).
138. D. R. White and R. C. Millikan, *J. Chem. Phys.*, **39**, 1807 (1963).
139. D. R. White, *J. Chem. Phys.*, **42**, 2028 (1965).
140. A. G. Gaydon and I. R. Hurle, *8th Symposium on Combustion*, Williams and Wilkins, Baltimore, 1962, p. 309.
141. I. R. Hurle, *Recent Advances in Aerothermochemistry*, Conference Proceedings, 12, AGARD, Paris, 1967, p. 125, 130.
142. D. L. Matthews, *J. Chem. Phys.*, **34**, 639 (1961).
143. R. C. Millikan, *J. Chem. Phys.*, **40**, 2594 (1964).
144. N. Basco, A. B. Callear and R. G. W. Norrish, *Proc. Roy. Soc. (London), Ser. A*, **260**, 459 (1961).
145. H. J. Bauer, H. O. Kneser and E. Sittig, *J. Chem. Phys.*, **30**, 1119 (1959).
146. K. L. Wray, *J. Chem. Phys.*, **36**, 2597 (1962).
147. W. Roth, *J. Chem. Phys.*, **31**, 720 (1959).
148. M. Camac, *Fundamental Phenomena in Hypersonic Flows*, (Ed. J. G. Hall), Cornell University Press, 1966, p. 195. Avco Report, **194** (1964).
149. J. G. Parker, *J. Chem. Phys.*, **45**, 3641 (1966).
150. P. W. Huber and A. Kantrowitz, *J. Chem. Phys.*, **15**, 275 (1947).
151. S. J. Lukasik and J. E. Young, *J. Chem. Phys.*, **27**, 1149 (1957).
152. H. Guenoche and R. Chareyre, *Recent Advances in Aerothermochemistry*, Conference Proceedings, 12, AGARD, Paris, 1967, p. 111.
153. A. P. Modica and J. E. La Graff, *J. Chem. Phys.*, **45**, 4729 (1966).
154. B. Cary, *Phys. Fluids*, **9**, 1047 (1966).
155. K. L. Wray and S. Byron, *Phys. Fluids*, **9**, 1046 (1966).
156. H. Miyama and R. Endoh, *J. Chem. Phys.*, **46**, 2011 (1967).
157. J. G. Parker, *J. Chem. Phys.*, **34**, 1763 (1961).
158. R. Holmes, F. A. Smith and W. Tempest, *Proc. Phys. Soc.*, **81**, 311 (1963).
159. D. R. White and R. C. Millikan, *J. Chem. Phys.*, **39**, 2107 (1963).
160. F. D. Shields and K. P. Lee, *J. Chem. Phys.*, **40**, 737 (1964).
161. U. E. Schnaus, *J. Acoust. Soc. Am.*, **37**, 1 (1965).
162. D. G. Jones, J. D. Lambert and J. L. Stretton, *Proc. Phys. Soc.*, **86**, 857 (1965).
163. M. C. Henderson, A. V. Clark and P. R. Lintz, *J. Acoust. Soc. Am.*, **37**, 457 (1965).
164. C. S. Tuesday and M. Boudart, *Princeton University Tech. Note*, AF33(038)-23976, 1955.
165. N. A. Generalov, *Dokl. Phys. Chem.*, **148**, 51 (1963).
166. D. R. White and R. C. Millikan, *AIAA Journal*, **2**, 1844 (1964).
167. H. J. Bauer and H. Roesler, *Chem. Soc. (London), Spec. Publ.*, **20**, 1966, p. 245.
168. R. C. Millikan, *J. Chem. Phys.*, **43**, 1439 (1965).
169. H. J. Bauer and K. F. Sahm, *J. Chem. Phys.*, **42**, 3400 (1965).

170. E. E. Nikitin, *Opt. Spectr. (USSR) (English Transl.)*, **9**, 8 (1960).
171. H. O. Kneser, H. J. Bauer and H. Kosche, *J. Acoust. Soc. Am.*, **41**, 1029 (1967).
172. F. D. Shields, *J. Acoust. Soc. Am.*, **32**, 180 (1960).
173. W. Griffith, D. Brickl and V. Blackman, *Phys. Rev.*, **102**, 1209 (1956).
174. A. Euken and Z. Numann, *Z. Physik. Chem.*, **36B**, 163 (1937).
175. J. W. Arnold, J. C. McCoubrey and A. R. Ubbelohde, *Proc. Roy. Soc. (London), Ser. A*, **248**, 445 (1958).
176. J. V. Martinez, J. G. Strauch and J. C. Decius, *J. Chem. Phys.*, **40**, 186 (1964).
177. W. D. Greenspan and V. H. Blackman, *Bull. Am. Phys. Soc.*, **2**, 217 (1957).
178. I. R. Hurle and A. G. Gaydon, *Nature*, **184**, 1858 (1959).
179. N. H. Johannesen, private communication.
180. P. V. Slobodskaya, *Opt. Spectr. (USSR) (English Transl.)*, **22**, 14 (1967).
181. P. V. Slobodskaya, *Opt. Spectr. (USSR) (English Transl.)*, **22**, 120 (1967).
182. P. K. Cheo, *Appl. Phys. Letters*, **11**, 38 (1967).

Addenda (*added in proof*)

3.A.1. Sections 3.2.2, 3.5.8, 3.6.4. *Expansions.* In these sections the relaxation of N_2 and CO in expansions was discussed. It had been found that the results could only be fitted with experimental Napier times much smaller than those obtained in normal shock experiments. Von Rosenburg, Taylor and Teare[A1] have now performed experiments with CO under similar conditions to those previously used and find a much smaller discrepancy. The measurements are very influenced by the presence of trace impurities and particularly by hydrogen atoms. The authors suggest that there may in fact be no discrepancy at all for systems free from impurities. Expansion experiments differ from normal shock experiments in that the nozzle flow is drawn from hot gas at the end of the tube which is stationary and has time to collect impurities from the boundary layers.

3.A.2. Sections 3.6.1, 3.6.9. The Napier time of HCl has been remeasured by Breshears and Bird[A2] and by Bowman and Seary[A3] who find it to be much shorter at 2000°K than previously suggested[A4]. The Napier time is given by

$$HCl(HCl): \quad \log_{10} \tau(\text{atm s}) = 27 \cdot 3 T^{-1/3} - 8 \cdot 87 \quad (3.152)$$

$$DCl(DCl): \quad \log_{10} \tau(\text{atm s}) = 33 \cdot 7 T^{-1/3} - 9 \cdot 17 \quad (3.153)$$

3.A.3. Section 3.2.4. There has been a review of the optical acoustic effect[A5].

3.A.4. Sections 3.5.7, 3.6.12, 3.6.13. Simpson, Bridgman and Chandler[A6,A7] have made careful measurements of relaxation in CO_2 and N_2O, and compared their results with those of Johannessen[15] and coworkers and Bhangu[14]. They concluded that curious change of Napier time with progress of relaxation observed previously (Figures 3.15, 3.16) was due to the assumption that the density change was exponential. If this was not done the results were more reasonable. They found that all three vibrational modes in CO_2 and N_2O relaxed together.

References

A1. C. W. von Rosenburg, R. L. Taylor and J. D. Teare, *J. Chem. Phys.*, **48**, 5731 (1968).
A2. W. D. Breshears and P. F. Bird, *J. Chem. Phys.*, **50**, 233 (1969).
A3. C. T. Bowman and D. J. Seary, *J. Chem. Phys.*, **50**, in press (1969).
A4. P. Borrell and R. Gutteridge, *J. Chem. Phys.*, **50**, in press (1969).
A5. A. W. Read, *Advances in Molecular Relaxation Processes*, **1**, 257 (1968).
A6. C. J. S. M. Simpson, K. B. Bridgman and T. R. D. Chandler, *J. Chem. Phys.*, **49**, 513 (1968).
A7. C. J. S. M. Simpson, K. B. Bridgman and T. R. D. Chandler, *J. Chem. Phys.*, **49**, 509 (1968).

4

Molecular Relaxation Processes in Liquids

W. J. Orville-Thomas
E. Wyn-Jones

4.1 GENERAL CONSIDERATIONS

4.1.1 Introduction

As is usual in such circumstances several branches of science made spectacular progress during the course of the last war. The application of ultrasonics to the study of liquid systems, e.g. to measure the draft of ships, was one of the fields so energized. As happened in the case of radar however, certain snags were struck which appeared to be inexplicable at the time. For example it was found that ultrasonic energy of certain frequencies was completely absorbed by sea water whilst these frequencies were not absorbed by artificial brine prepared in laboratories. It was subsequently discovered that the attenuation of the sound waves was caused by traces of magnesium sulphate present in the sea but not in the laboratory brine. It is now known that this is due to an extremely fast reaction involving the magnesium ion and the hydration sheath: the fast reaction occurs between the water molecules in the hydration sheath and those in the bulk of the solvent.

Scientists are continually on the lookout for new experimental methods which might give added knowledge about their more intransigent problems.

It has been found that the ultrasonic technique is peculiarly well adapted for the study of such very fast reactions and represents one of the most powerful new tools which have recently become available to the physical chemist. A whole new field, 'molecular acoustics', has thus been opened up. Amongst the problems already studied to some extent are the flexibility of molecules, a matter of great concern in the polymer field, aggregation processes in liquids such as the formation of hydration sheathes and hydrogen bonding, and the problem of vibrational relaxation

265

which is concerned with the mechanism of how energy is transferred from external translational degrees of freedom to internal vibrational modes.

It is now clear that in favourable cases molecular acoustic (ultrasonic) studies yield information on these and other problems, i.e. knowledge can be obtained about molecular structure and molecular processes (kinetic and thermodynamic data) as well as the structure of the liquid and solid states.

In molecular acoustic studies a sound wave of known characteristics is passed through the system.

The propagation of a sound wave through a liquid occurs adiabatically (over the available experimental frequencies). The pressure amplitude of the sound wave of frequency $f = \omega/2\pi$ travelling in the positive x direction is given by:

$$P = P_0 \exp[-\alpha x + i\omega(t - x/v)] \qquad (4.1)$$

where P = pressure in the liquid due to the impressed wave at x, P_0 = pressure amplitude at $x = 0$, α = absorption coefficient relating to excess pressure and v = phase velocity.

The attenuation of the sound wave passing through a fluid is therefore related exponentially to the distance travelled by the wave.

In practice one measures the velocity and the attenuation (i.e. extent of absorption) of the sound wave over the available frequency range; in addition the temperature of and sometimes the pressure on the system are varied as widely as possible.

Stokes, using classical theory, showed that the quantity α/f^2 should be constant for a given substance[1]. Experimentally, however, the information that becomes available is of the type shown in Figure 4.1, provided that the examination is carried out over a sufficiently wide frequency range. Examination of this diagram shows that as the frequency changes α/f^2 decreases from a constant value at low frequencies to a limiting value at high frequencies. At the same time the absorption/wavelength, $\alpha\lambda = \mu$, passes through a maximum value at a critical frequency, f_c, which is characteristic of the substance under investigation. This behaviour, in which a dispersion of acoustic energy happens, indicates that a relaxation process occurs. The velocity also changes, as shown in Figure 4.1—the variation of sound velocity with frequency being known as velocity dispersion.

The information available to the experimentalist is therefore the peak

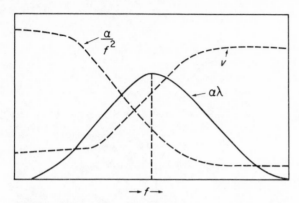

Figure 4.1 Variation of absorption of sound energy (α/f^2) and velocity (v) with frequency (f)

height of the loss curve $(\alpha\lambda \sim f)$ and the frequency f_c at which maximum loss occurs. It is these data that lead to structural information.

4.1.2 Energy of Liquids

The total energy of a liquid is the sum of many components. Some factors such as translational and vibrational energy are very familiar; others such as the energy due to the degree of order in a liquid, which is the result of a number of molecules clinging together in a quasi–crystalline assembly, are less so.

The topics dealt with in this chapter involve systems containing molecules in two or more energy states. An example of a two-state energy system would be an equilibrium mixture of rotational isomers (conformers) which differ in energy owing to their different shapes.

If we modify Litovitz's ingenious 'energy box' analogy[2] we can consider the separate energy components of a liquid to be contained in segments (Figure 4.2). Each segment is in contact with the others via a coupling mechanism so that a leakage (or transfer) of energy can occur from one energy segment to another.

For example if the temperature of an equilibrium system is raised this will increase the translational energy—part of this extra translational energy will be transferred and be transformed into the other types of energy present, e.g. vibrational energy, via a coupling mechanism which

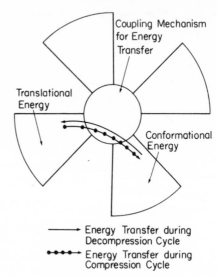

Figure 4.2   Transfer of energy

in this case involves molecular collisions, until a new equilibrium position is reached.

This will take a finite time depending upon the types of coupling process involved.

4.1.3   Relaxational Coupling

In liquids relaxational coupling is very important. Here if the energy of a particular segment is increased, part of this increase in energy will flow to the other segments at an exponential rate being transformed during the process. The exponential rate at which energy flows from segment (I) to another segment (II) involves a time constant, τ, known as the relaxational time. The relaxational time depends upon the nature of the particular relaxation process involved and is different for the various possible mechanisms of energy transfer.

Reverting to Figure 4.2, as the sound traverses the system the liquid is alternately compressed then decompressed at a frequency f. During the compression cycle, since the process is adiabatic, a rise in temperature occurs thus increasing the translational energy. If no leakage occurs to another energy segment the temperature is restored to the initial equilibrium value, T_e, at the end of the compression cycle.

When relaxational coupling exists, however, some of the translational energy is transformed into different types of energy such as vibrational or structural energy during the compression cycle. The rate at which this energy is returned to the translational modes depends upon the relaxation times of the coupling processes.

To simplify matters let us consider a two-stage process in which we imagine that energy is transferred from the translational segment to one other.

The energy-sharing process has a finite relaxation time, $\tau = 1/(2\pi f_c)$ and provided that the temperature is kept constant the rate at which energy is returned to the translational segment is independent of frequency. If conditions are such that $\omega\tau = f/f_c \ll 1$, i.e. one compresses and decompresses very slowly even though energy sharing occurs, the transferred energy has sufficient time to return in phase to the translational segment during the compression cycle, since $\omega \ll 1/\tau$.

When $\omega\tau = f/f_c \approx 1$ part of the returned energy will now enter the translational segment during the decompression cycle owing to the higher rate at which the liquid is being compressed and decompressed. In these circumstances the absorption per wavelength becomes finite and reaches its maximum value at $\omega\tau = 1$ (see Figure 4.1).

Finally when $\omega\tau = f/f_c \gg 1$, i.e. at very high frequencies, almost no energy is shared and the loss eventually becomes negligible.

Two factors determine the loss of energy per cycle. These are the amount of energy shared and returned out of phase and secondly the time constant of the transfer process. In a system where a single relaxation occurs the width of the loss curve is fixed and independent of the loss mechanism and takes the form,

$$\text{loss per cycle} \approx \omega\tau/(1+\omega^2\tau^2)$$

where α is the absorption coefficient at wavelength λ, τ is the relaxation time and $\omega = 2\pi f$.

4.1.4   The Interaction of Acoustic Waves with Liquids

4.1.4.1 *General Theory*

In order to fully understand the molecular nature of the relaxation process it is necessary to derive the relationships between the measured acoustic parameters α, μ and v and the physical properties of the liquid in which the relaxation process occurs. The theory has been discussed by many authors and in the present treatment only a resumé is given[3].

When developing theories for liquid systems the usual practice is to regard the liquid as a viscoelastic medium. Using this principle it is then possible to derive formulae which relate the measured acoustic parameters to the viscosity and modulus of the liquid medium. The relationship between the stress S and resultant strain, s, in any medium is given by,

$$S = Ks \tag{4.2}$$

where K is the modulus of the system. When the applied stress varies with time then Equation (4.2) is incomplete and one must turn to the general theory of relaxation where stress–strain relationships for liquid media have been derived. On application of a sinusoidal stress the modulus K becomes complex and,

$$K = K' + iK'' \tag{4.3}$$

A sound wave provides an example of a sinusoidal stress. When a relaxation process occurs in the medium, then the real and imaginary parts of the modulus will become frequency dependent and take the form,

$$K' = K'_0 + K'_r \frac{\omega^2 \tau^2}{1 + \omega^2 \tau^2} \tag{4.4}$$

and

$$K'' = \frac{K_r \omega \tau}{1 + \omega^2 \tau^2} + \omega \eta_\infty \tag{4.5}$$

where K_r is the relaxing (frequency dependent) part of the modulus given by

$$K_r = K'_\infty - K'_0 \tag{4.6}$$

where the subscripts ∞ and 0 refer to the values of the quantities at infinite and zero frequency respectively. It is also convenient to define a dynamic viscosity η by

$$\eta(\omega) = \frac{\eta_r}{1 + \omega^2 \tau^2} + \eta_\infty \tag{4.7}$$

where

$$\eta_r = \eta_0 - \eta_\infty \tag{4.8}$$

In the frequency regions which lie well outside the relaxation range it can be shown that

$$\eta_r = K_r \tau \tag{4.9a}$$

(At this stage it is well worth noting, as pointed out by Piercy, the similarity in the derivation of Equations (4.2) to (4.6) with those of dielectric relaxation theory where on dealing with electric stress and strain in an elastic medium (ether) the dielectric constant is defined by K^{-1} in Equation (4.2).)

4.1.4.2 *Relations between Acoustic Data and Relaxation Parameters*

The variation of K', K'' and η with frequency around the relaxation region is shown in Figure 4.3. We have seen that when acoustic relaxation occurs, the variation of the measured acoustic parameters, v, μ and α/f^2 with frequency is shown in Figure 4.1. If we compare the shapes of the curves in Figures 4.1 and 4.3 the similarity is obvious. An acoustic wave produces a sinusoidal stress in a liquid, and when conditions are such that $\tau \approx 2\pi f$ where f is the acoustic frequency, it becomes apparent from a comparison of Figures 4.1 and 4.3 that the acoustic parameters which are

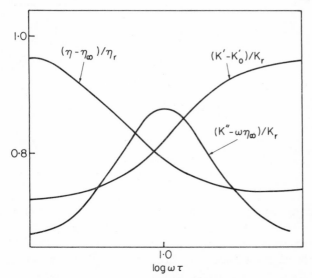

Figure 4.3 Variation of the normalized modulus and dynamic viscosity for a single-relaxation process

related to the physical properties of the medium are respectively v and K', μ and K'' and α/f^2 and η. This is a qualitative method of showing the relationship between the various quantities. The full relationships have been derived[4] and are:

$$v^2 = \rho^{-1}K'$$

$$\mu = \pi(\rho v^2)^{-1}K''$$

and

$$\alpha/f^2 = 2\pi^2(\rho v^3)^{-1}\eta$$

where ρ is the density of the medium.

In this work we are concerned with interpreting the results of measuring the properties of longitudinal sound waves in liquids. For an applied stress, such as the passage of a longitudinal sound wave through a liquid, the modulus K can be resolved into two components: K_c the compressional stress and K_s the shear stress as shown in Figure 4.4.

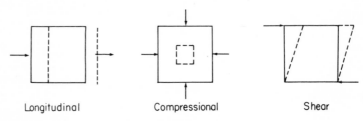

Longitudinal   Compressional   Shear

Figure 4.4   Effects of stress

The resultant modulus is now given by

$$K = K_c + \tfrac{4}{3}K_s$$

It, therefore, follows that in considering Equations (4.2)–(4.7) there should be contributions to K', K'' and η from both shear and compressional stress. In this work, however, we are only concerned with the results of the relaxation of the compressional modulus. These processes do not include any contributions from shear relaxation.

When dealing with liquid systems the usual practice is to discuss the compressibility, β, of the liquid rather than the modulus K. The relationship between the two quantities is

$$\beta = \frac{1}{K}$$

The passage of a longitudinal sound wave through a liquid is adiabatic and during compression there is a temperature rise in the liquid. This temperature rise, in turn, increases the translational energy of the liquid. Sound absorption in a liquid will occur because during compression some of the translational energy will be transferred to some of the internal degrees of freedom in the liquid via a relaxational coupling process, and this energy returns to the translational modes out of phase with the sound wave. In the relaxation region, i.e. when the relaxation time of the coupling process between the translational and internal degrees of freedom has a value of about $1 (2\pi f)$, the loss in sound absorption reaches a maximum. In this region the volume change accompanying the passage of the sound wave becomes out of phase with the pressure. Now the adiabatic compressibility of a liquid is defined as,

$$\beta = -\frac{1}{V}\left(\frac{\partial V}{\partial P}\right)_S \quad \text{where } S = \text{entropy}$$

and if ∂V is out of phase with ∂P as described above then β must become complex and also frequency dependent; it takes the form

$$\beta_\omega = \beta_0 - \frac{\Delta\beta\omega^2\tau^2}{1+\omega^2\tau^2} - \frac{i\Delta\beta\omega\tau}{1+\omega^2\tau^2}$$

(Compare with Equations (4.3), (4—4) and (4—5).)

The full relationships between the various β's in Equation (4.15) and the K's in Equations (4.3)–(4.6) are,

$$\beta_0 = \frac{1}{K_0}; \quad \beta_\infty = \frac{1}{K}, \quad \Delta\beta = \frac{K'_\infty - K'_0}{K'_\infty K'_0} \quad (4.9\text{b})$$

$\Delta\beta$ is the frequency independent value of the relaxing compressibility, and $\beta_0 = \beta_\infty + \Delta\beta$. Hence Equations (4.9) become

$$\eta = \frac{\Delta\beta\tau}{\beta_0^2} \quad (4.10)$$

The subscripts ∞ and 0 refer to the value of the quantities at infinity and zero frequency. In this work we assume in all the relaxational processes that $(\mu^2/4\pi^2) \ll 1$. This means that it is not necessary to distinguish between the various τ's in equations (4.8) and (4.10). The relationships between the acoustic parameters v, μ and α/f^2 and the adiabatic compressibilities are,

$$v^2 = \frac{1}{\rho\beta} \tag{4.11}$$

$$\mu = \frac{\pi\Delta\beta}{\beta_0}\left[\frac{\omega\tau}{1+\omega^2\tau^2}\right] \tag{4.12}$$

and

$$\alpha/f^2 = \frac{2\pi^2}{v}\frac{\Delta\beta}{\beta_0}\left[\frac{\tau}{1+\omega^2\tau^2}\right] \tag{4.13}$$

The expressions for μ and α/f^2 in Equations (4.12) and (4.13) refer only to the value of these quantities for the particular relaxation considered. Since we are dealing with a single relaxation process in a particular frequency region it is quite possible that some other relaxation process may occur at higher or lower frequencies. This means that the measured values of α/f^2 will contain, in addition to the relaxational contributions of Equations (4.12) and (4.13), a frequency independent contribution from other relaxation processes occurring at much higher or lower frequencies and in addition the classical viscosity contribution. This means that the overall measured absorption coefficient α may now be written as,

$$\frac{\alpha}{f^2} = J\left[\frac{1}{1+\omega^2\tau^2}\right] + B \tag{4.14}$$

Where B corresponds to α/f^2 at $\omega\tau \gg 1$ (see above) and J is a parameter of the relaxation process given by

$$J = \frac{2\pi^2}{v}\frac{\Delta\beta\tau}{\beta_0} \tag{4.15}$$

Most of the experimental work on ultrasonic absorption has been carried out by measuring the absorption coefficient α at different frequencies using temperature as an extra variable. In only very few cases have measurements been carried out at variable pressures. When a single relaxation occurs the pertinent equations that are always used to analyse the data and obtain molecular information are Equations (4.12) and (4.13). At this stage it is convenient to define a relaxational frequency f_c by

$$\tau = \frac{1}{2\pi f_c} \qquad (4.16)$$

In Equation (4.14) the value of α/f^2 represents contributions from the relaxation process $[J/(1+\omega^2\tau^2)]$ and also a frequency independent part B associated with classical viscous absorption and any other relaxational processes at frequencies much higher than f_c. The value of μ in Equation (4.12) represents purely the relaxational contribution. From Equations (4.14), (4.15) and (4.16) it follows that

$$\frac{\alpha}{f^2} = \frac{2\pi^2}{v}\frac{\Delta\beta}{\beta_0}\frac{1}{2\pi f_c}\left[\frac{1}{1+(f/f_c)^2}\right]+B \qquad (4.17)$$

The usual practice is to denote the quantity

$$\left[\frac{\pi}{v}\frac{\Delta\beta}{\beta_0}\frac{1}{f_c}\right]$$

by A when equation (4.17) becomes

$$\alpha/f^2 = A/[1+(f/f_c)^2]+B \qquad (4.18)$$

Substitution of A in equation (4.14) gives

$$\mu = Avf_c\frac{\omega\tau}{1+\omega^2\tau^2}$$

Which reduces to $\mu_m = \frac{1}{2}Avf_c$ when $\omega\tau = 1$. The value of μ_m, of course, represents the peak height of the curve.

From these equations it follows that the experimental value of μ, near the relaxation frequency, f_c, would be

$$\mu_{exp} = Avf_c\frac{(f/f_c)^2}{1+(f/f_c)^2}+Bvf$$

The velocity dispersion is given by

$$v^2-v_0^2 = \frac{\Delta\beta}{\beta_0}v_0v_\infty\frac{\omega^2\tau^2}{1+\omega^2\tau^2}$$

Where v is the velocity at frequency f and v_0 and v_∞ are the limiting low and high frequency values of the velocity.

4.1.4.3   *Theory for a Two-state Reaction*

An important example of a time-dependent molecular process is a two-state chemical equilibrium of the type:

$$A_1 \rightleftharpoons A_2$$

The Gibbs free energy ΔG of such a reaction is given by

$$\Delta G = \Delta H - T\Delta S = \Delta E + P\Delta V - T\Delta S$$

where the symbols have their usual meaning. The equilibrium constant is given by

$$K = \exp(-\Delta G/RT) = \frac{n_2}{n_1} = \frac{k_{12}}{k_{21}} \tag{4.19}$$

where n_1 and n_2 are the number of moles of A_1 and A_2, respectively and k_{12} and k_{21} are the appropriate forward and reverse rate constants. The variation of the Gibbs free energy of such a reaction is shown in Figure 4.5. Hence ΔG_{12} and ΔG_{21} refer to the free energies of activation for the

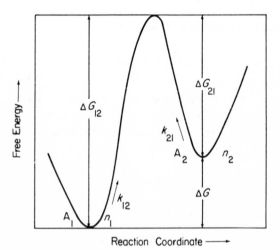

Reaction Coordinate ⟶

Figure 4.5   Gibbs free energy of a two-state equilibrium

forward and reverse reactions. What we are concerned with is interpreting the acoustic relaxation arising from the perturbation of such a reaction by a sound wave. Because of the chemical reaction present an additional degree of freedom has to be introduced. For convenience we call this extra variable, which describes the state of the chemical reaction, y. The equation of state of such a system now takes the form,

$$X = F(Y, Z, y)$$

where X, Y and Z are any three of the thermodynamic variable P, V, T and S, and F denotes the function which relates these properties. The passage of a sound wave through the liquid will cause small local fluctuations about the equilibrium in pressure, volume, temperature, entropy and also y; these fluctuations occur at the sound frequency, $\omega/2\pi$. For small displacements about the equilibrium it follows that

$$dX = \left(\frac{\partial X}{\partial Y}\right)_{Z,y} dY + \left(\frac{\partial X}{\partial Z}\right)_{Y,y} dZ + \left(\frac{\partial X}{\partial y}\right)_{Y,Z} dy$$

and

$$\left(\frac{\partial X}{\partial Y}\right)_{Z,\omega} = \left(\frac{\partial X}{\partial Y}\right)_{Z,y} + \left(\frac{\partial X}{\partial y}\right)_{Y,Z}\left(\frac{\partial y}{\partial Y}\right)_{Z,\omega} \tag{4.20}$$

In equation (4.20) the first term of the R.H.S. represents the condition in which the reaction is 'frozen'. This is achieved practically by raising the frequency, ω, until the reaction cannot follow the fluctuation in Y. This term is therefore called the infinite frequency component. Now, from thermodynamic definitions it can be shown that:

the adiabatic compressibility

$$\beta\beta_\omega = -\frac{1}{V}\left(\frac{\partial V}{\partial P}\right)_{S,\omega} \tag{4.21}$$

the specific heat at constant pressure

$$C_{p\omega} = T\left(\frac{\partial S}{\partial T}\right)_{P,\omega} \tag{4.22}$$

the isothermal compressibility

$$\beta_{T\omega} = \frac{-1}{V}\left(\frac{\partial V}{\partial P}\right)_{T,\omega} \tag{4.23}$$

and the thermal coefficient of expansion

$$\theta_\omega = \frac{1}{V}\left(\frac{\partial V}{\partial T}\right)_{P,\omega} \tag{4.24}$$

From equation (4.20) it follows that

$$\beta_\omega = \frac{-1}{V}\left(\frac{\partial V}{\partial P}\right)_{S,y} - \frac{1}{V}\left(\frac{\partial V}{\partial y}\right)_{P,S}\left(\frac{\partial y}{\partial P}\right)_{S,\omega} \tag{4.25}$$

$$= \beta_\infty + \Delta\beta_\omega \tag{4.26}$$

where

$$\beta_\infty = \frac{-1}{V}\left(\frac{\partial V}{\partial P}\right)_{S,y} \quad \text{and} \quad \Delta\beta_\omega = \frac{-1}{V}\left(\frac{\partial V}{\partial y}\right)_{P,S}\left(\frac{\partial y}{\partial P}\right)_{S,\omega}$$

$\Delta\beta_\omega$ is the frequency dependent part of the compressibility. Equations (4.22), (4.23) and (4.24) can also be expanded in a similar way.

The chemical reaction denoted by Equation (4.18) is reversible and if at any temperature the equilibrium is perturbed, the system will tend to return to equilibrium. The rate at which it returns to equilibrium will be a function of the displacement from equilibrium. If we use the variable y we can express this by the formula,

$$-\left(\frac{dy}{dt}\right) = f(y-y_0)$$

$$= \frac{1}{\tau}(y-y_0) + \frac{1}{\tau'}(y-y_0)^2 + \dots$$

where y_0 is the equilibrium value of y and τ is the relaxation time. For a small perturbation about the equilibrium we can safely reduce this Equation to

$$-\frac{dy}{dt} = \frac{1}{\tau}(y-y_0) \tag{4.27}$$

If the initial value of y at $t = 0$ is y_0 then integration of Equation (4.27) gives

$$y = \bar{y}[1 - e^{-t/\tau}] + y_0\, e^{-t/\tau}$$

If the final value of y is taken as zero (i.e. $\bar{y} = 0$) so that $y = y_0\, e^{-t/\tau}$, we see that the relaxation time τ is the time at which y decreases to $1/e$ time its original value y_0 at $t = 0$.

For sinusoidal perturbations of the equilibrium at frequency $\omega/2\pi\,(=f)$ we can write

$$y_0 = \bar{y}_0 + \mathrm{d}y_0\,e^{i\omega t}$$

and

$$y = \bar{y} + \mathrm{d}y\,e^{i\omega t}$$

where the $\bar{y}$'s denote frequency independent contribution.

It also follows that:

$$-\left(i\omega + \frac{1}{\tau}\right)\mathrm{d}y = \frac{1}{\tau}\mathrm{d}y$$

and

$$\mathrm{d}y = \frac{\mathrm{d}y_0}{1 + i\omega\tau} \tag{4.28}$$

From Equation (4.25) and (4.26) it follows that

$$\Delta\beta_\omega = \frac{-1}{V}\left(\frac{\partial V}{\partial y}\right)_{P,S}\left(\frac{\partial y}{\partial P}\right)_{S,\omega} \tag{4.29}$$

and from Equations (4.28) and (4.29)

$$\Delta\beta_\omega = \frac{-1}{V}\left(\frac{\partial V}{\partial y}\right)\left(\frac{\partial y}{\partial P}\right)_S\frac{1}{1+i\omega\tau} = \frac{\Delta\beta}{1+i\omega\tau} \tag{4.30}$$

where $\Delta\beta$ is the zero frequency value which is, of course, a real number. Now it can also be shown that

$$\beta_0 = \beta_\infty + \Delta\beta \tag{4.31}$$

when β_0 is the zero frequency value of β. From Equation (4.26)

$$\beta_\omega = \beta_\infty + \frac{\Delta\beta}{1+i\omega\tau} \tag{4.32}$$

and combining Equations (4.31) and (4.32) one obtains,

$$\beta_\omega = (\beta_0 - \Delta\beta) + \frac{\Delta\beta}{1+i\omega\tau}$$

or

$$\beta_\omega = \beta_0 - \Delta\beta\frac{i\omega\tau}{1+i\omega\tau} \tag{4.33}$$

If we multiply the top and bottom of the second term in the R.H.S. of Equation (4.33) by $(1 - i\omega t)$ we get

$$\beta_\omega = \beta_0 - \frac{\Delta\beta\omega^2\tau^2}{1+\omega^2\tau^2} - i\Delta\beta\frac{\omega\tau}{1+\omega^2\tau^2}$$

Also by considering the thermodynamics of the system and using Equations (4.20)–(4.26) it can be shown that,

$$\frac{\Delta\beta}{\beta} = (\gamma-1)\frac{\Delta C_p}{(C_p-\Delta C_p)}\left[1 - \frac{\Delta V}{\Delta H}\frac{C_p}{V\theta}\right]^2 \tag{4.34}$$

where ΔC_p is the relaxing part of the specific heat. When $\omega\tau = 1$, i.e. when $f = f_c$, (the relaxational frequency) Equation (4.12) reduces to

$$\mu_m = \frac{\pi}{2}\frac{\Delta\beta}{\beta_0} \tag{4.35}$$

where μ_m is the peak value of the loss curve in Figure 4.1.

In this work we are dealing with thermal relaxation. This is due to the rise in temperature accompanying an adiabatic compression and the fact that there is a time lag involved in the transfer of energy from external to internal degrees of freedom. This means that the equation determining the amount of energy shared between the states A_2 and A_1 is

$$\frac{\partial \ln K}{dT} = -\frac{\Delta H}{RT^2}$$

and this equation leads to a complex specific heat which in turn makes β complex.

The rate equation for the reaction denoted by Equation (4.18) is

$$\frac{dn_2}{dt} = k_{12}n_1 - k_{21}n_2 \tag{4.36}$$

If, for convenience, we define the variable y associated with the reaction as

$$y = \frac{n_2}{N} \tag{4.37}$$

where $N = n_1 + n_2 =$ constant then it follows that any change or perturbation of the equilibrium will involve a change in n_2 and hence y. From

Equations (4.36) and (4.37) it follows that

$$\frac{dy}{dt} = \frac{1}{N}[k_{12}n_1 - k_{21}n_2]$$

Now $K = k_{12}/k_{21}$ and also $(1-y) = n_1/N$, therefore

$$\frac{dy}{dt} = k_{21}[K(1-y)-y] \tag{4.38}$$

If we multiply the top and bottom of the right-hand side of Equation (4.38) by $(1+K)$ and rearrange we get

$$\frac{dy}{dt} = -k_{21}[1+K]\left[y - \frac{K}{K+1}\right] \tag{4.39}$$

From Equation (4.27) we have shown that:

$$\frac{-dy}{dt} = \frac{1}{\tau}(y - y_0)$$

and if we compare this with Equation (4.39) accepting the definition for y as

$$y = \frac{n_2}{N}$$

we find that

$$\frac{1}{\tau} = k_{21}[1+K]$$

or

$$= (k_{12} + k_{21})^{-1} \tag{4.40}$$

and also

$$y_0 = \frac{K}{1+K}$$

which is of course the equilibrium value of n_2/N.

Accepting the definition of y as n_2/N and also $y_0 = K/(1+K)$ Equation (4.28) now becomes

$$-dy = \frac{d[K/(1+K)]}{1+i\omega\tau}$$

and since

$$K = e^{-\Delta G/RT} \tag{4.41}$$

$$-dy = \frac{K}{(1+K)^2}\frac{d\ln K}{1+\omega\tau} \tag{4.42}$$

From Equations (4.20) and (4.22) and a consideration of the thermodynamics of the system it can be shown that:

$$\Delta C_{P_\omega} = \Delta H\left(\frac{dy}{dT}\right)_{P,\omega} \tag{4.43}$$

From Equations (4.19), (4.41), (4.42) and (4.43) it follows that

$$\Delta C_{P_\omega} = \Delta H\frac{K}{(1+K)^2}\frac{\Delta H}{RT^2}\frac{1}{1+i\omega\tau}$$

and by analogy with Equation (4.30) it follows that the real part of the relaxing specific heat ΔC_p is given by

$$\Delta C_p = R\left(\frac{\Delta H}{RT}\right)^2\frac{K}{(1+K)^2} \tag{4.44}$$

From equations (4.34), (4.35) and (4.44) the relationship between μ_m, the maximum absorption for wavelength, and the thermodynamics of the chemical equilibrium are derived. The kinetics of the equilibrium are related to the acoustic parameter through Equation (4.40) where $\tau = 1/2\pi f_c$.

4.2 EXPERIMENTAL METHODS

4.2.1 Introduction

When dealing with molecular relaxation processes in liquids using acoustic methods it is desirable to obtain information on both the absorption and velocity of sound over adequate frequency and temperature (or pressure) ranges. When relaxation occurs the dispersion in the quantity α/f^2 is appreciable and can be measured with reasonable accuracy. On the other hand, the dispersion in velocity, given by:

$$v^2 - v_0{}^2 = \frac{\Delta\beta}{\beta_0}v_0v_\infty\frac{\omega^2\tau^2}{1+\omega^2\tau^2} \tag{4.45}$$

is usually very small. At the relaxation frequency, when $\omega\tau = 1$, Equation (4.45) reduces to

$$v^2 - v_0^2 = \frac{\Delta\beta}{2\beta_0}v_0 v_\infty \tag{4.46}$$

It can also be shown that in these conditions Equation (4.12) becomes

$$\frac{\Delta\beta}{2\beta_0} = \frac{\mu_m}{\pi} \tag{4.47}$$

Substitution of (4.47) in (4.46) leads to

$$v^2 - v_0^2 = \frac{\mu_m}{\pi}v_0 v_\infty \tag{4.48}$$

and if

$$v = v_0 + \Delta v$$

where $\Delta v \ll v_0$ Equation (4.48) reduces to

$$\frac{\Delta v}{v} = \frac{\mu_m}{\pi}$$

For the typical relaxation process discussed in Section 4.3.3 on page 304 $\mu_m \approx 0{\cdot}03$ and hence $\Delta v/v \approx 0{\cdot}01$ which indicates that the velocity dispersion is very small and can be ignored. In practice, therefore, an ultrasonic relaxation process can be detected from measurements on sound absorption over a suitable frequency range but the magnitude of the velocity and data on thermodynamic quantities such as C_p, θ, and so forth are required in order to derive quantitative information from the experimental observations. The acoustic frequency range that would be desirable to investigate molecular processes in liquids ranges from a few cycles per second to 10^{13} c/s. Experimentally this range can be covered more or less successfully since there are methods available which can be used to measure sound absorption and velocity in the frequency region 10^2–10^{10} c/s. The accuracy of these measurements varies from qualitative estimates to values which are considered accurate to within a few per cent. The majority of acoustic measurements reported in the literature have been conducted by varying the frequency of the sound waves and temperature of the sample. Only in a few cases have acoustic relaxational parameters been measured at different pressures. Descriptions of different techniques which can be used to make absorption and velocity measurements over the acoustic spectrum are given below[5,6].

4.2.2   Resonance and Resonance-reverberation Methods

For the investigation of the acoustic properties of liquids in the frequency
region 5–800 kc/s the resonance (5–50 kc/s) and resonance-reverberation
(> 50 kc/s) methods have been used. A block diagram of a typical experimental arrangement is shown in Figure 4.6. The test liquid is kept in a
spherical glass container which is suspended in a low-pressure, tempera-
ture-controlled chamber. The principle involved is that the sphere is

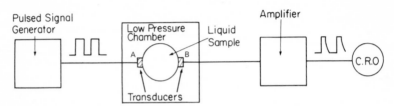

Figure 4.6   Resonance-reverberation apparatus

excited into resonance with an intermittent train of continuous sound
waves produced by means of the transducer A. The properties of the
resulting acoustic radial modes (eigenvibrations) set up in the spherical
container are then measured by using the second transducer B; this picks
up the mechanical reverberations of the sphere and converts them to
electrical waves, which in turn are suitably amplified and displayed on a
cathode-ray oscilloscope. At low frequencies the eigenvibrations are
reasonably well separated in frequency and can be excited preferentially
using single modes. In this case it is possible to measure the half-width
of the sharp resonance peaks or the decay time of the resonance modes.
The half-width of the resonance peak is a function of the Q factor and the
decay time τ is related to the absorption coefficient α by

$$\tau = 1/\alpha v$$

where v is the sound velocity.

At frequencies greater than 50 kc/s the eigenvibrations occurring during
reverberation are very close together and cannot be excited and detected
separately. In this case a large number of eigenvibrations are excited in a
narrow frequency band and their decay time measured. (It is assumed that
the decay times are constant within a narrow frequency band.)

There are several disadvantages in using resonators, the most serious
being the appreciable 'wall-losses'. These 'wall-losses' can be reduced by

making resonators from materials with low loss factors such as aluminium, glass or quartz. By means of calibration it is possible to estimate the wall losses, and in some cases they are found to be frequency dependent. Another disadvantage is the difficulty in differentiating between the radial resonant modes of vibration set up in the sphere and the other modes which are also excited. The former modes, which are used in measurements, usually have longer decay times. At the lower frequencies the resonating spheres have large volumes and therefore temperature control becomes difficult. At the higher frequencies the radial-mode decay times become smaller and comparable with the other modes set up in the system. Descriptions of suitable apparatus are given in references 7 to 12.

Recently Eggers[13] has developed a cylindrical resonator of the type shown in Figure 4.7. With this apparatus it is possible to measure absorption and velocity in the region 200 kc/s–2 Mc/s. A cylindrical resonator of this kind produces very sharp resonance peaks whose half bandwidths can easily be measured. At the lower frequencies allowance has to be made for the divergence of the sound beam due to the many paths that it travels during resonance. The dimensions of the resonator are such that only about 10–20 ml of liquid are needed. This means, of course, that this method has distinct advantages over other low-frequency methods and preliminary results are very promising.

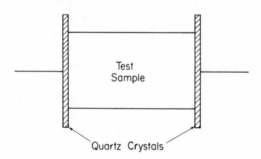

Figure 4.7   Eggers' cylindrical resonator

4.2.3   Tuning Fork

There are few reports available on a versatile technique which is capable of measuring the absorption of sound at very low frequencies ($10–10^2$ c/s). A proposed method[14] which gives qualitative results in the frequency

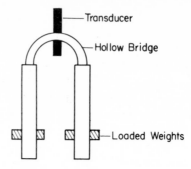

Figure 4.8   Tuning fork apparatus

region 10^2 c/s employs a tuning fork similar to that shown in Figure 4.8. The fork is made of fused quartz, has solid arms and a hollow bridge into which the test liquid is introduced. This hollow bridge rests on a transducer which is used to measure the resonance frequency of the tuning fork. The length of the solid arms can be varied and they can also be loaded with weights in order to change the resonant frequency of the tuning fork. The quality factor, Q, which is a function of the shape of the resonance peak of the tuning fork, can be estimated by measuring its decay time after excitation at the resonance frequency. When the tuning fork is excited, the resulting sound waves are detected by the transducer, which converts the acoustic to electrical energy, and this output is displayed on the screen of a cathode-ray oscilloscope. By measuring the Q factor and resonance frequency of the tuning fork with the hollow bridge alternatively empty and filled with the test liquid, and by varying the loading of the solid arms whose length can also be varied, it is possible to estimate the absorption coefficient and compressibility of the test liquid. The relatiohship between Q values, resonance frequencies and absorption coefficient has been derived by Andrae and others[14]. When the hollow bridge is filled with liquid an air bubble usually appears. The occurrence of such bubbles has to be taken into account in deriving acoustic data on the system. Qualitative measurements of this kind have confirmed an ultrasonic relaxation in beryllium sulphate solutions centered at 300 c/s.

4.2.4   Acoustic Streaming

This technique has been applied with reasonable success to measure the absorption of sound in liquids in the frequency region 100 kc/s–10 Mc/s.

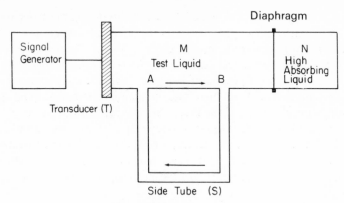

Figure 4.9 Acoustic streaming

The principle[15] of the method is based on measuring the pressure gradient formed along the direction of propagation of an acoustic wave. This pressure gradient causes the liquid to flow at a steady velocity which is related to the absorption of sound in the liquid. This principle was first applied by Lamb and his collaborators[16-18] to make quantitative measurements of the sound absorption in liquids. A simplified experimental arrangement is shown in Figure 4.9. The transducer, T, radiates sound waves directly into the main tube, M, which contains the liquid under test. This sound wave passes through the liquid and is absorbed in another tube, N, filled with a highly absorbing liquid. The liquids in tubes M and N are separated by a thin diaphragm. The pressure gradient set up by the acoustic wave causes the test liquid in tube M to flow through the side tube S in the direction shown in the diagram. The pressure gradient, Δp, between the points A and B is given by:

$$\Delta p = E \exp(-2\alpha l)$$

where l is the distance between A and B and E is the intensity of the sound wave in the liquid. The velocity of flow, v, of the liquid is related to the dimensions of the tube M and pressure gradient by

$$v = \Delta p r^2 / 4\eta l$$

where r and l are the radius and length of tube M respectively and η is the viscosity of the liquid. The intensity, E, of the sound wave in the liquid can be measured as described by Piercy and Lamb[16] or Hall and Lamb[18]. The velocity of streaming of the liquid is found by observing the movement of aluminium particles suspended in the test liquid. Using this method correction factors have to be introduced to allow for the end connections of the side tube S. Experimentally this is achieved by measuring v and E in liquids of known absorption and viscosity. The accuracy of absorption measurements derived from these measurements is estimated to be $\pm 5\%$. The disadvantages are that the measurements are time consuming and a fairly large amount of test liquid is needed: in addition cavitation due to non-Newtonian flow occurs. Details of experimental arrangements are given by Piercy and Lamb[16], Piercy[17] and Hall and Lamb[18] who also include a theoretical treatment on the principle of this travelling-wave technique.

4.2.5   Pulse Methods

The pulse technique has proved to be the most versatile for measurements of sound absorption and velocity in liquids in the frequency range 5–200 Mc/s. This technique was initially developed by Pellam and Galt[19]. A block diagram of a simplified 'pulse' apparatus is shown in Figure 4.10. A train of pulses produced by the pulse generator is fed into the transmitter, A, set at the desired frequency. These pulses are used to excite this oscillator, which, in turn, produces bursts of oscillations at the desired frequency, f. The resulting radio frequency pulses are fed to a transducer, T, which is acoustically coupled to a fused-quartz delay line, L_1. The acoustic pulse produced by this transducer passes through the lower fused-quartz rod, L_1, through the liquid under test and into the second fused-quartz delay line, L_2, the upper end of which is attached to the detecting transducer. This second transducer retransforms the acoustic pulse into an electric pulse which, after suitable amplification and demodulation, is finally displayed on the screen of a cathode-ray oscilloscope.

A second train of pulses, also produced by the pulse generator, is used to excite the 'comparison' oscillator, B, set at the same frequency as the transmitter A. These resulting 'comparison' radio frequency pulses, after suitable delay, are fed into an attenuator and subsequently amplified, demodulated and displayed at the side of the acoustic pulse on the screen of the cathode-ray oscilloscope.

The amplitude of the 'acoustic' pulse is changed by varying the distance between the launching and detecting transducers. This change of amplitude can be measured by visually comparing the height of the 'acoustic'

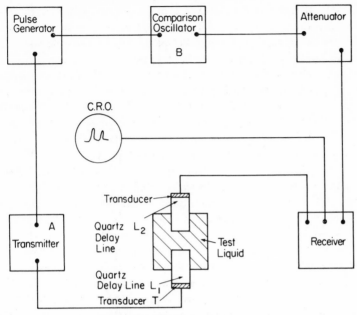

Figure 4.10 Pulse apparatus

pulse with that of the 'comparison' pulse which is attenuated by a known amount. By plotting the attenuation versus path length, the absorption coefficient, α, of the liquid follows from the equation:

$$P = P_0 \exp(-\alpha x) \exp(t - x/v)$$

where

$$\alpha = \ln(P_0/P)/x$$

The quantity $\ln(P_0/P)$ is directly proportional to the attenuator readings and x is the acoustic path in the liquid. The cell containing the liquid can be thermostated at temperatures between $-80 - +120°C$. An accuracy of $\pm 2\%$ is usually claimed for α/f^2 in the frequency range 10–120 Mc/s.

In general measurements are made with the pulse apparatus in the Fresnel region of the sound beam, that is, the region where the sound beam is assumed to be parallel. For measurements in this region the path length must not exceed $a^2/2\lambda$ where 'a' is the radius of the launching transducer and 'λ' the wavelength of the sound wave. Normally the lowest frequency at which pulse measurements can be made in the Fresnel region is around

4·5 Mc/s[20]. Below this frequency the maximum ultrasonic path length usually extends beyond the region within which the sound wave may be considered to propagate in a parallel beam. For measurements in this region, generally termed the Fraunhofer region, suitable corrections have to be applied[20,21] to take into account diffraction effects within the sound beam. By applying such corrections the pulse technique has been used to measure sound absorption at frequencies as low as 100 kc/s[22]. In very highly absorbing liquids (e.g. acetic acid) the Fresnel region extends to frequencies as low as 600 kc/s since the acoustic path in the liquid is relatively small. For measurements at frequencies higher than 250 Mc/s, the path length needed in the liquid to attenuate a 30 dB pulse becomes very small ($\sim 10^{-3}$ cm) and difficulty then arises in measuring this distance; the problem of the parallelism of the fused-quartz delay lines also becomes acute. A description of a mechanical system which has been used to measure absorption for frequencies as high as 800 Mc/s has been given by Hunter and Dardy[23]. When using the pulse technique spot frequencies, which are odd harmonics of the launching transducer, are used. Details of suitable apparatus are available in the literature[19-27]. Another difficulty, of course, at the higher frequencies, ~ 300 Mc/s, is that the acoustic frequency is produced as the harmonic of a 10 or 20 Mc/s quartz crystal. It follows that the higher the harmonic used then more power has to be supplied to excite the transducer.

Recently[28] it was found that by high frequency excitation of a quartz rod, acoustic vibration of frequency as high as 4×10^{10} c/s can be obtained. By using quartz resonators of this type instead of transducers in the mechanical unit, measurements have been carried out with a pulse apparatus to frequencies as high as 1000 Mc/s[29].

4.2.6   Optical Method

Debye and Sears[30] and also Lucas and Biquard[31] discovered that the alternate compression and decompression accompanying the passage of a sound wave through a liquid produces in the liquid a grating capable of diffracting light. If light is now passed through the liquid then the properties of the diffracted light will of course be dependent on the frequency and intensity of the sound beam. The basic experimental arrangements for such measurements are shown in the block diagram Figure 4.11. The sound wave, produced at the transducer A, passes through the liquid under test and is absorbed at B using an absorber. Monochromatic light from

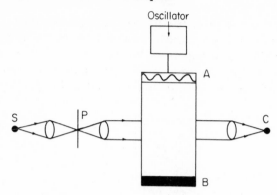

Figure 4.11 Debye–Sears optical method

the source S is then passed through the liquid, collected, and focussed at the detector, C, which can either be a photographic plate or a photocell. The resulting light at C will, of course, be the image of the slit, P, in the various diffracted orders. The diffraction angle of the grating set up in the liquid, θ, can be measured experimentally and is related to the sound and light wavelength, λ_s and λ_l, respectively, by

$$\sin \theta = n\lambda_l/\lambda_s$$

where n is an integer. If the frequency, f, of the transducer is known then the velocity of sound in the liquid follows from the relation,

$$v = f\lambda_s$$

The diffraction angle θ can be measured conveniently using a photographic plate.

The sound-absorption coefficient, α, can be found by measuring the intensity of one of the diffracted orders. This is done by using a photocell or a detector. In practice it is found that in the first order of the diffracted spectrum there is a linear relationship between the sound and light intensity. The sound intensity is then also proportional to the attenuation constant, α. Experimental arrangements are described in references 32 to 34.

4.2.7   Radiation Pressure

This method[35–41] involves measuring the small pressure that is set up on an absorbing or reflecting surface due to an incident sound beam. This small pressure, known as radiation pressure, is proportional to the energy density of the sound beam, which in turn is proportional to the square of the wave amplitude. A simplified version of an experimental apparatus is shown in Figure 4.12. The transducer, A, produces the sound beam which passes through the test liquid on to the movable absorber or detector, B.

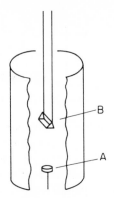

Figure 4.12
Radiation pressure
apparatus

The radiation pressure set up at the detector can then be measured by means of a sensitive balance. The absorption coefficient α then follows from the relationship

$$\Delta W_x / \Delta W_0 = e^{-2\alpha x}$$

where W_0 is that part of the measured weight of the detector due to the radiation pressure at the initial position where $x = 0$ and ΔW_x is the corresponding value at x. This method can be used in the frequency region 3–100 Mc/s. The disadvantages in this method are in the mechanical difficulties of measuring ΔW.

4.2.8 Ultrasonic Interferometer

Sound velocities in liquids are usually determined using an interferometric technique. A diagram of a simple arrangement is shown in Figure 4.13. The circular tube A, closed at the bottom by a transducer B, holds the test liquid. This transducer is driven by an oscillator and the resulting sound wave passes through the test liquid and is reflected off the movable reflector C. This reflected sound wave can then be detected by the transducer and can be investigated visually by displaying the resulting electrical wave on the screen of a cathode-ray oscilloscope or it can be measured by audio-frequency comparison with a crystal oscillator. The instrument is initially set by adjusting the frequency of the oscillator and position of the reflector to give a zero beat corresponding to standing-wave conditions in

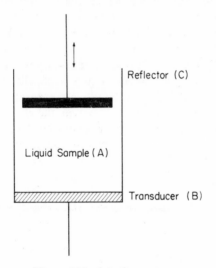

Figure 4.13   Interferometer

the test liquid. The reflector is then moved until the position of about 20 zero beats have been measured. The mean distance between standing-wave conditions is equal to one half sound wavelength, $\lambda/2$, and, therefore, if l is the distance travelled by the reflector for n zero beats

$$l = n(\lambda/2)$$

The sound velocity, v, can then follow from the relation $v = f\lambda$ where f is the frequency of the oscillator. It is worth noting that velocities can also be measured in a similar fashion using the pulse apparatus described in Section 4.2.5. There are several references available on descriptions of systems which can be used to determine sound velocities[42–45].

A description of how an interferometer can be used to measure the absorption of sound in liquids has recently been given by Ilgunas and

Paulauskas[46]. In the frequency region 1–15 Mc/s the results were satisfactory, although the corrections due to the diffraction of the sound
beam were quite high.

4.2.9 Comparison Measurements

Using an arrangement of the kind shown in Figure 4.14, Carstensen[47] was
able to make some comparison velocity measurements using progressive
waves to frequencies as low as 100 kc/s. The mechanical arrangement
consists of the launching and detecting transducers which are mounted at a
fixed distance apart so that the whole system can be moved in the directions
indicated. The launching transducer is immersed in the test liquid and the
detecting transducer in a reference liquid where the velocity of sound is
known. The test and reference liquids are separated by a diaphragm.
When moving the whole transducer system, the effective path lengths in
the two liquids change and the difference in sound velocities can be found
from the phase difference between the transmitter and receiver voltage.

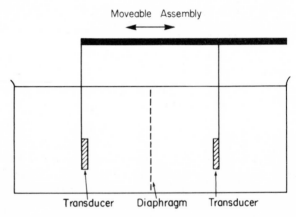

Figure 4.14 Comparison method

As the distance between the launching and detecting transducers is
fixed the divergence of the sound beam is constant. Siegert[48], using a
mechanical arrangement of this kind, was able to make very accurate
absorption measurements in the frequency region 200 kc/s–2 Mc/s using
the pulse technique.

In effect the movement of the transducer system produces an effective path-length change in the test liquid and, therefore, the absorption coefficient can be found. The divergence of the sound beam, which is constant for all measurements, can be accounted for by calculation using the dimensions of the transducer or by suitable calibration.

4.2.10 Brillouin Spectroscopy

A technique which has recently been used to measure the velocity and absorption of sound waves in the Gigacycle region uses the principle of light scattering first predicted by Brillouin. In a liquid medium an assembly of thermally excited sound waves (sometimes termed Debye waves) are present with a range of wavelengths varying from the dimensions of the container to interatomic distances. The motion of the molecules in the medium can, therefore, be regarded as being distributed amongst a set of acoustic modes. These thermally excited acoustic modes produce periodic fluctuations in the refractive index of the medium and Brillouin first showed that light energy could be used to determine the properties of these modes. If monochromatic light is incident on the medium, the Rayleigh scattered light contains, in addition to the incident frequency, the Brillouin components which arise from light scattered as a result of 'reflection' (or more correctly Doppler shifted) off these sound waves. The resulting spectrum, when examined, will consist of a central component which has the same frequency as the incident light and the Stokes and anti-Stokes Brillouin components on each side. The scattering, or reflection of these sound waves, will only occur with appreciable intensity when the Bragg equation,

$$\lambda_e = 2\lambda_s \sin(\theta/2)$$

where λ_e is the wavelength of the incident light, λ_s the wavelength of the sonic wave and θ the scattering angle, is satisfied. It follows, therefore, that the frequency separation, Δv, of the Brillouin components from the central line, given by

$$\Delta v = \pm 2v_0(c/v) \sin \theta/2$$

is identical to the frequency of the sound waves. In this latter equation v_0 = frequency source, c = velocity of light and v = velocity of sound at frequency Δv. By varying, θ, it is possible to obtain sound velocity measurements simply by measuring $\Delta v(10^9-10^{10} \text{ c/s})$. The relative intensities of the Brillouin components have been theoretically investigated by Landau and Placzek[49]. The width of the Brillouin components is related to the lifetime of the sound wave involved in the scattering process. If the exciting light source is monochromatic it is possible to measure Δv, the broadening of the Brillouin component over that of the source. The absorption coefficient, α, can then be determined from the relationship

$$\alpha = (\delta v/\Delta v)2\pi/\lambda_s$$

A block diagram of a simplified experimental arrangement is shown in Figure 4.15. Monochromatic light from a laser source is used to radiate the liquid sample kept in a glass container. The resulting scattered light is then collected at the desired angle and analysed by a Fabry–Perot interferometer. The resulting spectrum can either be detected photographically or recorded using a photomultiplier. In order to record the spectrum, the interferometer must be pressure scanned. The velocity of the acoustic waves can be calculated from the frequency of the Brillouin components and the absorption follows from the shape of these bands. This technique is still at a fairly early state of development and should prove useful for high-frequency measurements. Descriptions of various experimental arrangements are given in the literature[50–56].

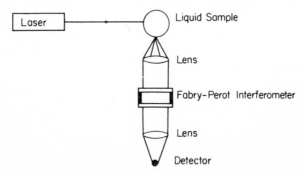

Figure 4.15   Brillouin spectrometer

4.3   ANALYSIS OF EXPERIMENTAL DATA

4.3.1   Calculation of Thermodynamic Data

Two important aspects of a reaction are its thermodynamic properties and its kinetic behaviour. Ultrasonic data yield information in both areas.

In this section the procedure used to derive thermodynamic data is described. The necessary theory has been briefly derived in Section 4.1, and has been outlined in full in references 1, 4 and 48. Consequently the fundamental equations used here are quoted without derivation.

4.3.1.1 *Basic Equations*

When many systems are perturbed by acoustic energy it is found that the experimental data fit a single relaxation time. An important example of a chemical reaction where behaviour is of this type is the two-state energy equilibrium,

$$A_1 \rightleftharpoons A_2$$

Provided that the experimentally measured velocity dispersion is negligible the ultrasonic quantity μ_m and the thermodynamic parameters are related as follows (see Section 4.1.4.3).

$$2\mu_m/\pi = (\gamma - 1)[\Delta C_p/(C_p - \Delta C_p)][1 - (\Delta V/V)(C_p/\Delta H \cdot \theta)]^2 \qquad (4.49)$$

$$v^2 = (\gamma - 1)JC_p/\theta^2 TM \qquad (4.50)$$

$$\Delta C_p = R(\Delta H/RT)^2 \exp(-\Delta G/RT)/[1 + \exp(-\Delta G/RT)]^2 \qquad (4.51)$$

where ΔC_p is the 'relaxing' contribution to the specific heat, ΔG is the Gibbs free energy for the reaction related to the enthalpy and entropy differences by

$$\Delta G = \Delta H - T\Delta S$$

ΔV is the volume change in the total volume V, θ is the coefficient of thermal expansion, γ is the specific heat ratio, J the Joule and M the molecular weight.

Hence we have the situation in which, in general, there are three unknowns ΔH, ΔS and ΔV expressed in terms of the variation of μ_m with temperature—the quantity obtained experimentally. Provided $\Delta C_p \ll C_p$ Equation (4.49) can be rearranged to give,

$$\Delta C_p = 2\mu_m C_p/\pi(\gamma - 1)[1 - (\Delta V/V)(C_p/\Delta H\theta)]^2 \qquad (4.52)$$

Combining Equation (4.51) and Equation (4.52) gives:

$$\frac{2\mu_m C_p}{\pi(\gamma - 1)R} = \left[\frac{\Delta H}{RT} - \frac{\Delta V}{V}\frac{C_p}{\theta RT}\right]^2 \frac{\exp(-\Delta G/RT)}{[1 + \exp(-\Delta G/RT)]^2} \qquad (4.53)$$

$(\gamma - 1)$ can be eliminated from Equation (4.53) by using Equation (4.50) giving

$$\frac{2\mu_m C_p{}^2 J}{\pi R v^2 \theta^2 TM} = \left[\frac{\Delta H}{RT} - \frac{\Delta V}{V}\frac{C_p}{\theta RT}\right]^2 \frac{\exp(-\Delta G/RT)}{[1+\exp(-\Delta G/RT)]^2} \quad (4.54)$$

From Equation (4.54) the only piece of experimental data available is μ_m whereas the unknown thermodynamic parameters are ΔV, ΔH and ΔS (ΔG is related to ΔH and ΔS). In order to solve Equation (4.54) the acoustic relaxation is assumed to be a thermal process, and the assumption $(\Delta V/V)(C_p/\theta) \ll \Delta H$ is made. This is a major assumption and has to be applied in order to derive any thermodynamic data. When this assumption is made Equation (4.54) reduces to

$$\frac{2\mu_m C_p{}^2 J}{\pi R v^2 \theta^2 TM} = \left(\frac{\Delta H}{RT}\right)^2 \frac{\exp(-\Delta G/RT)}{[1+\exp(-\Delta G/RT)]^2} \quad (4.55)$$

On rearrangement and substitution of equation (4.51) one obtains the relation,

$$\Delta C_p = \frac{\mu_m}{Tv^2}\left[\frac{C_p}{\theta}\right]^2\left(\frac{2J}{\pi M}\right) \quad (4.56)$$

The way in which ΔC_p varies with $(\Delta H/RT)$ can be computed from Equation (4.51) provided a value is assumed for the entropy change, ΔS, of the reaction. A typical plot, assuming that $\Delta S = 0$, is shown in Figure 4.16.

Examination of Figure 4.16, which is known as Schottky's function, shows that at low temperatures, when $\Delta H/RT$ is large, ΔC_p increases with increasing temperature until a maximum value is reached when $\Delta H = 2\cdot 4RT$ for $\Delta S = 0$. After this point ΔC_p decreases with increasing temperature. The general shape of the $\Delta C_p \sim (\Delta H/RT)$ curve remains the same as ΔS takes a range of values. All that occurs is that the maximum value of ΔC_p and the point on the $(\Delta H/RT)$ scale at which ΔC_p reaches its maximum value varies. Corresponding values of these quantities are given in Table 4.1.

4.3.1.2   Lamb's Method

Molecular acoustic studies provide experimental data on the variation of μ_m and v with temperature. Equation (4.56) shows how these quantities are related to ΔH via Equation (4.51): it is clear that one proceeds to quantitative thermodynamic data pertaining to the reaction via Equation

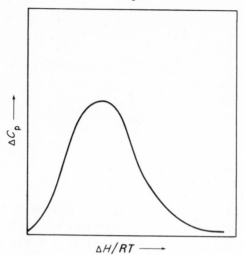

Figure 4.16 Relation between ΔC_p and $\Delta H/RT$
with $\Delta S = 0$

(4.56) in which the third term on the right-hand side is constant for a given substance. If C_p and θ are known over the appropriate temperature range Equation (4.55) can be manipulated as follows to yield a value for the enthalpy difference.

Table 4.1

Calculation of ΔC_p values

ΔS e.u.	ΔC_p (max) cal/mole/deg	$\Delta H/RT$ at ΔC_p (max)
4	3·61	3·2
3	2·66	3·0
2	1·88	2·75
1·5	1·57	2·6
1	1·29	2·4
0·5	1·06	2·4
0	0·87	2·4
−0·5	0·71	2·4
−1	0·57	2·3
−1·5	0·45	2·3
−2	0·36	2·3
−3	0·23	2·2

If $\Delta G \gg 3RT$ then the last term on the right-hand side of Equation (4.55) is very nearly unity and the relation becomes

$$\left[\frac{T\mu_m}{v^2} \cdot \frac{C_p^2}{\theta^2}\right]\left[\frac{2J}{\pi MR}\right] = \left[\frac{\Delta H}{R}\right]^2 \exp(-\Delta G/RT)$$

$$= \left[\frac{\Delta H}{R}\right] \exp(-\Delta H/RT)\exp(\Delta S/R)$$

and therefore

$$\ln\left[\frac{T\mu_m}{v^2}\frac{C_p^2}{\theta^2}\right] = -\frac{\Delta H}{RT} + \text{constant}$$

This expression corresponds to a straight-line plot whose slope provides a value for $-\Delta H/R$, and whose intercept is proportional to ΔS.

For many substances the quantities C_p and θ have not yet been measured experimentally. In these cases it is necessary to make the further assumption that the ratio (C_p/θ) is independent of temperature. Equation (4.55) then reduces to

$$\left[\frac{T\mu_m}{v^2}\right]\text{constant} = \left[\frac{\Delta H}{R}\right]^2 \exp(-\Delta G/RT)$$

and the slope of a plot of

$$\log\left[\frac{T\mu_m}{v^2}\right] \text{ against } 1/T \text{ yields a value for } \Delta H.$$

When $\Delta G < 3RT$ and the magnitude of C_p/θ is unknown a different procedure has to be followed. A further assumption that $\Delta S = 0$ is introduced and Equation (4.51) reduces to

$$\Delta C_p = R\left[\frac{\Delta H}{RT}\right]^2 \exp(-\Delta H/RT)[1+\exp(-\Delta H/RT)]^{-2}$$

A plot of ΔC_p against $\Delta H/RT$ is shown in Figure 4.16 and it can be seen that ΔC_p reaches a maximum when $\Delta H = 2\cdot4RT$.

The manner in which ΔC_p is related to the experimentally measured quantities μ_m and c (provided that $\Delta H \gg \Delta V C_p/V\theta$) is summarized by Equation (4.56).

If the temperature dependence of C_p/θ is neglected then from Equation (4.56) it follows that the temperature variation of ΔC_p is governed by the quantity (μ_m/Tc^2). An analysis of how the experimental values of (μ_m/Tc^2) change with temperature gives a clear indication of which part of the Schottky function (Figure 4.16) fit the experimental data. If $\mu_m Tc^2$ increases with temperature then obviously $\Delta H > 2.4RT$ which means that the assumption $\Delta G > 3RT$ is reasonably safe to use (see previous section). If μ_m/Tc^2 decreases with increasing temperature, then $\Delta H < 2.4RT$. In this case the measurements should be extended to lower temperatures so that μ_m/Tc^2 passes through a maximum.

An approximate value for the enthalpy difference then follows from the relation

$$\Delta H \approx 2.4RT_{max}$$

It should be pointed out, however, that if C_p/θ is known then ΔC_p (maximum) can be found via Equation (4.56). Corresponding, more accurate values for ΔH and ΔS, follow from the procedure that leads to the values quoted in Table 4.1.

4.3.1.3 Piercy's Method

In a different approach to determine the thermodynamic parameters of the equilibrium, Piercy[57] has used Equation (4.55) and taken the square root of both sides giving:

$$(2\mu_m J C_p^2/\pi R^2\theta^2 Tv^2 M)^{1/2} = \frac{\Delta H}{RT}\frac{\exp(-\Delta G/2RT)}{1+\exp(-\Delta G/RT)} \tag{4.57}$$

(It should be noted there that the transformation:

$$\frac{\exp(-\Delta G/2RT)}{1+\exp(-\Delta G/RT)} = \frac{\exp(\Delta G/RT)}{1+\exp(\Delta G/RT)}$$

has been used and also the basic assumption that $\Delta H \gg \Delta V C_p/V\theta$ is applied.)

Denoting the l.h.s. of Equation (4.57) by F it follows that

$$\frac{\Delta G}{RT} = 2\ln\left[\frac{\zeta_1+1+\{F_1^2+(\zeta_1+1)^2\}^{1/2}}{F_1}\right] \tag{4.58}$$

and

$$\frac{\Delta H}{RT} = 2[F_1^2+(\zeta_1+1)^2]^{1/2} \tag{4.59}$$

where

$$\zeta_1 = (T_1/F_1)(\mathrm{d}F/\mathrm{d}T)_1 \qquad (4.60)$$

and the subscript 1 refers to measurements at a particular temperature.

From its definition it follows that F can be determined from experimental data; the value of ζ_1 then follows from Equation (4.60). From Equations (4.58) and (4.59) it is then possible to obtain ΔH and ΔS (via Equation 4.51) directly from the experimental data without making any *a priori* assumptions regarding the value of ΔG.

This method requires knowledge of the thermal data on C_p and θ and, therefore, is not often applicable to rotational isomers of ethane derivatives since very little thermal data are available. However, Piercy has pointed out that one big advantage of this method is that the work can be carried out in solution (~ 5 mole %). In this case the quantities C_p, θ and $(\gamma - 1)$ are essentially those of the solvent, which in most cases are accurately known over a wide temperature range.

4.3.2 Calculation of Kinetic Parameters

The molecular systems dealt with subsequently in this section are of two main types. In the first (e.g. in methylcyclohexane) the absorption of sound energy senses a change in the equilibrium between two of the possible isomers and the overall reaction can be represented as

$$\mathrm{A}_1 \underset{k_{21}}{\overset{k_{12}}{\rightleftharpoons}} \mathrm{A}_2$$

(see Figure 4.5).

For this reaction the relaxation frequency f_c, related to the relaxational time by $\tau = 1/(2\pi f_c)$, is given by Equation (4.40).

$$f_c = (k_{12} + k_{21})/2\pi \qquad (4.61)$$

Since in addition

$$K = k_{12}/k_{21} = \exp(-\Delta G/RT) \qquad (4.62)$$

substitution of Equation (4.61) in (4.62) yields

$$2\pi f_c = k_{12} + k_{21} = k_{21}[1 + \exp(-\Delta G/RT)]$$

or

$$2\pi f_c/[1 + \exp(-\Delta G/RT)] = k_{21} \qquad (4.63)$$

If the reaction is assumed to be a first-order rate process, the rate constant k_{21} can be expressed as

$$k_{21} = \frac{kT}{h} \exp(-\Delta H_{21}^+/RT) \exp(\Delta S_{21}^+/R) \qquad (4.64)$$

substitution of Equation (4.63) in (4.64) yields

$$2\pi f_c/[1 + \exp(-\Delta G/RT)] = \frac{kT}{h} \exp(-\Delta H_{21}^+/RT) \exp(\Delta S_{21}^+/R) \qquad (4.65)$$

where k and h are Boltzmann's and Planck's constants respectively. ΔS_{21}^+ is the entropy and ΔH_{21}^+ the enthalpy of activation (or barrier height)—see Figure 4.5.

Hitherto the usual procedure [58-61] used in deriving a value for the barrier height hindering rotation, ΔH_{21}^+, has been to assume that ΔG is sufficiently large ($> 3RT$) so that the denominator on the l.h.s. of Equation (4.65) equals unity and,

$$f_c/T = (k/2\pi h) \exp(-\Delta H_{21}^+/RT) \exp(\Delta S_{21}^+/R) \qquad (4.66)$$

A plot of $\log(f_c/T)$ against $1/T$ then is a straight line of slope $(-\Delta H^+/R)$ and intercept ΔS^+, provided that ΔS^+ does not vary with temperature.

In the case of rotational isomerism in ethane derivatives the assumption that $\Delta G > 3RT$ need not necessarily be true. However this approach is still applicable as has been shown by Wyn-Jones and Orville-Thomas[62] who argued as follows.

Manipulation of Equation (4.62) shows that

$$k_{21} = 2\pi f_c/(1 + K) \qquad (4.67)$$

The assumption that ΔG is large is tantamount to saying that K is very small and so Equation (4.67) reduces to

$$k_{21} = 2\pi f_c$$

It has been shown[62] for a number of rotational isomers that when K values are computed from enthalpy differences obtained from infrared studies†, within the range $-2 \leqslant \Delta S \leqslant 2$, the quantity $2\pi f_c$ changes by a factor of 20–30, whereas $(1 + K)$ only changes by some 10% over an 80°C range and therefore remains sensibly constant and approximately unity as demanded by the assumption leading to Equation (4.66). This

† These are generally accepted to be accurate.

means that to within an accuracy of 5–10% or so a plot of $\log(f_c/T)$ against $1/T$ can still be used to evaluate ΔH_{21}^{+}.

4.3.3 Numerical Example

At this stage it is useful to illustrate the way in which relaxation parameters are derived from experimental acoustic data by a definite example; the rotational isomeric relaxation observed for 1,2-dibromo-2-methyl propane will be considered.

Experimentally the quantity α/f^2 for this liquid was measured over a range of temperature over the relaxation region. The experimental results are given in Table 4.2. The α/f^2 values are then plotted against temperature for each frequency and smooth curves are drawn through the experimental points as shown in Figure 4.17. Smoothed values of α/f^2 are then read at 5–10°C temperature intervals. These averaged values are then substituted in the basic equation

$$\alpha/f^2 = A/[1+(f/f_c)^2]+B \tag{4.68}$$

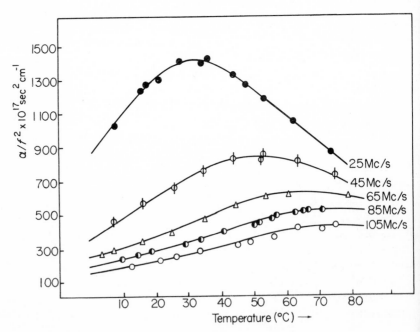

Figure 4.17 Ultrasonic absorption data for 1,2-dibromo-2-methyl propane

Table 4.2

Experimental data for 1,2-dibromo-2-methyl propane

Frequency 25.4 Mc/s		Frequency 44.9 Mc/s		Frequency 65.6 Mc/s		Frequency 83.3 Mc/s		Frequency 105.5 Mc/s	
Temp. °C	$\alpha/f^2 \times 10^{17}$ sec^2 cm^{-1}	Temp. °C	$\alpha/f^2 \times 10^{17}$ sec^2 cm^{-1}	Temp. °C	$\alpha/f^2 \times 10^{17}$ sec^2 cm^{-1}	Temp. °C	$\alpha/f^2 \times 10^{17}$ sec^2 cm^{-1}	Temp. °C	$\alpha/f^2 \times 10^{17}$ sec^2 cm^{-1}
69.0	863	75.0	720	79.9	582	71.0	517	71.0	409
62.2	1056	63.9	807	61.3	602	67.0	515	64.2	408
54.0	1185	53.2	850	55.0	595	66.0	520	56.7	375
47.8	1268	53.0	807	54.1	587	62.5	507	49.3	342
44.0	1321	44.3	820	44.8	540	57.0	488	45.2	316
36.6	1424	35.3	767	35.2	467	52.5	472	39.8	303
34.6	1393	26.2	657	25.4	405	50.5	430	33.6	282
28.6	1422	16.3	559	16.5	344	41.5	396	26.9	257
22.2	1289	8.0	466	7.0	292	34.0	343	21.5	237
17.8	1273			3.7	267	29.2	312	12.5	210
16.5	1234					18.5	275		
7.0	1029					10.0	245		

to yield values for the relaxation parameters A, B and f_c. This is best done either (1) graphically or (2) by computation.

4.3.3.1 Graphical Method

Graphical solutions of Equation (4.68) have been discussed by Lamb and Pinkerton[63], Heasell and Lamb[58] and Andrae and others[64]. In brief a value for B is chosen such that a plot of $(\alpha/f^2 - B)^{-1}$ against f^2 is linear giving a slope of $(1/f_c)^2$ and intercept $1/A$. This value of A can then be used to plot $[A/(\alpha/f^2 - B) - 1]^{1/2}$ against f to obtain a better estimate for f_c. Finally the values of A and f_c are then substituted in the above Equation (4.68) and the value of B adjusted until the best fit is obtained with experimental results. Normally a fit of about 5% or less is aimed for.

An alternative graphical solution is to choose a value of B until a plot of $(\alpha/f^2 - B)$ against $(\alpha - Bf^2)$ gives a straight line. The intercepts of this line with the coordinate axes are equal to A and Af_c^2.

4.3.3.2 Computer

When the computer facilities are available it is easy to manipulate Equation (4.68) and write a computer programme to solve for A, B and f_c so that the quantity

$$\sum_i \left[\left(\frac{\alpha}{f_i^2} \right)_{obs} - \frac{A}{1 + (f_i/f_c)^2} - B \right]^2$$

over all the experimental frequencies is minimized. In this way A, B and f_c are evaluated using a least mean-squares method.

In Table 4.3 the values of A, B and f_c derived from analysing Equation (4.68) using an I.B.M. 1620 computer are listed for 1,2-dibromo-2-methyl propane. These relaxation parameters can now be substituted in Equation (4.68) to provide 'theoretical' values for α/f^2, when compared with the experimental data summarized in Figure 4.17 excellent agreement is found. The variation of the velocity of sound with temperature is shown in Figure 4.18. The experimental points were found to fit the equation $v = 1035 - 2.64\,G$ where t is the temperature in °C. The absorption per wavelength μ of the sound waves is given by

$$\mu = 2\lambda = \alpha v/f \qquad (4.69)$$

From Equations (4.68) and (4.69) it follows that:

$$\mu = Avf/[1 + (f/f_c)^2] + Bvf$$

Table 4.3

Relaxational parameters for 1,2-dibromo-2-methyl propane

Temp. °C	$A \times 10^{17}$ sec^2 cm^{-1}	$B \times 10^{17}$ sec^2 cm^{-1}	f_c Mc/s	μ'_m
5	7066	143	9·31	0·0336
10	6122	150	10·6	0·0344
15	5063	143	13·2	0·0332
20	4092	138	16·1	0·0324
25	3621	144	18·3	0·0336
30	3082	137	21·5	0·0317
35	2552	138	25·6	0·0308
40	2104	137	30·8	0·0301
45	1763	132	36·9	0·0298
50	1502	138	42·9	0·0291
55	1320	147	48·1	0·0283
60	1145	143	56·0	0·0282
65	1001	149	63·2	0·0274
70	882	140	72·6	0·0273
75	753	149	82·9	0·0263

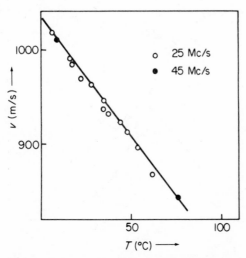

Figure 4.18 Variation of sound velocity with temperature for 1,2-dibromo-2-methyl propane

and the absorption per wavelength (μ') relating to the relaxation is given by

$$\mu' = \frac{Avf}{1+(f/f_c)^2} \qquad (4.70)$$

When $f = f_c$ μ' reaches a maximum value of $\mu'_{max} = \frac{1}{2}Avf_c$. The values of μ'_{max} for 1,2-dibromo-2-methyl propane are also listed in Table 4.3. The variation of μ' and α/f^2 with frequency is shown in Figures 4.19 and 4.20 for different temperatures. In these graphs the solid lines are obtained from the relaxational parameters via Equations (4.68) and (4.70) and the points are experimental.

In Figure 4.21 a plot of $\log(T\mu'_m/v^2)$ against $1/T$ for 1,2-dibromo-2-methyl propane is shown. From the slope of this graph a value of $\Delta H = 0.97\,\mathrm{Kcal/mole}$ is obtained. In a similar way a value for the barrier height of $\Delta H_{21}^+ = 5.5\,\mathrm{Kcal/mole}$ is found from a plot of $\log(f_c/T)$ against $1/T$ as shown in Figure 4.22.

In Figure 4.23 a typical plot of μ'_m/Tv^2 against temperature is shown for 1,2-dibromobutane. This reaches a maximum corresponding to the

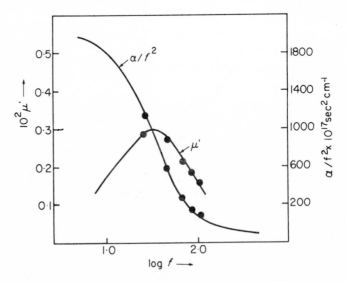

Figure 4.19   Variation of μ' and α/f^2 with $\log f$ at 40°C for 1,2-dibromo-2-methyl propane

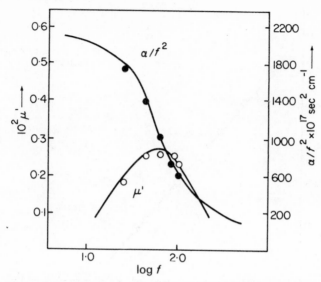

Figure 4.20 Variation of μ' and α/f^2 with $\log f$ at 60°C for 1,2-dibromo-2-methyl propane

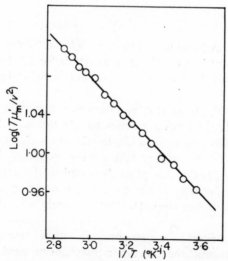

Figure 4.21 Determination of enthalpy difference

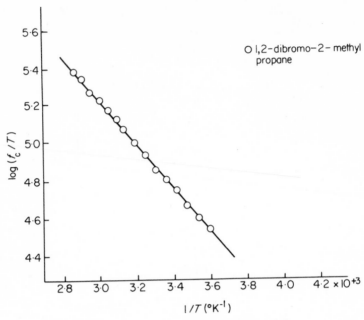

Figure 4.22   Determination of barrier height

same maximum as Schottky's function. This substance was chosen to illustrate this point because the data available for 1,2-dibromo-2-methyl propane cover only the descending portion of the Schottky function where $\Delta H > 3RT$.

In most cases the relaxation parameters of Equation (4.68) are more conveniently solved by using a computer. In this way the experimental results are conditioned to fit a single relaxation using the least-mean-squares method. It is useful to take a cross-section of the experimental results to check whether the plots discussed in the graphical solutions are linear. The linearity of these plots is a useful guide to determine whether the results are consistent with a single relaxation mechanism or not.

Even though a good fit can be obtained with a computer owing to the conditioning process, the relaxation mechanism need not strictly be governed by a single relaxation time[65] and care and judgement must be exercised. To emphasize this point an example of a process which can be

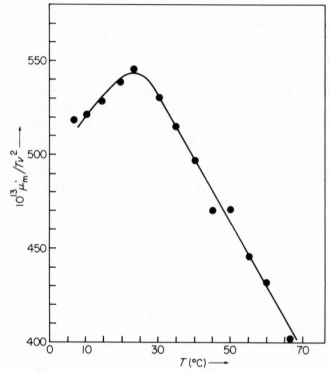

Figure 4.23 Application of Shottky's function for 1,2-dibromo-
butane

analysed in terms of a single *or* a double relaxation mechanism will be
considered.

4.3.3.3 *Difficult Analyses*

In a large number of cases the acoustic data can be analysed adequately
in terms of a single relaxation. However cases are known of systems
having more than one relaxation time of comparable magnitude; i.e.
multiple relaxation occurs.

For a process with multiple relaxations Equation (4.1) is modified to

$$\alpha/f^2 = \sum_{i}^{n}\left[\frac{A_i}{1+(f/f_c)^2}\right]+B \tag{4.71}$$

A few systems involving multiple relaxation have been studied. For example Tamm[66] has demonstrated that this phenomenon exists in a number of electrolytic solutions. In effect it was shown that a plot of the absorption per wavelength, μ, against frequency possessed more than one maximum.

In the example given below of the binary system triethylamine/H_2O, Andrae and his collaborators[20,24,67] found that they were able to analyse the experimental data in terms of both a single and a double relaxation mechanism using Equation (4.71) with $n = 1$ and $n = 2$ respectively.

4.3.3.3.1   *Triethylamine/water system.* For an aqueous solution (mole fraction of triethylamine 0·125) Andrae and his workers found that the single relaxation ultrasonic parameters were

$$A = 6200 \times 10^{-17} \sec^2 cm^{-1}$$

$$B = 390 \times 10^{-17} \sec^2 cm^{-1}$$

$$f_c = 25 \, Mc/s$$

These data, with $f = 4·5$ and $7·5$ Mc/s in turn, lead to calculated values of 6400 and $6078 \times 10^{-17} \sec^2 cm^{-1}$ for α/f^2 as compared with experimental values of 6200 and $5400 \times 10^{-17} \sec^2 cm^{-1}$.

The corresponding figures assuming a double relaxation were (same units),

$$A(1) = 4936, \qquad A(2) = 1993$$

$$f_c(1) = 9·7, \qquad f_c(2) = 62$$

$$B = 284$$

These parameters lead to values of α/f^2, with $f = 4·5$ and $7·5$ Mc/s, of 6337 and 5348 as compared with observed values of 6200 and 5400.

It is clear, therefore, that within the experimental error of $\pm 5\%$ it is difficult to differentiate between a single and a multiple relaxation process from computed results. In these cases care should be exercised when dealing with the results and the plots suggested in the graphical methods should also be tried. Typical plots of $(\alpha/f^2 - B)^{-1}$ against f^2 for 2-iodo-butane and 2,3-dichlorobutane are shown in Figures 4.24 and 4.25. These plots are reasonably straight indicating that the experimental results fit a single relaxation mechanism. The molecules exist as three separate rotational isomers which means that more than one isomeric relaxation

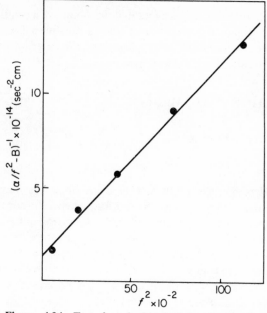

Figure 4.24 Test for single-relaxation mechanism (2-iodobutane, −10°C)

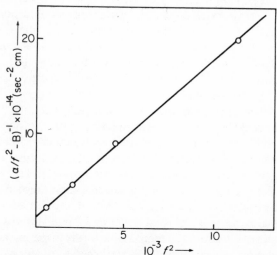

Figure 4.25 Test for a single-relaxation mechanism (2,3-dichlorobutane, −10°C)

is expected. Since the experimental data point to a single relaxation this indicates either a distribution of relaxation times over a short frequency range or that only one relaxation has been observed. These possibilities are discussed in Section 4.4.1.

4.4   ROTATIONAL ISOMERISM

4.4.1   Substituted Ethanes

4.4.1.1   *Introduction*

Rotational isomerism about a C–C single bond has been extensively and successfully studied by infrared and Raman spectroscopy[68,69]. These techniques provide the enthalpy difference between stable isomers from a temperature study of the intensities of two corresponding bands—one from each isomer. These data supplement those obtained from ultrasonic studies which in addition yield barrier heights. In this section the energetic parameters associated with internal rotation in liquid substituted ethanes and other simple molecules are considered.

The rotation of one part of a molecule with respect to another about an internal chemical bond is, in general, accompanied by a change in the potential energy, V, of the molecule. In the case of the simplest molecule, ethane, and also saturated derivatives where one of the carbon atoms has three identical substituents, internal rotation about the C–C bond is associated with a potential-energy curve whose shape is similar to that shown in Figure 4.26. This type of curve is obtained by plotting V against the azimuthal angle θ. The minima in the potential-energy curve represent a molecular conformation corresponding to the staggered form of the molecule and the maxima of the curve to the eclipsed form. The various conformations of the molecule corresponding to the rotation of one part of the molecule with respect to the other is also indicated in Figure 4.26. The potential barrier hindering rotation in these molecules is the difference in energy between the maxima and the minima of this energy diagram. The value of this energy parameter is of extreme importance in determining the physical properties of the molecules. In this discussion we are concerned with potential barriers of the same order as kT implying that a dynamic equilibrium exists between the more stable forms of the molecule.

For molecules which have potential-energy diagrams similar to Figure 4.26 it is possible to estimate the barrier hindering rotation from

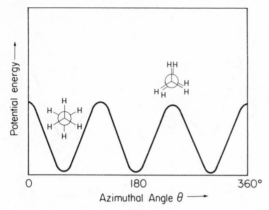

Figure 4.26   Potential-energy diagram for ethane

microwave and far infrared spectroscopy. Ultrasonic relaxation due to internal rotation will not occur in these systems because all the stable forms have the same energy, i.e. there is no enthalpy difference.

The energy barrier hindering rotation in these molecules undoubtedly arises from interactions amongst the electrons and nuclei in the molecule. The contributions to the barrier from the various interactions are not fully understood although several empirical attempts have been carried out to try and account for the barriers. Even for the simple molecule, ethane, sophisticated theoretical treatments arrive only at semi-quantitative conclusions regarding the cause of the potential barrier.

4.4.1.2   *Barrier Heights to Rotation*

4.4.1.2.1   *Two-energy states, one of which is degenerate.* In derivatives of ethane which have two or three different substituents on each carbon atom the potential-energy curves are modified since the various conformers have different energies. In the simplest molecules, i.e. those which have only two different substituents attached to each carbon atom e.g. CX_2Y-CX_2Y or CX_2Y-CXY_2, there can be two types of potential-energy curves depending on the physical properties of the molecule. The shapes of these curves are indicated in Figures 4.27 and 4.28. In molecules of this kind two of the staggered forms are optical isomers having the same energy. Figure 4.27 represents an energy diagram where these optical isomers are the higher energy stable forms and Figure 4.28 represents the reverse case.

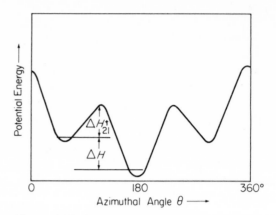

Figure 4.27 Potential-energy diagram for 1,2-dibromo-2-methyl propane

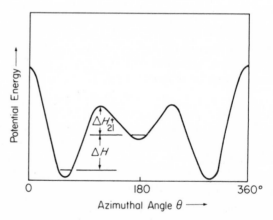

Figure 4.28 Potential-energy diagram for isobutyl bromide

The properties of these molecules depend on the energy differences between the stable isomers and also the barrier height hindering rotation. On the potential-energy curves the barrier height is represented by the differences in energy between neighbouring maxima and minima.

The enthalpy difference between the isomers can be determined both from infrared and Raman spectroscopy and also from ultrasonic absorption studies. Enthalpy differences derived from spectroscopic methods have been reviewed by Mizushima[68] and Sheppard[69] whereas ultrasonic values are liable to error and are discussed in Section 4.4.1.3.

The experimental information about the potential barrier hindering rotation in these molecules has been very limited until the recent development of the ultrasonic technique. It has been found that ultrasonic waves perturb the equilibrium between the stable forms that differ in energy. (No ultrasonic relaxation will occur due to the dynamic equilibrium between forms having the same enthalpy.) In order to derive the energy parameters of rotation from ultrasonic data it is necessary to take into account the statistical weight of the optical isomers in these systems. This means that Equations (4.51) and (4.61) for the two state $A_1 \rightleftharpoons A_2$ process are modified as follows:

When the two optical isomers have the higher energy:

(i) $\quad \Delta C_p = 2R(\Delta H/RT)^2 \exp(-\Delta G/RT)/[1 + 2 \exp(-\Delta G/RT)]^2$

$\quad \tau = (2k_{12} + k_{21})^{-1}$

(ii) When the two optical isomers have the lower energy:

$\quad \Delta C_p = (R/2)(\Delta H/RT)^2 \exp(-\Delta G/RT)/[1 + \tfrac{1}{2} \exp(-\Delta G/RT)]^2$

$\quad \tau = (k_{12} + 2k_{21})^{-1}$

In the following discussion the potential barriers hindering rotation in these molecules are discussed. This quantity derived from ultrasonic data represents the difference in energy between the highest energy stable form and the eclipsed form which can be regarded as a transition state between the two stable rotational isomers. The barrier height is indicated as ΔH_{21}^{+} in the potential-energy diagrams. Listed in Table 4.4 are the energy barriers for a number of molecules derived from ultrasonic data together with the molecular conformation of the more stable isomer and also the type of potential-energy curve associated with internal rotation in these molecules; these can usually be determined from spectroscopic techniques.

For the molecules listed in Table 4.4 we can investigate how different substituents affect the value of the potential barrier hindering rotation. The evidence presented in this discussion will be concerned with the effect of substitution which can be directly correlated with steric and electrostatic non-bonding forces in the molecule. Non-bonding interac-

Table 4.4
Potential barrier for molecules with two isomers having the same energy

Molecule	Stable Conformation	Potential Energy Curve	Barrier ΔH_{21}^{+} (kcal/mole)
1,2-dichloroethane $ClCH_2CH_2Cl$	trans	Fig. 2	$3 \cdot 2^{a,h}$
1,2-dibromoethane $BrCH_2CH_2Br$	trans	Fig. 2	$4 \cdot 2^{a,h}$
n-propylbromide $CH_3CH_2CH_2Br$	gauche	Fig. 1	$3 \cdot 6^{b}$
n-butane $CH_3CH_2CH_2CH_3$	trans	Fig. 2	$3 \cdot 4^{c,h}$
1,1,2-trichloroethane Cl_2HCCH_2Cl	Cl–Cl trans	Fig. 1	$5 \cdot 8^{b}$
1,1,2-tribromoethane Br_2HCCH_2Br	Br–Br trans	Fig. 1	$6 \cdot 4^{b}$
2-methyl butane $(CH_3)_2HCCH_2CH_3$	$CH_3–CH_3$ trans	Fig. 1	$4 \cdot 7^{d}$
isobutylchloride $(CH_3)_2HCCH_2Cl$	$CH_3–Cl$ trans	Fig. 1	$3 \cdot 8^{e}$
isobutylbromide $(CH_3)_2HCCH_2Br$	$CH_3–Br$ trans	Fig. 1	$4 \cdot 7^{e}$
isobutyliodide $(CH_3)_2HCCH_2I$	$CH_3–I$ trans	Fig. 1	$5 \cdot 4^{e}$
1,2-dichloro-2-methyl propane $(CH_3)_2ClCCH_2Cl$	Cl–Cl trans	Fig. 2	$4 \cdot 5^{e}$
1,2-dibromo-2-methyl propane $(CH_3)_2BrCCH_2Br$	Br–Br trans	Fig. 2	$5 \cdot 5^{e}$
1,1,2,2-tetrabromoethane $Br_2HCCHBr_2$	trans	Fig. 1	$4 \cdot 3^{f}$
2,3-dimethylbutane $(CH_3)_2HCCH(CH_3)_2$	trans	Fig. 1	$2 \cdot 8^{d}$
1,2-dibromo, 1,1,2,2-tetrafluoroethane BrF_2CCF_2Br	trans	Fig. 1	$6 \cdot 1^{g}$

[a] J. E. Piercy, *J. Chem. Phys.*, **43**, 4066 (1965).
[b] R. A. Padmanabhan, *J. Sci. Ind. Res.* (*India*), **19B**, 336 (1960).
[c] J. E. Piercy and M. G. Seshagiri Rao, *J. Chem. Phys.*, **46**, 3951 (1967).
[d] J. H. Chen and A. A. Petrauskas, *J. Chem. Phys.*, **30**, 304 (1959).
[e] E. Wyn-Jones, *Ph.D. Thesis*, University of Wales, 1963.
[f] K. Krebs and J. Lamb, *Proc. Roy. Soc.* (*London*), *Ser. A*, **244**, 558, (1957).
[g] A. A. Petrauskas quoted by J. Lamb in *Physical Acoustics*, (Ed. W. P. Mason), Vol. II, Part A, 1965, p. 203.
[h] These values represent Arrhenius Activation Energies, $E = \Delta H_{21}^{+} + RT$.

tions obey a Lennard-Jones type potential and are proportional to a term in r^{-12} where r is the internuclear distance. On the other hand electrostatic forces obey Coulomb's law and vary with the reciprocal of the appropriate length. It therefore follows that the closer the non-bonding atoms are the greater the steric terms become. Using the same argument that Mizushima[68] developed in discussing enthalpy differences between rotational isomers, the non-bonded steric forces for substituents are estimated in terms of van der Waals radii. For the substituents considered in this discussion steric forces increase in the following order

$$H, Cl, Br, CH_3, I$$

Electrostatic contributions to non-bonded forces, estimated from electronegativity of substituents and bond moments decrease in the following order

$$CH_3 < I < Br < Cl$$

In all the molecules listed in Table 4.4 the barrier for the bromo-substituted compounds are higher than the corresponding chloro-compounds. This indicates that steric interactions are very important in determining these barriers. The steric contributions in the bromo-compounds are greater than the chloro-compounds even though the eclipsed non-bonded $Cl \cdots Cl$ (2·69 Å), $H \cdots Cl$ (2·55 Å) and $CH_3 \cdots Cl$ (2·62 Å) lengths are less than the corresponding $Br \cdots Br$ (2·82 Å), $H \cdots Br$ (2·63 Å) and $CH_3 \cdots Br$ (2·68 Å) distances. In addition the electrostatic contributions from the chloro-compounds are greater. This argument also applies to the isobutyl halide series where the barrier for the bulkier iodo-compound is greater than the bromo-derivative which, in turn, is greater than the chloro-compound. However, the electrostatic contribution must not be neglected. For instance, the steric contributions for the bromine and methyl groups are approximately the same since the van der Waals radii are similar. Let us consider the series n-butane $(CH_3CH_2CH_2CH_2)$, n-propyl bromide $(CH_3CH_2CH_2Br)$ and 1,2-di-bromoethane $(BrCH_2CH_2Br)$; 2-methylbutane $[(CH_3)_2CHCH_2CH_3]$, isobutyl bromide $[(CH_3)_2CHCH_2Br]$ and 1,1,2-tribromoethane (Br_2CHCH_2Br) and also 2,3-dimethylbutane $[(CH_3)_2CHCH(CH_3)_2]$ and 1,1,2,2-tetrabromoethane $(Br_2CHCHBr_2)$. The successive substitution of bromine for methyl groups in the molecular skeleton increases the barrier considerably. This demonstrates the effect of the electrostatic forces associated with the bromine atom which become appreciable when more than one halogen atom is present in the molecule. One can go from 2-methyl butane to 1,1,2-trichloroethane and from n-butane to 1,2-dichloroethane by exchanging the methyl groups for chlorines. Examination of Table 4.4 shows that this results in an increase in barrier height thus indicating that even though the steric forces are reduced the electrostatic forces become appreciable and the value of the barrier is increased. On the other hand substituting one methyl group for one chlorine atom as in 2-methyl butane and isobutyl chloride decreases the barrier. This demonstrates that the electrostatic forces become appreciable only when more than one of the electronegative atoms are present in the molecule.

In conclusion, for the molecules considered in this work, the major part of the barrier hindering rotation is caused by non-bonded steric and electrostatic forces. The individual contributions from these forces depend

Table 4.5

Molecules with the three isomers having different energies

Molecule	Barrier ΔH_{21}^{+} kcal/mole
1,2-dichloropropane $\Big\}$ $ClCH_2CHClCH_3$	$4 \cdot 7^a$
1,2-dibromopropane $\Big\}$ $BrCH_2CHBrCH_3$	$4 \cdot 9^a$
1-chloro-2-bromopropane $\Big\}$ $ClCH_2CHBrCH_3$	$5 \cdot 5^b$
sec-butylchloride $\Big\}$ $CH_3CH_2CHClCH_3$	$4 \cdot 4^b$
sec-butylbromide $\Big\}$ $CH_3CH_2CHBrCH_3$	$4 \cdot 8^b$
sec-butyliodide $\Big\}$ $CH_3CH_2CHICH_3$	$5 \cdot 3^b$
d,l,2,3-dichlorobutane $\Big\}$ $CH_3HClCCClHCH_3$	$5 \cdot 0^b$
d,l,2,3-dibromobutane $\Big\}$ $CH_3HBrCCBrHCH_3$	$3 \cdot 7^b$

[a] R. A. Padmanabhan, *J. Sci. Ind. Res. (India)*, **19B**, 336 (1960).

[b] E. Wyn-Jones, *Ph.D. Thesis*, University of Wales, 1963.

on the nature of the substituents. The above conclusions are very similar to those derived by Mizushima[68].

4.4.1.2.2 *Three-isomer system, all differing in energy.* Other molecules exist as rotational isomers having three different substituents on at least one carbon atom. The shape of the potential-energy curve associated with internal rotation in these molecules is as shown in Figure 4.28 with all the maxima and minima at different energy levels.

Since all the stable isomers have different energies it is theoretically possible to have more than one ultrasonic relaxation for these systems. In practice, for molecules of this kind, the acoustic data have been analysed in terms of a single relaxation. This implies that either one of the equilibria has been perturbed by sound waves and that the remaining processes have a frequency dependence outside the present experimental range or that the observed parameters derived from the acoustic data refer to an average of the relaxation processes occurring. In practice, it is very difficult to differentiate between these possibilities. In Table 4.5 a list of energy barriers derived from analysing the acoustic data for a single relaxation time for molecules of this type is given.

The simplest arguments used in the previous section concerning non-bonded interactions can now be applied to the secondary butyl halide series to show that for this substance

the equilibrium between isomers A and B is that most probably perturbed by the sound wave. The following reasons are suggested to support this contention.

(i) Since only one halogen atom is present in the molecule the electro-static forces play only a very secondary rôle in determining the potential barrier hindering rotation. The major rôle played by the steric forces is seen as the barrier increases from the chloride derivative to the iodide. This suggests that the size of the halogen atom plays a major rôle in determining the barrier.

(ii) In the related series of molecules, the isobutyl halides, the molecular conformations of the stable isomers are

$$
\begin{array}{ccc}
\text{(E)} & \text{(F)} & \text{(F')}
\end{array}
$$

with F and F′ being the optical isomers having the same energy. In both series of molecules the barrier heights for the corresponding isomers are very similar. For the secondary butyl halides the eclipsed form, corresponding to the equilibrium between A and B, involves a direct methyl–halogen interaction similar to that in the isobutyl halides (E–F).

(iii) The energy differences between the stable isomers in the secondary butyl halides are (in kcal/mole)[70]:

	sec₋butyl chloride	sec/butyl bromide
ΔH (A–B)	0·65±0·3	0·6±0·3
ΔH (C–B)	0·4±0·2	0·5±0·2
ΔH (A–C)	0·25±0·1	0·1±0·1

and for the isobutyl halides:

ΔH (E–F)	0·4±0·2	0·3±0·1

In isomer B of the secondary halides and isomer F(=F′) of the isobutyl halides the nearest non-bonded interactions occur between the following pairs of atoms and groups:

$$CH_3-X, H-H, H-X \quad \text{and} \quad \text{three } (CH_3-H) \text{ interactions}$$

In the eclipsed conformation of the equilibria $A \rightleftharpoons B$ and $E \rightleftharpoons F$ the nearest non-bonded interactions occur between:

$$CH_3-X, CH_3-H \quad \text{and} \quad H-H$$

If we assume that the potential energy of these molecules arises mainly from non-bonded interactions, it should follow, therefore, to a first approximation, that the sum

$$\Delta H_{21}^{+} + \Delta H = \Delta H_{12}^{+}$$

for corresponding molecules in each series should be the same.

For secondary butyl chloride and isobutyl chloride $\Delta H_{12}^{+} = 5·05\pm0·4$ and $4·2\pm0·8$ kcal/mole respectively and for the bromo-compounds $\Delta H_{12}^{+} = 5·4\pm0·4$ and $5·01\pm0·4$ kcal/mole respectively. This agreement is rather good considering the simplicity of the model used and is a good indication that the ultrasonic relaxation observed in the secondary butyl halides refers to the equilibrium between isomers A and B.

4.4.1.2.3 *Longer chain molecules.* The energy barriers for molecules with longer carbon-atom chains are listed in Table 4.6. In these systems

Table 4.6

Molecule	ΔH_{21}^{+} Barrier kcal/mole
2-chloropentane $\left.\begin{array}{l}\\ CH_3HClCCH_2CH_2CH_3\end{array}\right\}$	4·4[a]
2-bromopentane $\left.\begin{array}{l}\\ CH_3HBrCCH_2CH_2CH_3\end{array}\right\}$	4·80[b]
2-methylpentane $\left.\begin{array}{l}\\ (CH_3)_2HCCH_2CH_2CH_3\end{array}\right\}$	3·9[c]
3-chloropentane $\left.\begin{array}{l}\\ CH_3CH_2CHClCH_2CH_3\end{array}\right\}$	3·4[a]
3-bromopentane $\left.\begin{array}{l}\\ CH_3CH_2CHBrCH_2CH_3\end{array}\right\}$	3·6[a]
3-methylpentane $\left.\begin{array}{l}\\ CH_3CH_2CHCH_3CH_2CH_3\end{array}\right\}$	4·1[c]
1,2-dibromopentane $\left.\begin{array}{l}\\ BrCH_2CHBrCH_2CH_2CH_3\end{array}\right\}$	3·5[a]
1,4-dibromopentane $\left.\begin{array}{l}\\ BrCH_2CH_2CH_2CHBrCH_3\end{array}\right\}$	4·7[a]
2-bromohexane $\left.\begin{array}{l}\\ CH_3CHBrCH_2CH_2CH_2CH_3\end{array}\right\}$	4·2[a]
3-bromohexane $\left.\begin{array}{l}\\ CH_3CH_2CHBrCH_2CH_2CH_3\end{array}\right\}$	3·6[a]
1,2-dibromobutane $\left.\begin{array}{l}\\ BrCH_2CHBrCH_2CH_3\end{array}\right\}$	3·4[b]
n-pentane $\left.\begin{array}{l}\\ CH_3CH_2CH_2CH_2CH_3\end{array}\right\}$	3·9[d,e]
n-hexane $\left.\begin{array}{l}\\ CH_3CH_2CH_2CH_2CH_2CH_3\end{array}\right\}$	2·6[d,e]

[a] T. H. Thomas, *Ph.D. Thesis*, University of Wales, 1966.

[b] E. Wyn-Jones, *Ph.D. Thesis*, University of Wales, 1963.

[c] J. H. Chen and A. A. Petrauskas, *J. Chem. Phys.*, **30**, 304 (1959).

[d] J. E. Piercy and M. G. Seshagiri Rao, *J. Chem. Phys.*, **46**, 3951 (1967).

[e] See note (*h*) Table 4.4.

internal rotation is possible about more than one C–C bond. For example, the possible conformations of 2-bromo pentane are:

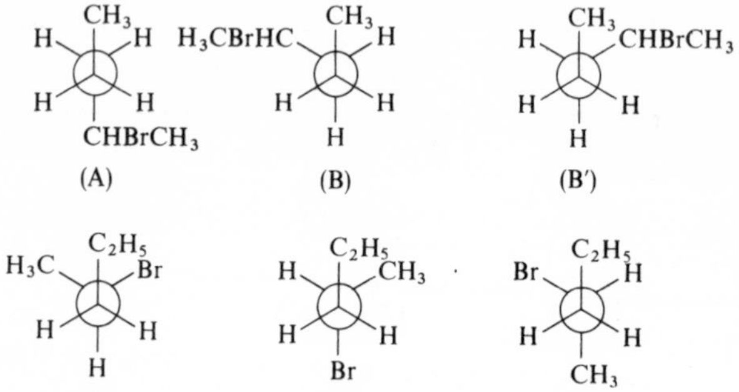

CH3
H
H
H
H
CHBrCH3
(A)
H3CBrHC
CH3
H
H
H
H
(B)
CH3
H
CHBrCH3
H
H
H
(B′)
C2H5
H3C
Br
H
H
H
(C)
C2H5
H
CH3
H
H
Br
(D)
C2H5
Br
H
H
H
CH3
(E)

The acoustic data for these systems has been treated for a single relaxation time. These energy barriers therefore most likely refer to an average of part or all the possible rotational mechanisms.

4.4.1.3 *Enthalpy Differences*

It must be realized that in discussing enthalpy differences in the liquid phase the actual numerical values are very small (0–1 kcal) for these compounds[68,69]. This means that uncertainties of up to ± 400 cals are to be expected owing to experimental errors, inherent in estimating the slopes of straight-line graphs whose gradients are small. The ultrasonic and spectroscopic values of the enthalpy differences for a number of ethane derivatives are listed in Table 4.7. In the majority of these molecules where the enthalpy difference has been found from both techniques, the ultrasonic values are much larger than the infrared values. In a few cases there is reasonable agreement to within ± 400 cals/mole.

In general it is accepted that enthalpy differences obtained by spectroscopic methods are reliable. This view is supported by the fact that for the simple molecules in Table 4.8 the liquid-state values of ΔH obtained from infrared, Raman and dipole moment data are in good agreement. In addition for the vapour state the spectroscopic values agree with those obtained from microwave, dipole-moment and heat-capacity data. Another supporting feature is that the relationship between the spectroscopic vapour and liquid-state ΔH values can often be explained in terms of Onsager's dielectric theory.

Table 4.7

Enthalpy differences between rotational isomers

Molecule	kcal/mole ΔH (Ultrasonic)	kcal/mole ΔH (Infrared)
n-propylbromide	1·3[a]	0, 0·44, 0·5[i]
1,1,2-trichloroethane	2·1[a]	0·1, 0·22, 0·29[i]
1,1,2-tribromoethane	1·6[a]	0·5[i]
2-methylbutane	0·9[b]	0[i]
isobutylchloride	1·31[c]	0·36[e]
isobutylbromide	0·70[d]	0·26[e]
1,2-dichloro-2-methylpropane	2·01[e]	0[e]
1,2-dibromo-2-methylpropane	0·97[e]	0·74[e]
1,1,2,2-tetrabromoethane	0·9[f]	0·9, 0·75[i]
2,3-dimethylbutane	1·0[b]	0·1[i]
1,2-dibromo-1,1,2,2-tetrafluoroethane	1·4[g]	0·93[i]
1,2-dichloropropane	1·1[a]	
1,2-dibromopropane	0·9[a]	1·02, 0·33, 0·67[j,k]
1-chloro-2-bromopropane	2·88[e]	
sec-butylchloride	1·3[e]	0·65, 0·40, 0·25[j,l]
sec-butylbromide	1·8[e]	0·6, 0·5, 0·1[j,l]
sec-butyliodide	1·01[e]	
d,l,2,3-dichlorobutane	1·30[e]	
d,l,2,3-dibromobutane	2·0[e]	0·5, 0·7, 0·3[j,l]
2-methylpentane	0·9[b]	
3-chloropentane	1·2[h]	
3-bromopentane	1·1[h]	
3-methylpentane	0·9[h]	
1,2-dibromopentane	1·3[h]	
1,4-dibromopentane	1·3[h]	0·17[h]
2-bromohexane	1·5[h]	0·44[h]
1,2-dibromobutane	1·5[e]	

[a] R. A. Padmanabhan, *J. Sci. Ind. Res. (India)*, **19B**, 336 (1960).

[b] J. H. Chen and A. A. Petrauskas, *J. Chem. Phys.*, **30**, 304 (1959).

[c] E. Wyn-Jones and W. J. Orville-Thomas, *Chem. Soc. (London), Spec. Publ.* **20**, 209 (1966).

[d] A. E. Clark and T. A. Litovitz, *J. Acoust. Soc. Am.*, **32**, 1221 (1960).

[e] E. Wyn-Jones, *Ph.D. Thesis*, University of Wales, 1963.

[f] K. Krebs and J. Lamb, *Proc. Roy. Soc. (London), Ser A*, **244**, 558 (1957).

[g] A. A. Petrauskas quoted by J. Lamb in *Physical Acoustics*, (Ed. W. P. Mason), Vol. II, Part A, 1965, p. 203.

[h] T. H. Thomas, *Ph.D. Thesis*, University of Wales, 1966.

[i] Values quoted in N. Sheppard, *Advances in Spectroscopy*, Vol. 1, 1959, p. 288.

[j] Values for a three-isomer system.

[k] Y. A. Pentin and U. M. Tatevski, *Dokl. Akad. Nauk. USSR*, **108**, 290 (1966).

[l] L. P. Melikhova, Yu. A. Pentin and O. D. Ul'yanova, *Zh. Strukt. Khim.*, **4**, 535 (1963).

Table 4.8

Enthalpy differences found from different techniques

Molecule	ΔH (liq.) kcal/mole	Method
1,2-dichloroethane	0	Infrared
	0	Raman
	0	Dielectrics
1,2-dibromoethane	0·74	Infrared
	0·65	Raman
	1·0	Dielectrics

The discrepancies revealed by Table 4.7 are a clear indication that the assumptions made using the molecular acoustic method need to be scrutinized carefully. Summarized they amount to assuming that the stoichiometric equation is correct, that the kinetics are first order, the systems are ideal and that

$$\frac{\Delta V}{V} \cdot \frac{C_p}{\theta} \ll \Delta H; \quad \text{in addition it is taken that}$$

$$\Delta C_p \ll C_p$$

These assumptions have to be used because the only experimental piece of information usually available is the magnitude of the maximum absorption per wavelength μ_m at different temperatures. In many cases there is also insufficient information available on the thermal parameters C_p and θ in which case their dependence on temperature is neglected. A further assumption, that $\Delta S = 0$, has also been used in applying Schottky's function. It is, therefore, not surprising that the enthalpy differences obtained from molecular acoustic measurements are incompatible with the spectroscopic data for many substances.

A number of attempts have been made to test these various approximations. Davies and Lamb[59] pointed out that in order to obtain accurate thermodynamic data for an isomeric equilibrium the temperature and pressure dependence of the acoustic absorption should be studied. When only temperature measurements are available the critical assumption is that $\Delta V/V \cdot C_p/\theta \ll \Delta H$. Since they did not have sufficient data on a pair of ethane conformers to test this point they carried out the necessary tests on the related molecules cis- and trans-1,2-dichloroethylene in which no rotation about the double bond occurs. For these molecules, which are very similar in size to dihalogenated ethanes, they showed that

$$\frac{\Delta V}{V} \cdot \frac{C_p}{\theta} = 0.41 RT$$

whilst

$$\Delta H = 0.74 RT$$

In this case therefore the assumption is completely wrong. This example provides a vivid illustration of the care that should be taken in using this approximation.

Andrae and his coworkers[71] using the ultrasonic data of Krebs and Lamb[72] plotted the possible values of ΔH both as a function of ΔS and ΔV at different temperatures for 1,1,2,2-tetrabromoethane. None of the possible values on these curves equalled the spectroscopic value of ΔH, emphasizing once more the basic weakness in the theory used.

An attempt has been made by Wyn-Jones and Orville-Thomas[62] to estimate values for $\Delta V/V$ using relations (4.49), (4.50) and (4.51). A consideration of the entropy difference between rotational isomers of ethane derivatives shows that this quantity should lie in the region $2 \geqslant \Delta S \geqslant -2$. Within these limits ΔC_p values were then estimated using spectroscopic values for ΔH via Equation (4.51). The experimentally determined velocity of sound was then used, together with C_p and θ values found in the literature, to estimate $(\gamma - 1)$. Finally substitution of values for the experimentally determined parameter μ_m together with the known, or estimated, values for the other quantities in Equation (4.49) led to estimates for $\Delta V/V$.

For four ethane derivatives $\Delta V/V$ was found to lie between 0.02 and 0.05. These values, though small, indicate that the inequality

$$(\Delta V/V)(C_p/\theta) \ll \Delta H$$

is unlikely to hold since ΔH is very small ($0 \sim 300$ cals/g mole) in these compounds.

These conclusions indicate therefore that the enthalpy differences for ethane derivatives estimated from ultrasonic studies can be inaccurate. It would certainly help if some of the systems already studied were remeasured in a variety of inert solvents. In this way the ratio of the pressure to temperature dependence of the sound waves would be obtained. According to Piercy this procedure is comparable to varying the pressure on the pure liquid rotamer—another advantage is that the static parameters C_p

and θ, in dilute solution, would be those of the solvent which in general are known. As an independent check similar studies could be carried out spectroscopically. One of the greatest difficulties in this type of study is that the magnitude of ΔH is very small in most ethane derivatives. In dilute solutions, however, it is found that the position is improved since ΔH is usually bigger owing to the dielectric effect of the solvent[68,69].

4.4.1.4 Kinetic Aspects

Since $\Delta G = \Delta H - T\Delta S$ and $K = \exp(-\Delta G/RT)$ the infrared ΔH values can be used to calculate equilibrium constant values provided that an assumed value is substituted for ΔS.

When ultrasonic data (f_c values) are available the analysis can be carried further to yield rate constants since

$$2\pi f_c = k_{12} + k_{21}$$
$$= k_{21}(1 + K)$$

or

$$k_{21} = 2\pi f_c/(1 + K)$$

That is when f_c and K are known it is possible to calculate values for k_{21} over a range of temperature assuming that both ΔH and ΔS are temperature independent. Taking the Arrhenius equation

$$k_{21} = A \exp(-E/RT)$$

as our basis it is seen that a plot of $\log k_{21}$ against $1/T$ leads to values for A and E. Also from Equation (4.64) a plot of $\log k_{21}/T$ against $1/T$ will yield a value for ΔH_{21}^+.

In Table 4.9 are given values for A, E and ΔH_{21}^+ calculated in this fashion assuming that $-2 \leqslant \Delta S \leqslant 2$ e.u. These results show that the difference between the Arrhenius 'E' and ΔH_{21}^+ is approximately RT as expected. The Arrhenius 'A' factors are of the magnitude expected for a first-order reaction. They are also very close to the torsional frequency[73,74] of the higher energy isomer. This is reasonable since the pertinent motion involved in the transition from the higher to the lower energy rotamer is the torsional mode which must be involved in the following factor governing the backward reaction. The values of ΔH_{21}^+ listed in Table 4.9 are, of course, very close to the ΔH_{21}^+ values derived from the $\log(f_c/T)$ against $1/T$ plots discussed in Section 4.3.2 and listed in Tables 4.4 and 4.5.

Table 4.9

Calculated values for barrier heights and Arrhenius parameters

Molecules	ΔS e.u.	ΔH_{12}^+ kcal/mole	E kcal/mole	'A' factor $\times 10^{-12}$ c.p.s.	Torsional frequencies $\times 10^{-12}$ c.p.s.
1,2-dichloro-2-	−2	4·5	5·1	0·2	
methylpropane	0	4·5	5·1	0·4	3·33[a]
	2	4·5	5·1	0·8	
1,2-dibromo-2-	−2	5·6	6·1	0·7	
methylpropane	0	5·2	5·9	1·5	3·96[b]
	2	5·1	5·7	2·8	
isobutylchloride	−2	3·7	4·5	0·4	
	0	3·8	4·7	0·5	3·72[c]
	2	3·7	4·5	0·4	4·98[c]
isobutylbromide	−2	4·7	5·2	0·5	3·51[c]
	0	4·8	5·2	1·0	5·37[c]
	2	4·7	5·3	0·7	
sec-butylchloride	−2	4·4	4·9	3	
	0	4·4	4·8	2·5	3·96[d]
	2	4·2	4·7	1·4	
sec-butylbromide	−2	4·8	5·3	6	
	0	4·7	5·2	4	3·39[d]
	2	4·8	5·1	2	

[a] M. Hayashi, T. Shimanouti and S. Mozushima, *Spectrochim. Acta*, **10**, 1 (1967).

[b] M. Hayashi, *Nippon Kagaku Zasshi*, **78**, 1749 (1957).

[c] N. I. McDevitt, A. L. Rozek, F. F. Bentley and A. D. Davison, *J. Chem. Phys.*, **42**, 1173 (1965).

[d] Kiyoshi Nakamura, *Nippon Kagaku Zasshi*, **78**, 1164 (1957).

4.4.2 Substituted Cyclohexanes

Isomerism in cyclohexane and its derivatives has been reviewed in excellent articles by Mizushima[68], Dauben and Pitzer[75], and by Sheppard[69]. Owing to the tetrahedral distribution of valence bonds around the carbon atoms a number of conformers are possible for cyclohexane. The two most commonly known isomeric forms are the chair (I) and boat (II) forms.

(I) (II) (III)

It is now realized, however, that the above forms represent only two of the possible flexible conformations. Amongst other possible conformations are also the twist (III) and half-boat forms. Calculation indicates that the free-energy diagram for inversion in the cyclohexane molecule is as shown in Figure 4.29 in which the order of magnitude of the enthalpies between the various forms is indicated. In the monosubstituted cyclohexanes,

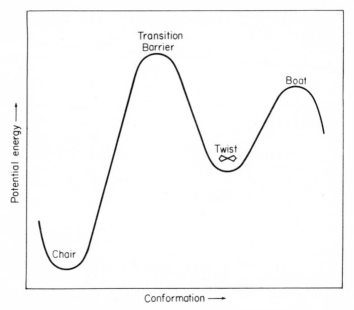

Figure 4.29   Potential-energy diagram for cyclohexane

however, it is possible to get two isomeric forms for each of the above flexible conformers due to the substituent taking up an axial (a) or equatorial (e) positions. For instance the two isomeric chair forms of methylcyclohexane are as shown below

Ultrasonic relaxation due to isomerization in cyclohexane molecules could therefore arise either from a perturbation of the equilibrium between, say, the chair and twist forms or between the various isomeric forms due to axial and equatorial isomers. Experimentally ultrasonic relaxation was not observed in cyclohexane from 200 kc–200 Mc/s but at low frequencies an isomeric relaxation was observed in the monosubstituted cyclohexanes (Table 4.10). This was attributed[76-78] to the perturbation of the equilibrium between the axial and equatorial isomers of the more energetically favourable chair form.

Isomeric chair forms in the disubstituted cyclohexanes are also possible. For instance in 1,4-dimethylcyclohexane four conformations of the $C–CH_3$ bonds are possible viz (1a, 4a), (1a, 4e), (1e, 4a) and (1e, 4e) where 'a' denotes axial and 'e' equatorial substitution. This means that in this molecule two separate isomers *cis*- and *trans*-1,2-dimethylcyclohexane are possible, whose conformational structure in terms of chair configuration may be expressed as *trans* (1a, 4a), (1e, 4e) and *cis* (1a, 4e), (1e, 4a). In *trans*-1,4-dimethylcyclohexane the two conformational isomers will have different energies and therefore ultrasonic relaxation will occur as shown in Table 4.11. On the other hand in the *cis*-1,4-dimethyl isomer the rotation of the 1-methyl group from axial to equatorial is accompanied by the rotation of the 2-methyl group in the reverse direction during the transition from one equilibrium state to another as shown in Figure 4.30. Since these two conformers have equal energy no ultrasonic relaxation is expected and no absorption of acoustic energy was found experimentally[77].

Piercy has made a complete study on the energetics of axial to equatorial isomerism in monosubstituted methyl-, chloro- and bromocyclohexanes in various solvents. The results are listed in Table 4.10. For the methyl derivative Piercy argues that if the solvents used are sufficiently inert the mixing will be ideal and the quantity $(\gamma - 1)/C_p$ is that of the solvent.

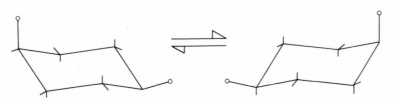

Figure 4.30 Equilibrium between the identical chair forms of *cis*-1,4-dimethylcyclohexane

Table 4.10

Energy parameters for ring inversion in some cyclohexane derivatives

Molecule	Phase	$\Delta H^{\ddagger}_{21}$ kcal/mole	Ultrasonics			N.M.R. $\Delta H^{\ddagger}$ kcal/mole	I.R. $\Delta H^{\ddagger}$ kcal/mole
			$\Delta S^{\ddagger}_{21}$ e.u.	ΔH kcal/mole	ΔS e.u.		
Methylcyclohexane { xylene solution		10·9		3·5	4·0[a]		
nitrobenzene solution		10·8		3·6	4·0[a]		
liquid		6·36	11·0	1·9	2·22[b]		
liquid		10·3[c,d]		2·9[c]	3·2[c]		
Chlorocyclohexane	xylene solution	12±2[a]				10·8[e]	15±4·5[f]
Bromocyclohexane	xylene solution	12±2[a]				10·8[e]	15±4·5[f]

[a] J. E. Piercy, *J. Acoust. Soc. Am.*, **33**, 198 (1961).

[b] M. E. Pedinoff, *J. Chem. Phys.*, **36**, 777 (1962).

[c] J. E. Piercy and S. V. Subrahmanyam, *J. Chem. Phys.*, **42**, 4071 (1965).

[d] The more reliable values for methylcyclohexane are given in references (*a*) and (*c*).

[e] L. W. Reeves and K. O. Stromme, *Can. J. Chem.*, **38**, 1241 (1960).

[f] Yr. A. Pentin, Z. Sharipov, G. G. Kotova, A. V. Kamemitzku and A. A. Akhrem, *Zh. Strukt. Khim.*, **4**, 194 (1963).

Hence, by varying the solvent, the ratio of the pressure to the temperature dependence ($dp/dt = C_p/\rho\theta T$) of the sound wave is also varied, where ρ is the density of the system, and this is equivalent to varying the pressure on the system. The volume of change during isomerization was estimated by comparing the densities of the separate *cis* and *trans* isomers of 1,4-dimethylcyclohexane. The *cis* molecule has one axial and one equatorial group whereas the majority of the *trans* molecules are believed to be in a state with two equatorial groups. The change of density from *cis* to *trans* is, therefore, an indication of the change in density when one methyl group is changed from an equatorial to an axial position and leads to an acceptably low figure of 0·4 % for $\Delta V/V$.

Values of $\Delta H = 3·50 \pm 0·05$ kcal/mole were obtained for both solutions. These seem to indicate that when dilute solutions are used for systems in which $\Delta V/V$ is known to be small the assumptions necessary to evaluate ΔH values from ultrasonic studies are not drastically invalid. Following the work of Piercy on methylcyclohexane in solution, Pedinoff[22] made a complete study of the molecule in the pure liquid and his results are quoted in Table 4.10. As shown in this table the energy values for the isomerization derived by these authors do not agree. These discrepancies were later clarified in a further study on liquid methylcyclohexane by Piercy and Subrahmanyam. Following this work Verma and his collaborators[79,80] have made some comments on $\Delta V/V$ the volume change during isomerization. Verma and his collaborators used Piercy's results on methylcyclohexane and computed values for ΔH and ΔS for different values of $\Delta V/V$. It was shown that when $\Delta V/V$ changes from 0 to 10% the corresponding change in ΔH was from 2·9 to 6·0 kcal/mole and ΔS changes from 3·20 e.u. to 15·77 e.u. In his earlier publication on methylcyclohexane, Piercy[57] estimated $\Delta V/V$ to be about 0·4% from density measurements. Verma and his collaborators conclude therefore that the thermodynamic parameters evaluated by Piercy assuming that $\Delta V/V = 0$ are sensibly unaffected when using a value of $\Delta V/V = 0·4\%$.

From the published ultrasonic work in this field it is clear that the characteristic frequencies of cyclohexane derivatives lie in the kilocycle region. This means that absorption measurements at these frequencies are difficult and the accuracy of the results is very low (see Section 4.2). Due to this difficulty only the few complete studies listed in Table 4.1 have been conducted on these molecules using acoustic methods. One way of overcoming this problem has been applied by Piercy and his collaborators. This is done by measuring the acoustic parameters at

Table 4.11

Relaxation parameters for some cyclohexane derivatives

Molecule	Temp. °C	f_c Mc/s	A $10^{17} \sec^2 cm^{-1}$	B $10^{17} \sec^2 cm^{-1}$
cyclohexanol[a]	32	0·22	351,000	540
cyclohexylamine[a]	19	0·11	210,000	100
dicyclohexylamine[a]	19	0·22	260,000	
methylcyclohexanone[a]	18	0·22	260,000	
ethylcyclohexane[b]	16	0·06	500,000	
trans-1,2-dimethylcyclohexane[b]	16	0·12	55,000	35
trans-1,4-dimethylcyclohexane[b]	16	0·18	26,000	87

[a] J. Karpovich, J. Chem. Phys., 22, 1767 (1954).
[b] J. Lamb and J. Sherwood, Trans. Faraday Soc., 51, 1674 (1955).

high temperature ($\sim 150°C$) in solution, with the sample under a small pressure (~ 20 atm). At the higher temperatures the characteristic frequencies of these molecules lie in the frequency range 1–5 Mc/s. This means that absorption measurements can be obtained relatively accurately with the pulse technique. The potential barriers hindering the inversion of the two chair forms in chloro- and bromocyclohexane have also been determined from N.M.R.[81] and infrared studies[82]. These values are also listed in Table 4.10. Within experimental error the agreement between the different techniques is very good.

4.4.3   Tertiary Amines

Ultrasonic relaxation in pure tertiary amines was first observed by Heasell and Lamb[58]. They argued that the mechanism of the relaxation process could be due to one or more of the following:

(i) there might be a time delay in adjustment of the molecules to a state of closer packing as observed in water and ethyl alcohol,

(ii) an intermolecular equilibrium between monomeric and dimeric forms of the amine arising from hydrogen bonding might exist,

(iii) a vibrational relaxation due to time delay in establishing equilibrium amongst the internal vibrational modes in the molecule is possible, and

(iv) internal rotation about C–N bonds as shown in Figure 4.31.

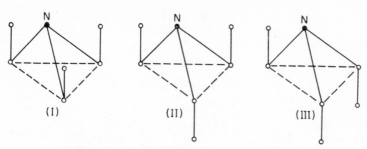

Figure 4.31 Rotational isomers of triethylamine

A consideration of the evidence available on the properties of tertiary amines led Heasell and Lamb to conclude that mechanism (iv) was most likely to be solely responsible for the observed ultrasonic relaxation. Using molecular models they considered all possible rotamers and concluded that the following conformations are most likely, Figure 4.31.

This view has been confirmed in an interesting and valuable study carried out by Litovitz and Carnevale[83]. They found that the characteristic frequency, f_c, of triethylamine was independent of pressure up to 3000 atmospheres. This result confirms the intramolecular mechanism proposed by Heasell and Lamb[58]. These authors argue further that since isomeric rotational relaxation is thermal in nature then there should be a zero volume change in going from either stable state to the activated state. The experimental result supports this argument.

The observed relaxation in all the amines listed in Table 4.12 appears to be consistent with a single relaxation mechanism. This would seem to indicate that only two of the isomeric forms shown in Figure 4.31 take part in the relaxation: it is difficult to decide which are the two isomers since the physical evidence on the structure of these molecules is incomplete. The evidence available is summarized below:

A study of molecular models shows that there is a great deal of steric hindrance present in conformer (III), Figure 4.31. This decreases as the number of alkyl groups that are folded back towards the nitrogen lone-pair increases. This means that from a steric point of view the energetically more favoured conformers are (I) and (II).

There is evidence[87,88] from both spectroscopic and thermochemical experiments that triethylamine is a weaker base than trimethylamine and also primary and secondary amines. In order to explain this fact it was

Table 4.12
Kinetic and thermodynamic data on amines

Molecule	ΔH_{21}^{+}	ΔS_{21}^{+}	ΔH	ΔS
triethylamine	$6\cdot8^{a}$	44	$3\cdot4$	$4\cdot7$
	$10\cdot8^{b}\}$	$17\}$	$5\cdot9\}$	11
	$10\cdot1\ \}$	$19\}$	$5\cdot9\}$	16
tripropylamine	$4\cdot5^{c}$	$-3\cdot2$	$1\cdot46$	$1\cdot1$
tributylamine	$4\cdot3^{d}$		$0\cdot85$	$-0\cdot9$
tripentylamine	$4\cdot8^{c}$	$-3\cdot3$	$1\cdot16$	$-0\cdot9$
triisopenyltamine	$3\cdot0^{c}$	$-9\cdot7$	$1\cdot11$	$-1\cdot1$
trihexylamine	$4\cdot3^{c}$	$-5\cdot6$	$1\cdot13$	$-1\cdot3$
triallylamine	$5\cdot9^{c}$	$2\cdot2$	$1\cdot51$	$2\cdot1$
trimethallylamine	$4\cdot3^{c}$	$-3\cdot9$	$1\cdot33$	$1\cdot9$
3,N-diethylaminopropiononitrile	$4\cdot5^{c}$	$-4\cdot1$	$1\cdot16$	$0\cdot7$
3,N-dipropylaminopropiononitrile	$5\cdot9^{c}$	$-0\cdot84$	$1\cdot19$	$0\cdot7$
3,N-diethylaminobutane-2-one	$6\cdot5^{c}$	$2\cdot7$	$1\cdot36$	$2\cdot0$
3,N-diethylaminopropylamine	$6\cdot3^{c}$	$2\cdot2$	$1\cdot31$	$1\cdot8$
N,N-diethylcyclohexylamine	$3\cdot68^{c}$	$-4\cdot1$	$1\cdot36$	$1\cdot7$
2-diethylaminoethylamine	$5\cdot88^{c}$	$-2\cdot7$	$1\cdot6$	$1\cdot9$
1-diethylamino-2-chloropropane	$5\cdot80^{c}$	$3\cdot4$	$1\cdot1$	$-0\cdot4$

[a] E. L. Heasell and J. Lamb, *Proc. Roy. Soc. (London)*, Ser A, **237**, 233 (1956).
[b] R. A. Padmanabham and E. L. Heasell, *Proc. Phys. Soc. (London)*, **76**, 321 (1960).
[c] E. J. Williams, E. Wyn-Jones, T. H. Thomas and W. J. Orville-Thomas, unpublished work.
[d] K. Krebs and J. Lamb, *Proc. Roy. Soc. (London)*, Ser A, **244**, 558 (1958).

proposed that at least one of the alkyl groups in triethylamine is folded back so that it projects towards the lone-pair of the nitrogen atom. This, of course, hinders the approach of any electron-deficient group and thus weakens the amine as a base.

On this basis Heasell and Lamb[58] concluded that the most likely conformers taking part in the equilibrium are (I) and (II) of Figure 4.31. However, from the ultrasonic data they could not deduce which is the more stable isomer since in theory the *a priori* assumption is always made that ΔH is positive. There are no assumptions made, however, regarding the sign of ΔS. These authors considered the molecular models (I) and (II) and concluded that since there is more freedom of movement of the methyl groups in (II) then the entropy of (II) is greater than (I). From the experimental data on triethylamine ΔS was positive which suggests, therefore, that isomer (I) is more stable. In a further study on tri-n-butylamine Krebs and Lamb[72] found ΔS to be negative which suggests that isomer (II) is more stable. This conclusion was supported by parachor calculations[84] on the higher tri-n-alkyl amines. These calculations showed that good agreement was found between experiment and theoretical parachors on assuming conformation (II) to be the most stable. This work was then followed by a further study on triethylamine by Padmanabhan and Heasell[85]. The energy parameters derived from this work are listed in Table 4.12. Following this work Williams and coworkers[86] have carried out some more studies on some tertiary amines and their results are listed in Table 4.12. The entropy difference between the isomers in the tri-n-alkylamines changes sign in going from triethylamine to tri-n-butylamine. For the higher tri-alkylamine ΔS is negative which suggests that by comparison with molecular models isomer (II) is more stable. This is in complete agreement with the earlier views of Lamb and his collaborators and also the parachor calculations. It is also worth noting that the energy barrier is fairly constant around 4·5 kcal/mole for the tri-n-alkyl amines propyl to hexyl. This suggests that the forces giving rise to the potential barriers in these molecules are roughly the same in this series; this would be true if rotation occurs around the CN bond as postulated. The magnitude of the energy barrier listed in Table 4.12 is very similar to that found for substituted ethanes. Padmanabhan and Heasell[85] extended the earlier work of Heasell and Lamb[58] on triethylamine. In order to explain both sets of experimental results which covered the temperature range -60 to $40°C$ and the frequency range 10–250 Mc/s the former authors proposed that the entropy difference between the isomers and also the entropy of activation were temperature dependent. These temperature-dependent entropy terms were introduced in order to account for the non-linear plots obtained when $\log[(T\mu_m/v^2)(C_p/\theta)^2]$ and $\log(f_c/T)$ were plotted against $(1/T)$. These authors concluded that the entropy of the higher energy isomer (II) in Figure 4.31 decreases with decreasing temperature, presumably as the rotation of the methyl group becomes restricted.

In the work of Krebs and Lamb[72] and Williams and coworkers[86] all the kinetic plots (log f_c/T against $1/T$) were linear, and therefore indicate that both the entropies and energies of activation are independent of temperature. Thermodynamic plots were not used in this work because the necessary conditions regarding the magnitude of ΔG were not obeyed. However, since the entropy of activation is independent of temperature it is safe to assume that the entropy difference is also independent of temperature.

4.4.4   Unsaturated Molecules

4.4.4.1   *Aldehydes and Ketones*

Lamb and de Groot[61] carried out extensive ultrasonic studies on
internal rotation in unsaturated molecules. Their results are listed in
Tables 4.13 and 4.14 for unsaturated aldehydes and ketones. Although

Table 4.13

Relaxation parameters and energy values for some unsaturated aldehydes

Molecule	Temp. °C	Relaxation parameters			ΔH_{21}^{+} kcal/mole	ΔH kcal/mole
		A $\times 10^{-17}$ sec^2 cm^{-1}	B $\times 10^{-17}$ sec^2 cm^{-1}	f_c Mc/s		
acrolein[a]	−24·6	790	38	29·1		
	−0·3	380	37	77·8	4·96	2·06
	25·2	208	36	176		
crotonaldehyde[a]	0	1397	24	12·0		
	25	674	29	30·3	5·51	1·93
	50·1	355	34	70·0		
cinnamaldehyde[a]	25·1	722	78	15·7		
	50·2	331	63	36·0	5·62	1·5
	75·8	154	74	65·2		
methacrolein[a]	−24·8	282	29	22·2		
	0	150	27	65·2	5·31	3·07
	24·9	94·7	25	174		
furacrolein[a]	59·9	1840	61	13·5		
	79·4	1150	66	23·0	5·10	1·2
	99·8	617	132	31·4		
α-methyl- cinnamaldehyde[b]	35	216	59	18·3		
	45	158	57	27·6	6·30	—
	65	112	60	47·5		

[a] M. S. de Groot and J. Lamb, *Proc. Roy. Soc.* (*London*), *Ser. A*, **242**, 36 (1957).
[b] R. A. Pethrick and E. Wyn-Jones, unpublished work.

complete data were only obtained for five molecules this work has proved
valuable in understanding the problem of internal rotation in these
molecules. In brief, molecules with a general skeleton of the type:

$$\begin{array}{c} X \\ \diagdown \\ H \diagup \end{array} C_1 = C_2 \begin{array}{c} \diagup Y \\ \diagdown \\ C_3 \diagdown Z \end{array} {=} O$$

Table 4.14

Relaxation parameters for a number of aldehydes and ketones[a]

Molecule	T °C	Velocity $\times 10^{-5}$ cm sec^{-1}	A $\times 10^{-17}$ sec^2 cm^{-1}	B $\times 10^{-17}$ sec^2 cm^{-1}	f_c Mc/s	μ'_m $\times 10^2$
2-ethyl-3-propyl acrolein	24·9	1·306	159	39	24·6	0·255
	50·1	1·211	87	44	56·0	0·295
hexylcinnamic aldehyde	25·0	1·448	129	150	18·2	0·170
	50·2	1·362	61	103	35·8	0·148
methylvinyl ketone	−24·8	1·451	107	23	80·6	0·626
	−0·1	1·342	34	25	178	0·406
methylisopropenyl	−24·8	1·459	96·3	59	26·4	0·186
ketone	−0·1	1·347	28·2	51	59·6	0·113
thiophene-2-aldehyde	75·0	1·298	2495	71	6·73	0·436
furfural acetone	50·1	1·383	1371	67	9·78	0·93
	74·7	1·299	564	67	23·2	0·85
propionaldehyde	−24·7	1·379	69	25	189	0·899
n-butyraldehyde	−24·8	1·406	29	25	165	0·336

[a] M. S. de Groot and J. Lamb, *Proc. Roy. Soc. (London), Ser. A*, **242**, 36 (1957).

were investigated. In these molecules physicochemical evidence has shown

that the X and $-C \overset{\displaystyle O}{\underset{\displaystyle Z}{\big\Vert}}$ groups are usually *trans* with respect to the

$C_1=C_2$ double bond. Due to the presence of conjugated double bonds it is thought that delocalization of π-electrons occurs. This delocalization would clearly stabilize the planar structures

(I)   (II)

where I and II are conformers obtained by rotation around the C_2C_3 bond. These two forms are rapidly interconvertible since the energy barrier is small; they are known as the s-*trans* and s-*cis* forms respectively.

For the aldehydes dipole moment studies[89] have shown that the s-*trans* conformation is the more stable. Lamb and de Groot argued that the potential barrier hindering rotation about the C_2–C_3 bond in these molecules is a function of non-bonded steric forces, electrostatic forces, conjugation between double bonds and repulsive forces between bonds. From a molecular-orbital calculation of acrolein (X, Y, Z=H) Lamb and de Groot showed that the s-*cis* form was 2 kcal/mole more stable than the s-*trans* form. Steric forces were considered to be unimportant in acrolein. Hence, to explain the relatively greater stability of the s-*trans* isomer these authors concluded that the repulsive forces between bonds were much higher in the s-*cis* than in the s-*trans* form.

In acrolein, crotonaldehyde (X=CH_3, Y, Z=H) and cinnamaldehyde (X=C_6H_5, Y, Z=H) the energy barrier hindering rotation about the C_2–C_3 bond increases respectively from 4·96 through 5·52 to 5·62 kcal/ mole. This has been attributed to the substitution of electron-donating groups in the X-positions thus strengthening conjugation. The electron-donating power of the phenyl group is greater than that of the methyl which in turn is greater than that of hydrogen. This is reflected in the energy barriers found for these molecules. The energy difference between the s-*cis* and s-*trans* form in these molecules decreases with conjugation. This is explained as follows. As conjugation increases, the electrostatic forces are also increased; this in turn stabilizes the s-*cis* isomer thus decreasing the energy barrier. It was also found that as the energy barrier decreases in these molecules the dipole moment increases as expected showing that more of the s-*cis* form is present.

In these molecules the relative ease of rotation about the C_2–C_3 bond is therefore determined by the amount of conjugation present which in turn gives a double-bond character to this bond. The characteristic frequencies of these molecules are also a measure of the rate of rotation. At 25°C the characteristic frequencies, f_c, of acrolein and crotonaldehyde are respectively 176, 30·3 and 15·7 Mc/s. This shows that f_c decreases as the rotation becomes more difficult.

Substitution of groups in the Y position should not alter conjugation appreciably since in the extreme resonance hybrid a positive charge is associated with the atom C_1. This is found to be so as is shown by a comparison of the data (table 4.13) presented for methacrolein, 2-ethyl-3-propyl acrolein and hexyl cinnamic aldehyde. For instance the characteristic frequencies of acrolein and methacrolein are very similar. This argument is also true for cinnamaldehyde and hexyl cinnamaldehyde. A complete ultrasonic study was also carried out on methacrolein ($X=H$, $Y=CH_3$, $Z=H$). The high value for the enthalpy difference in this molecule ($\Delta H = 3.07$ kcal/mole) compared to acrolein can be explained as being due to electrostatic attraction in the s-*trans* form between the methyl group and oxygen atom.

In the unsaturated ketones where $Z=CH_3$, the presence of the methyl group introduces steric forces which will play an additional rôle in the potential energy barrier hindering rotation about the C_2–C_3 bond. Complete data on these molecules are not available but for methylvinyl ketone (X, $Y=H$, $Z=CH_3$) and isopropenyl ketone ($X=H$, Y, $Z=CH_3$) the maximum absorption per wavelength, μ_m, decreases with temperature. An examination of Schottky's function shows that, ΔH, the enthalpy difference between the isomers is less than 1.5 kcal/mole. This means that the s-*cis* conformation is more stable which is in accord with dipole-moment measurements. In the extreme case of mesityl oxide, $(CH_3)_2C:CHCOCH_3$, the s-*trans* configuration is completely blocked due to steric hindrance and consequently no ultrasomic relaxation was found in this molecule.

In thiophene-2-aldehyde and 5-methyl-2-acetyl furane the characteristic frequencies have low values even at high temperature. These systems contain very mobile π-electrons and therefore increased conjugation restricts the rotation of the $-C{\Large\diagup}^{O}_{\diagdown}{}_H$ group. In furacrolein (I) and furfural acetone (II),

(I)   (II)

there are two possibilities of internal rotation as indicated by the arrows shown in the diagram. The experimental results appear to be consistent with this interpretation.

On the other hand in propionaldehyde and n-butyraldehyde the rotation of the aldehyde group is governed mainly by steric and electrostatic forces. The amount of conjugation in these molecules, compared to acrolein, is very small. The characteristic frequencies for these molecules are very high showing that the rotation is less restricted than in the corresponding unsaturated aldehydes.

4.4.4.2 Vinyl Ethers

Absorption of ultrasonic energy attributable to a relaxation process has been found for methyl, ethyl and 2-chloroethyl vinyl ethers. The experimental results are listed in Table 4.15. It is believed that owing to

Table 4.15

Relaxation parameters for the vinyl ethers[a]

Molecule	Temp.	A 10^{17} sec^2 cm^{-1}	B[b] 10^{17} sec^2 cm^{-1}	f_c Mc/s	μ'_m 10^{-2}
Methylvinyl ether	-24.8	272	25	217	3.64
Ethylvinyl ether	-24.8	299	25	200	3.75
2-chloroethylvinyl ether	-24.7	130	20	305	2.85

[a] M. S. de Groot and J. Lamb, *Proc. Roy. Soc.* (*London*), *Ser. A*, **242**, 36 (1957).
[b] Assumed values.

stabilization by the presence of delocalized π-bonding the planar forms (I) and (II) are the most likely conformers.

(I) s-*cis*   (II) s-*trans*

Rotational isomerism in methyl and ethyl vinyl ethers has been studied recently by infrared spectroscopy[90,91]. From this work it was concluded that the s-*cis* form is more stable and also the s-*trans* form is possibly non-planar due to steric interactions.

4.4.4.3 Esters

Similar behaviour is found for esters where rotation about the ester CO 'single' bond has been suggested as the mechanism leading to the possibility of rotational isomerism. For these molecules two planar forms are possible. For example in methyl acetate the equilibrium system perturbed by the passage of acoustic energy is:

Historically esters were amongst the first molecules studied by the ultrasonic technique. Of the earlier work[92-95], only that of Pinkerton appears to be complete. Although Tabuchi[96] appears to have made a complete study on ethyl formate, Piercy[97] has pointed out that since only velocity dispersion was measured the results are not considered accurate enough for quantitative treatment.

In Table 4.16 the energy parameters for a number of esters, for which complete ultrasonic studies have been carried out, are listed. Other ultrasonic work on esters has been described by Hall and Lamb[98] and also by Nozdrev and his colleagues[99,100].

Table 4.16

Energy parameters for internal rotation in esters

Molecule	State	ΔH_{21}^{+} kcal/mole	ΔS_{21}^{+} e.u.	ΔH kcal/mole	ΔS e.u.
Methyl formate[a]	pure liquid	7·8	8·7	2·3	—
	xylene solution	7·0	6·0	—	—
	nitrobenzene solution	—	—	2·3	—
Ethyl formate[a]	pure liquid	8·0	8·8	2·3	—
	xylene solution	7·8	9·5	2·3	—
	nitrobenzene solution	7·5	8·0	2·3	—
isopropyl formate[b]	xylene solution } heptane solution }	5·8	4·3	3·7	1·5
ethyl acetate[c]	pure liquid	5·66	—	2·92	—
	pure liquid	5·70	—	1·61	—

[a] S. V. Subrahmanyam and J. E. Piercy, *J. Acoust. Soc. Am.*, **37**, 340 (1965).

[b] J. E. Piercy and S. V. Subrahmanyam, *J. Chem. Phys.*, **42**, 1475 (1965).

[c] W. M. Slie and T. A. Litovitz, *J. Chem. Phys.*, **39**, 1638 (1963); J. M. M. Pinkerton, 'Ultrasonic Conference, Brussels,' *Mededel Koninkl. Vlaam. Acad. Wetenshap, Belg., Kl. Welenshap*, 1951, p. 177.

Slie and Litovitz[101] carried out a study on liquid ethyl acetate over the temperature range 11–80°C and up to 1000 atmospheres pressure (Table 4.17). The characteristic frequency of the isomeric relaxation was

Table 4.17

Relaxation parameters of ethyl acetates[a]

Temp. °C	Pressure atm	f_c Mc/sec
	1	22·7
40	381	23·1
	760	23·7
	1000	23·1
	1	24·1
80	351	20·0
	700	16·7
	1000	14·3

[a] W. A. Slie and T. A. Litovitz, *J. Chem. Phys.*, **39**, 1538 (1963).

found to increase with temperature, as expected, but was independent of pressure. These results were also in good agreement with the earlier work of Pinkerton[94] (Table 4.16). This experimental work proves that for the isomerism in ethyl acetate the assumption that $\Delta V/V = 0$, used in the ultrasonic theory to evaluate ΔH, is correct.

Piercy has suggested that the main cause of the potential barrier to rotation in esters is the partial double-bond character of the CO bond adjacent to the carbonyl link. This arises because of the delocalization of π-electrons in the system.

$$\underset{O}{\overset{H_3C}{\diagdown}} C - OX \longrightarrow \underset{O}{\overset{H_3C}{\diagdown}} C \cdots OX$$

That is, the bond order of the carbonyl link decreases from two whilst that of the C–O 'single' bond becomes greater than unity.

The work described above has provided evidence that in esters rotation occurs around the ester grouping C–O bond. In isopropyl formate an examination of the structure reveals that rotation about the methoxyl C–O bond is also possible leading to the possibility of rotamers of the form,

$$\underset{H}{\overset{H_3C \quad CH_3}{\diagdown \diagup}} \underset{O-C}{\overset{C}{\diagdown}} \underset{H}{\overset{O}{\diagup}} \rightleftharpoons \underset{H_3C}{\overset{H_3C \quad H}{\diagdown \diagup}} \underset{O-C}{\overset{C}{\diagdown}} \underset{H}{\overset{O}{\diagup}}$$

This possibility has been confirmed by Piercy and Subrahmanyam[102] who detected a second relaxation region (centred about 45 Mc/s). This has been attributed to the type of isomerism described above.

This mechanism, first put forward by Piercy and Subrahmanyam, has been confirmed by a recent microwave study.

Microwave spectroscopy has provided a powerful tool so far as the study of comparatively small rotamers is concerned. In a recent study Riveros and Bright Wilson Jr.[103] have detected the presence of at least two rotational isomers of ethyl formate in the vapour phase. These isomers have the conformations (I) and (II).

That is rotation has occurred around the $C'O$ bond. An enthalpy difference of 186 ± 60 cal/mole and a potential barrier of 1100 ± 250 cal/mole were found for the system. In the *trans* isomer (I) the heavy atom skeleton is coplanar whilst in the *cis* isomer (II) the methyl group is 95° out of the plane.

4.5   CARBOXYLIC ACIDS

It is interesting to note that the first quantitative study using molecular acoustic techniques was carried out on liquid acetic acid. Lamb and Pinkerton[63] measured the ultrasonic parameters for pure acetic acid in the frequency range 7·5–65 Mc/s and over the temperature range 10–60°C. Their ultrasonic parameters were found to fit a single relaxation around 1 Mc/s. The results were treated for the perturbation of an $A_1 \leftrightarrow A_2$ type equilibrium and the thermodynamic parameters derived from the experimental data are listed in Table 4.18. These data were analysed to provide kinetic information as well. In order to account for their results these authors suggested that the absorption of sound energy was caused by the perturbation of a monomer ⇌ dimer equilibrium,

Table 4.18

Thermodynamic energy parameters derived for pure acids
from ultrasonic studies

Acid	ΔH	ΔS	ΔH_{21}^{+}
acetic	2·32	3·89	8·46[a]
			8·1[b]
formic	—	—	6·2[b]
propionic	3·86	4·4	7·51[c]
			7·3[b]
butyric	—	—	1·11[d]
valeric	—	—	5·4[b]

[a] J. Lamb and J. M. M. Pinkerton, *Proc. Roy. Soc. (London)*, *Ser. A*, **199**, 114 (1949).

[b] C. Moriamiez, M. Moriamiez and A. Moreaux, *J. Chim. Phys.*, **63**, 615 (1966).

[c] J. A. Lamb and D. H. A. Huddart, *Trans. Faraday Soc.*, **46**, 540 (1950).

[d] P. White and G. C. Benson, *Can. J. Chem.*, **30**, 1135 (1958).

but the actual mechanism was not deduced. They did, however, point out that this reaction may take place in two steps. In the first step they envisaged the joining together of two molecules by a *single* hydrogen bond,

(ii)
$$2\,CH_3.C\!\!\begin{array}{c}O\\ \\OH\end{array} \qquad CH_3.C\!\!\begin{array}{c}O\cdots HO\\ \\OH \quad O\end{array}\!\!C.CH_3$$

representing a $2A \rightleftharpoons B$ reaction followed by the formation of a *second* hydrogen bond to complete the dimer.

(iii)
$$CH_3.C\!\!\begin{array}{c}O\cdots HO\\ \\OH \quad O\end{array}\!\!CCH_3 \rightleftharpoons CH_3.C\!\!\begin{array}{c}O\cdots HO\\ \\OH\cdots O\end{array}\!\!C.CH_3$$

This latter process, an $A_1 \rightleftharpoons A_2$ type reaction, was thought to be the rate-determining step and thus related to the observed relaxation.

These measurements were extended to propionic acid by Lamb and Huddart[105] who found that the thermodynamic-energy parameters were very similar to those derived for acetic acid (see Table 4.18).

The thermodynamic parameter $\Delta H°$ is known for the monomer $\rightleftharpoons$ dimer equilibrium in acetic acid both in solution and in the vapour phase. The change in $\Delta H°$ with change of phase can be explained in terms of Onsager's theory of electrostatic interaction between polar molecules.

Freedman[106] estimated a value for $\Delta H°$ for the pure liquid acetic and propionic acids from vapour-state data. From these values of $\Delta H°$, a theoretical value for μ_m, the maximum sound absorption per wavelength, was calculated on the basis of a $2A \rightleftharpoons B$ (monomer $\rightleftharpoons$ dimer) reaction. Freedman also recalculated experimental values for μ_m from the ultrasonic data on the basis of a $2A \rightleftharpoons B$ reaction and the values are listed in Table 4.19.

Table 4.19

Freedman's data on acetic and propionic acid[a]

	Acetic acid			Propionic acid	
$T°C$	$10^2 \times \mu_m(\text{exp})$	$10^2 \times \mu'_m(\text{calc})$	Temp.	$10^2 \times \mu'_m(\text{exp})$	$10^2 \times \mu'_m(\text{calc})$
20	5·10	5·10	8	1·07	1·08
30	5·45	5·40	21	1·34	1·36
40	5·83	5·74	31	1·54	1·55
50	6·10	6·10	41	1·78	1·78
60	6·41	6·48	51	2·07	1·99

[a] E. Freedman, *J. Chem. Phys.*, **21**, 1784 (1953).

The agreement between the experimental and calculated values of μ_m is remarkable and is consistent with the view that the excess ultrasonic absorption in acetic and propionic acid is caused by a perturbation of the monomer $\rightleftharpoons$ dimer equilibrium. In order to test this theory Piercy and Lamb[107] carried out extensive measurements on the ultrasonic absorption of solutions of acetic acid in non-associated solvents where the concentrations of monomer and dimers have been determined experimentally using other techniques.

They came to the conclusion that two relaxation processes occurred in the solutions studied, and an analysis of the results was carried out using the equation

$$\alpha/f^2 = B + A_1/[1 + (f/f_{c1})^2] + A_2/[1 + (f/f_{c2})^2]$$

It was found that one relaxing mechanism predominates at low frequencies ($\leqslant 7$ Mc/s) whilst the second was predominant at frequencies greater than 15 Mc/s when the concentration was low. No explanation was put forward to account for the low frequency relaxation but it was thought that this was a continuation into the region of low concentration of the relaxation already found in pure acetic acid (i.e. the $A_1 \rightleftharpoons A_2$ type reaction described at the start of this section).

The analysis of the high-frequency data was found to be possible in terms of a monomer $\rightleftharpoons$ dimer equilibrium and this possibility was offered as the most likely explanation for the high-frequency relaxation.

The work done by Piercy and Lamb[107], therefore, does not confirm the theoretical expectations of Freedman[106]. Further doubt on this work has been cast by Tabuchi[108] who found that Freedman's conclusions are not unique since the relaxation parameters in pure acetic acid can also be reproduced from a consideration of equilibrium (iii) above.

Carnevale and Litovitz[83] found that the characteristic frequency of pure acetic acid was independent of pressure. This showed that the sound absorption at lower frequencies (~ 1 Mc/s) is caused by an intramolecular reaction and excludes such possible causes as vibrational relaxation or viscosity effects. Other data derived from measurements[109,110] on pure acids are also listed in Table 4.18. It is clear that a great deal of further work is required on ultrasonic relaxation in acids in order to provide a satisfactory interpretation. Recent work[111] done in aqueous solution shows that the acids formic, acetic, propionic and butyric form associated complexes with water at mole fractions of 0·75, 0·50, 0·67 and 0·11 respectively. These correspond to molecular association ratios of 3, 1, 2 and 1 to 8 respectively. These results also indicate that part of the excess absorption found in these systems is caused by equilibria involving hydrogen-bonded forms.

It is also worth noting that a number of other possible effects could cause anomalous sound absorption in carboxylic acids. For example, in propionic acid and higher homologues rotation about a C–C bond is possible,

$$CH_3-CH_2 \diagdown \diagdown \begin{array}{c} O \\ \diagup\!\!\!\!\diagup \\ C \\ \diagdown \\ OH \end{array}$$

In addition internal rotation about the C–O bond would lead to the equilibrium system,

$$R.C \diagup\!\!\!\!\diagdown \begin{array}{c} O \cdots \\ \diagdown H \\ \diagup \\ O \end{array} \rightleftharpoons R.C \diagup\!\!\!\!\diagdown \begin{array}{c} O \\ \diagdown \\ O \\ | \\ H \end{array}$$

Piercy[112] believes that this is the process that occurs in carboxylic acids. As evidence he cites his work on formic, butyric and octanic acids in the pure liquid state and in solution. This argument appears to hinge on the value of $\sim 8 \cdot 2$ kcal/mole found for the activation energy for the reaction occurring (whatever its nature) in carboxylic acids. Since spectroscopic evidence[112] points to a similar value ($\sim 8 \cdot 9$ kcal/mole) for the barrier height to rotation about the C–O bond in acids, Piercy concludes that the same mechanism is responsible for the ultrasonic relaxation. As additional evidence Piercy points out that in formates the barrier height to internal rotation about the C–O bond is of the same order of magnitude.

Recent measurements of the Brillouin spectrum of acetic acid show that a third relaxation probably occurs around 4 Gc/s. This relaxation was detected when the hypersonic and ultrasonic velocities were compared[113]. This relaxation might be associated with the frequency dependence of shear viscosity or possibly with vibrational relaxation.

4.6 VIBRATIONAL RELAXATION

4.6.1 Introduction

The redistribution of energy between translational and vibrational modes in a molecule involves the relaxation of the vibrational specific heat and gives rise to thermal ultrasonic relaxation[1]. Vibrational relaxation of this kind has been observed several times in the vapour or gas phase but only a few times in liquids. The reason for the relatively few measurements in the liquid phase is that the relaxation times associated with vibrational specific heat are relatively short except in a few cases. Recent developments in Brillouin spectroscopy (see Section 4.2) should provide enough information to measure these fast relaxation times. In some liquids it is found that the ultrasonic relaxation involves the deactivation of all the normal modes[114] whereas in others only a fraction of the normal vibration modes are involved[115]. In the latter case a second relaxation involving the remaining mode or modes is expected at higher frequencies. For the same molecule this pattern of single or multiple dispersions is identical in the liquid and gas phase. In general when multiple relaxation occurs the vibrational mode or modes which deactivate with the same relaxation time, τ, are usually close in frequency.

4.6.2 Basic Theory

Several complete treatments of the methods of deriving relaxation parameters both from experimental data and from theoretical models are contained in the literature[1,114–116]. Hence only a brief outline is included here. The usual procedure is to determine the value for a certain parameter (e.g. μ_m) directly from experimental data and then to compare the value so obtained with an estimate obtained from a theoretical treatment involving the vibrational specific heat of the molecule computed from the usual Planck–Einstein formula. This is done as follows.

4.6.2.1 *Experimental Determination of Relaxation Parameters*

When considering the phenomenon of vibrational relaxation the left-hand side of the usual expression for a single relaxation

$$\alpha/f^2 = \{A/[1+(f/f_c)^2]+B\}$$

is incomplete and has to be modified to $(\alpha/f^2)(c/c_0)$, where c is the sound velocity at the driving frequency f, and c_0 is the low-frequency velocity. The term (c/c_0) is introduced to take account of velocity dispersion which can be appreciable when vibrational relaxation occurs[114,115]. It follows therefore that the maximum absorption per wavelength is given by

$$\mu'_m = Ac_0 f_c/2$$

These equations can be used with the experimental data to yield values for the relaxation parameters.

4.6.2.2 *Theoretical Determination of Relaxation Parameters*

Vibrational relaxation is a thermal effect and, therefore, this leads to a complex or relaxing specific heat. When relaxation occurs the specific heat, C_p, becomes frequency dependent and is given by:

$$C_p = C_{p\infty} + \frac{\Delta C_p}{1+ij\omega\tau}$$

where $C_{p\infty}$ is the specific heat at infinite frequency, ΔC_p is the relaxing specific heat, τ is the relaxation time and ω is the angular velocity given by $2\pi f$.

Vibrational relaxation involves the deactivation of normal modes of vibration of the molecule. The vibrational frequencies, v_i, are related to ΔC_p by the Planck–Einstein formula,

$$\Delta C_p = \sum n_i R x_i^2 \exp(-x_i)/[1-\exp(-x_i)]^2$$

where $x_i = hv_i/kT$, n_i is the degeneracy of the ith vibrational mode, and k is Boltzmann's constant.

A quantity, r, known as the relaxation strength has been defined[58,115] as

$$r = \Delta\beta/\beta$$

and this is useful in discussing the problem of vibrational relaxation.

During vibrational relaxation a group of vibrational modes relax with a single relaxation time; this implies a many-state system. Since vibrational relaxation is a thermal process there is no volume change between the upper and lower vibration states. For a two-state system the relevant expressions involving μ'_m, the maximum absorption per wavelength, and the thermodynamic parameters of the system are given by Equation (4.35).

$$\mu'_m = (\pi/2)(\Delta\beta/\beta) = r(\pi/2)$$

For vibrational relaxation Equation (4.34) reduces to

$$r = \Delta\beta/\beta = (\gamma - 1)(\Delta C_p/C_p) \tag{4.72}$$

In addition the relaxation strength is related to the relaxation parameters A, B and f_c by

$$r = Av_0 f_c/\pi - (1 - 2\beta/A)(Av_0 f_c/\pi)^2/4$$

and

$$f_c = 2\pi\tau(1 - r)^{1/4} \tag{4.73}$$

In most cases $r \ll 1$ (see Table 4.20), the factor $(1 - r)^{1/4} \approx 1$ and hence Equation (4.73) reduces to $f_c \approx 1/(2\pi\tau)$.

When sound absorption is interpreted as arising from vibrational relaxation, i.e. the deactivation of normal modes of vibration, the relations given above can be used to calculate theoretical values for μ'_m, r or ΔC_p. These can then be compared with those obtained experimentally. In cases where only a fraction of the normal modes is involved the contribution of the vibrational modes to ΔC_p is found by trial and error until agreement is found between experiment and theory. This can then be checked by carrying out the ultrasonic measurements at different temperatures when the type of agreement indicated above should be maintained.

Vibrational relaxation studies have been carried out on a number of liquids. Some of these are discussed below.

Table 4.20

Summary of experimental data on vibrational relaxation in small molecules

Molecule	Temp. °C	Pressure (atm)	f_c Mc/s	μ'_m Exptl	μ'_m Theor.	r Exptl	r Theor.	ΔC_p cal/mole/deg Exptl	ΔC_p cal/mole/deg Theor.
CS_2[a]	25		78	0·260	0·262				
	−63		31	0·134	0·133				
CO_2[b]	0	70	10·9			0·1088	0·1026		
	0	37·5	10·5			0·1163	0·1100		
	25	97	11·2			0·1379	0·1308		
	25	69	9·65			0·1454	0·1363		
	30	86	9·85			0·1428	0·1375		
SO_2[c]	0		23					0·291	0·259
	25		23					0·383	0·373
	50		24					0·461	0·502
	25		170	0·102	0·104				
CH_2Cl_2[d,e,f]	−21		171	~0·064	0·021				
	25		147	0·109	0·104				
	0		117	0·087	0·086				
	−60			0·035	0·034				
CH_2Br_2[g]	25							3·24	3·15
	−10							2·60	2·55
	−40							1·98	2·02

[a] J. H. Andrae, E. L. Heasell and J. Lamb, *Proc. Phys. Soc. (London), Ser. B*, **69**, 625 (1956).
[b] R. Bass and J. Lamb, *Proc. Roy. Soc. (London), Ser. A*, **247**, 168 (1958).
[c] R. Bass and J. Lamb, *Proc. Roy. Soc. (London), Ser. A*, **243**, 94 (1957).
[d] J. H. Andrae, *Proc. Phys. Soc. (London), Ser. B*, **70**, 71 (1957).
[e] J. H. Andrae, P. L. Joyce and R. J. Oliver, *Proc. Phys. Soc. (London)*, **75**, 82 (1960).
[f] J. L. Hunter and H. D. Dardy, *J. Chem. Phys.*, **42**, 2961 (1965).
[g] J. L. Hunter and H. D. Dardy, *J. Chem. Phys.*, **42**, 3637 (1965).

4.6.3 Specific Systems

4.6.3.1 *Carbon Disulphide*

The first molecular acoustic study to be analysed in terms of vibrational relaxation of the specific heat was carried out on carbon disulphide[114] (Table 4.20).

The linear triatomic molecule CS_2 has three fundamental modes of vibration; two of these are bond stretching in nature whilst the third, which is doubly degenerate, involves the bending of the S-C-S framework. When *all* of these normal modes are taken into account the experimental and theoretical estimates for μ'_m are sensibly equal at different temperatures. That is it is plausible to assume that the relaxation process is entirely vibrational in character and involves the deactivation of all normal modes of vibration. As will be shown later the mechanism of energy transfer between vibrational and translational degrees of freedom can be explained in terms of binary collisions. This viewpoint is strengthened by some hypersonic-velocity data[117–119] which indicates that the velocity dispersion can also be accounted for by postulating vibrational relaxation.

Measurements have recently been extended[120] to 500 Mc/s and the resulting data are in good agreement with the earlier work.

4.6.3.2 *Carbon Dioxide*

The analysis of molecular acoustic data (Table 4.20) for carbon dioxide[127] over the temperature range 0–30°C and 32·5 to 100 atmospheres pressure shows that the deactivation of all normal vibrational modes takes place with a single relaxation time. Further work[128] will be discussed more fully in Section 4.6.4.

4.6.3.3 *Sulphur Dioxide*

Ultrasonic measurements (Table 4.20) on liquid sulphur dioxide at ten atmospheres pressure have been carried out by Bass and Lamb[115]. Sulphur dioxide is a non-linear triatomic molecule with three vibrational degrees of freedom. The fundamental modes of vibration can be described as antisymmetrical and symmetrical bond-stretching modes, $v_a(SO_2)$ and $v_s(SO_2)$, and a bond-bending vibration, $b(SO_2)$ involving changes in the OSO angle. It was found that the single relaxation found in the megacycle region could be associated with the deactivation of the two bond-stretching modes whose frequencies are 1378 and 1148 cm^{-1}. The experimental value quoted in Table 4.20 for ΔC_p was found by substituting the experimental value for the relaxation strength in Equation (4.72). It was estimated that the lower frequency bond-bending mode relaxes at a frequency of about 1500 Mc/s—well outside the range available to these workers. Two commercial samples of sulphur dioxide were used. The second sample had characteristic frequencies about 8% higher than those included in Table 4.20. This discrepancy was attributed to impurities. This point is discussed in Section 4.6.4.

4.6.3.4   Methylene Halides

Several molecular acoustic studies have been carried out on methylene chloride[119–125] and methylene bromide[126]. For methylene chloride a single relaxation in the ultrasonic region (Table 4.20) was found to be consistent with the deactivation of all normal vibrational modes except the low frequency ClCCl bending mode, b(CCl$_2$), at 283 cm^{-1}.

Recent hypersonic measurements suggest that the velocity dispersion is associated with a double relaxation. The indications are therefore that the relaxation associated with the b(CCl$_2$) mode occurs in the gigacycle region. This view is strengthened by an absorption measurement carried out at a frequency of 3·52 Gc/s[125]. This datum together with ultrasonic-absorption and viscosity data shows that the relaxation associated with the b(CCl$_2$) mode occurs at a frequency slightly less than 10 Gc/s.

For the corresponding bromoderivative[126] molecular acoustic measurements up to 500 Mc/s show that a single relaxation occurs accountable in terms of the vibrational relaxation of all normal vibrational modes except the b(CBr$_2$) mode at 174 cm^{-1}.

4.6.3.5   Benzene

Benzene has been extensively studied by Carome and his coworkers[127]. The project included ultrasonic-absorption (7–500 Mc/s) and velocity measurements as well as hypersonic-velocity determinations. In considering the ultrasonic data Carome compared the experimentally determined f_c values with theoretical estimates obtained from the relation

$$f_c \,(\text{theor.}) = 2\mu'_m \,(\text{theor.})/Av_0$$

where A is the usual relaxation parameter and v_0 is the velocity of sound at low frequency. With this equation values for f_c (theor.) were calculated assuming (i) that all normal modes were deactivated and (ii) that the lowest vibrational mode was not involved in the relaxation process. These data are given in Table 4.21 and it is clear that the assumption (ii) makes the better fit. In order to fit the hypersonic velocity-dispersion data it was found necessary to invoke a second relaxation process involving the low-frequency vibrational mode. This second relaxation was estimated to occur in the 6–15 Gc/s region. Very recently[128] in describing new laser light-scattering work on sound-velocity measurements in the frequency range from 1·5 to about 5 kMc/s, Carome has again claimed that the observed ultrasonic-absorption and hypersonic-velocity data may be interpreted in terms of double relaxation of the vibrational specific heat.

Table 4.21
Ultrasonic relaxation data for benzene[a]

	f_c Mc/s		
T °C	Observed	Calculated from ΔC_p (total)	Calculated from ΔC_p (total-lowest mode)
25	560 ± 50	748	470
6	510 ± 50	830	520

[a] J. L. Hunter, E. F. Carome, H. D. Dardy and J. A. Bucaro, *J. Acoust. Soc. Am.*, **40**, 313 (1966).

This interpretation has been challenged by O'Connor, Schlupf and Stakelon[129] who carried out velocity measurements over a substantial portion of the dispersion region, namely, from about 0·6 to 6·0 Gc/s. In their work these authors measured the Doppler shift of the Brillouin components in the scattered radiation in liquid benzene using a gas laser source. The experimental results were compared with those predicted on the basis of Mountain's theory[130]. The conclusion was that the thermal relaxation of vibrational states in benzene can be described in terms of a single-relaxation mechanism. It was felt that the conflicting interpretations are due principally to the use of thermodynamic data based on inaccurate assignments of frequency values to the fundamental modes of vibration of benzene.

4.6.3.6 *Other Molecules*

There have been several other hypersonic and ultrasonic measurements of relaxation associated with the vibrational specific heat. At present these results are not complete and no definite conclusions can be made. This work has shown, however, that acoustic relaxation in carbon tetrachloride, methyl chloride, cyclopropane, nitrous oxide and sulphur hexafluoride is associated with vibrational relaxation[115,131,132].

4.6.4   General Discussion

The work done so far shows that the pattern of vibrational relaxation in gases is very similar to that in the corresponding liquid.

A comparison of gas- and liquid-phase results for a number of fluids shows that the density is the main factor governing the change in relaxation time. It has been shown that to a first approximation the product of density (ρ) and relaxation time (τ) is approximately constant even though the relaxation time in the gas and liquid phase may differ by several orders of magnitude.

In view of the similarity between vibrational relaxation in gas and liquids Litovitz[2,133,134] has come to the conclusion that binary collisions are primarily responsible for energy transfer in the liquid phase. This means that, as in the gas phase, the interaction of colliding molecules leads to energy transfer in favourable cases. No energy transfer is assumed to take place when the molecule collides with the cell wall as a whole or when a molecule collides with a group of molecules forming a complex.

The relaxation time, τ, for a binary collision is given by:

$$\tau = [\tilde{N}\bar{P}]^{-1} \tag{4.74}$$

where $\tilde{N}$ is the number of collisions per second and $\bar{P}$ is the probability that during a collision the vibrational-energy state of the molecule, which is assumed to be simple harmonic, will change.

In order to proceed further the results of the application of kinetic theory to liquids have to be used. From kinetic theory it can be shown that $\tilde{N}$ is given by

$$\tilde{N} = \bar{v}/L_f \tag{4.75}$$

where

$$\bar{v} = (8RT/\pi m)^{1/2} \tag{4.76}$$

is the average velocity of the molecules as defined from kinetic theory and L_f is the mean free path. If the mechanism of vibrational relaxation is the same for liquid and gas then to a first approximation:

$$\bar{P}_{liq} = \bar{P}_{gas} \tag{4.77}$$

In the liquid state the evaluation of $\tilde{N}$ and $\bar{P}$ is rather difficult. In accounting for the pressure and temperature dependence of the relaxation times in carbon disulphide and also carbon dioxide both as dense gas and liquid, Litovitz has used different methods for the evaluation of L_f together with the above theory. In the gas phase the actual volume of the molecules present is only a small fraction of the total volume available. However, in the liquid phase these two quantities are comparable owing to the close proximity of the molecules. This is the most critical factor which must be taken into account in applying gas-phase kinetic theory to liquid-state systems. Therefore, in order to calculate mean free paths one must calculate a value for the 'free volume' or some expression for the distances between the edges of the molecules. The value one obtains for free volume depends critically on the type of theory used. Litovitz has used a number of theories to calculate values for L_f: these include Eyring's free volume theory[135], Kittel's available volume model[136], the kinetic-theory hard-sphere model due to Sutherland, and Eyring and Hirschfelder's cell model[135]. In addition Litovitz has used Enskog's theory[138] to predict $\tilde{N}$ in the gas and liquid phase from p.v.t data.

A critical examination of the earlier work done on carbon dioxide[139–141] and carbon disulphide[2,83,114,133], showed that the original gas-phase data on carbon dioxide could be accounted for from Sutherland's kinetic-theory hard-sphere model, whilst the temperature dependence of the relaxation time in carbon disulphide could be accounted for on the basis of Eyring and Hirschfelder's cell-model theory.

Madigosky and Litovitz[137] were able to extend the previous measurements on carbon dioxide both as a gas and liquid at various pressures. The relaxation parameters associated with vibrational relaxation were found from measurements at $50.2°$ where carbon dioxide is a gas up to 600 amagat density and also at $25°$ where up to 600 amagat density carbon dioxide is a liquid. The Sutherland kinetic-theory hard-sphere model predicts a value of L_f given by the relation

$$L_f = [\sqrt{2}\pi\sigma^2\rho/(1+C/T)]^{1/2} \tag{4.78}$$

where σ is the hard-sphere molecular diameter, ρ is the density and C the Sutherland constant. From Equations (4.74), (4.75), (4.76) and (4.78) it therefore follows from kinetic-theory considerations that (f_c/ρ) is a constant (f_c is related to τ by Equation 4.73). For gaseous and liquid carbon dioxide, plots of f_c against density are shown in Figures 4.32 and 4.33 and these data indicate that there is large departure from the gas

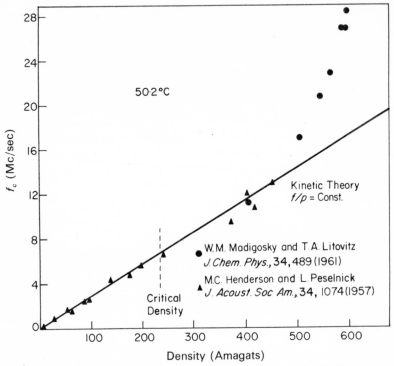

Figure 4.32   Plot of relaxation frequency against amagat density for gaseous
carbon dioxide

kinetic theory in the high-density region. As shown in Figure 4.32 the
original work by Henderson and Peselnick[139] was not extended to this
region of high density.

The next step carried out by Madigosky and Litovitz was to apply the
cell model proposed by Eyring and Hirschfelder to evaluate L_f. This
theory predicts that L_f is given by:

$$L_f = 2(1/\rho)^{1/3} - 2\sigma \tag{4.79}$$

From Equations (4.74), (4.75) and (4.76) and (4.79) it follows that

$$\tau = (2/\bar{v}\bar{P})[(1/\rho)^{1/3} - \sigma] \tag{4.80}$$

A plot of τ against $(1/\rho)^{1/3}$ shown in Figure 4.34 for both the dense gas and
liquid carbon dioxide, is an excellent straight line which on extrapolation

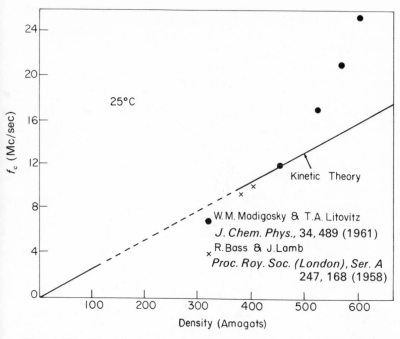

Figure 4.33   Plot of relaxation frequency against amagat density for liquid carbon dioxide

gives a value of 3·56 Å for σ in both phases. This value of σ is also in excellent agreement with that of 3·54 Å predicted from the hard-sphere kinetic theory of Sutherland.

A further test of the experimental data was to plot experimental and theoretical values of (L_f/σ) against density in the high-density region (350–600 amagats). The experimental values of L_f were found from Equations (4.74) and (4.75) which give

$$\tau(\text{expl}) = \frac{L_f}{\bar{v}}[\bar{P}]^{-1} \tag{4.81}$$

For substitution in this equation $\bar{v}$ was calculated from Equation (4.76) and the gas-phase value of $\bar{P}$ was used[1] assuming Equation (4.81). A second set of experimental points were calculated by correcting $\bar{P}$ to give a value for $\bar{P}_{\text{liq}}$ as follows,

$$\bar{P}_{\text{liq}} = \bar{P}_{\text{gas}}(1 + C/T)/\exp(\varepsilon/kT) \tag{4.82}$$

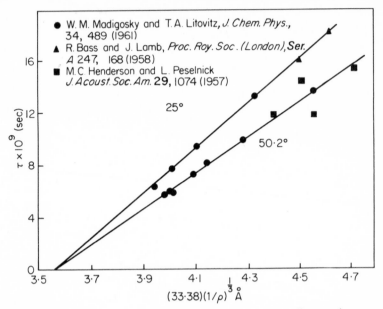

Figure 4.34 Plot of relaxation time against the mean distance between molecular centres

This equation was derived by Litovitz using a theoretical treatment for gas-phase vibrational relaxation derived by Schwartz and Herzfeld[1]. This correction was introduced by Litovitz in the belief that long-range attractive forces probably do not play a rôle in determining $\bar{P}_{liq}$ and therefore should account for the differences between $\bar{P}_{gas}$ and $\bar{P}_{liq}$. In Equation (4.82) C is the Sutherland constant and ε is the depth of the Lennard-Jones potential. For comparison the theoretical values of (L_f/σ) were calculated from Equation (4.78) for the Sutherland kinetic theory and Equation (4.79) from the cell-model method. As shown in Figure 4.35 neither theoretical model accounts satisfactorily for the experimental points. Madigosky and Litovitz then modified the cell model for a liquid so that the wandering molecule was trapped not by a fixed but by a moving wall. This meant that L_f is now given

$$L_f = \Sigma[(1/\rho)^{1/3} - \sigma] \qquad (4.83)$$

This is one half the value proposed by the cell model. Theoretical values of (L_f/σ) from Equation (4.83) were then plotted in Figure 4.35. It can be

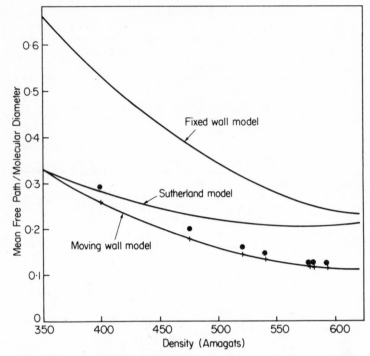

Figure 4.35 Comparison of theoretical and experimental data using different mean-free-path models

● Experimental using $\bar{P}(\text{liq}) = \bar{P}(\text{gas})$
× Experimental using $\bar{P}(\text{liq})/\bar{P}(\text{gas}) < 1{\cdot}0$

seen now that excellent agreement is found between the 'moving wall cell model' and the experimental points when $\bar{P}$ has been corrected for attractive forces. It is also seen that the value of σ calculated from Equation (4.83) as shown in Figure 4.34 will not alter when Equation (4.83) is used.

Furthermore experimental values of $\bar{P}_{\text{liq}}$ and $\bar{P}_{\text{gas}}$ can be found from Equations (4.74), (4.75) and (4.83) followed by a plot of $\tau(\text{exptl})$ against $(1/\rho)^{1/3}$. These values can be checked with the theoretical values given by Equation (4.82). For both carbon dioxide and carbon disulphide these results are shown in Table 4.22.

The theoretical expression for $\bar{P}$ derived from the gas-phase results by Schwartz and Herzfeld indicates that $\bar{P}$ is independent of pressure.

Table 4.22

Comparison data from vibrational relaxation in gas and liquid phase

Molecule	Temp. °C	$\bar{P}$ (exp) dilute gas	$\bar{P}$ (exp) dense gas	$\bar{P}$ (exp) liquid	$\dfrac{\bar{P}\text{ (dil gas)}}{\bar{P}\text{ (liq)}}$ exp.	$\dfrac{\bar{P}\text{ (dil gas)}}{\bar{P}\text{ (liq)}}$ calc.	$\dfrac{\bar{P}\text{ (dense gas)}}{\bar{P}\text{ (liq)}}$ exp.
CO_2	25	$1\cdot75^{a,b}$	$1\cdot69^{a,e}$	$1\cdot55^{a,e}$	$1\cdot14$	$1\cdot13^{g}$	$1\cdot1$
CO_2	50·2	$2\cdot08^{c}$	$2\cdot07^{e}$	$1\cdot87^{e}$	$1\cdot11$	$1\cdot11^{g}$	$1\cdot1$
CS_2	25	$10\cdot9^{d}$		$5\cdot4^{f}$	$2\cdot5$	$2\cdot03^{g}$	

[a] All values multiplied by 10^{-5} sec^{-1}.
[b] F. D. Shields, *J. Acoust. Soc. Am.*, **29**, 450 (1956).
[c] C. Henderson and L. Peselnick, *J. Acoust. Soc. Am.*, **29**, 1074 (1957).
[d] F. A. Angona, *J. Acoust. Soc. Am.*, **25**, 1110 (1954).
[e] W. M. Madigosky and T. A. Litovitz, *J. Chem. Phys.*, **34**, 489 (1961).
[f] J. H. Andrae, E. L. Heasell and J. Lamb, *Proc. Phys. Soc.* (*London*), Ser. B, **69**, 625 (1956).
[g] Calculated using Equation (4.82).

This is shown in the linearity of the plots in Figure 4.34. This means that

$$\frac{\tau_p}{\tau_0} = \frac{\Sigma[(1/\rho_p)^{1/3} - \sigma]}{\Sigma[(1/\rho_0)^{1/3} - \sigma]} \tag{4.84}$$

where the subscripts p and 0 refer to pressure p and atmospheric pressure respectively. This expression is derived using Equations (4.74), (4.75), (4.76) and either (4.79) or (4.83). Using a value of $\sigma = 4\cdot32$ Å for carbon disulphide Litovitz was able to check the pressure dependence of the relaxation times[2,83,128,133]. His results are summarized in Table 4.23. Again excellent agreement is obtained when using the cell model or moving cell model. The value of σ used in this experiment was 4·32 Å which is slightly less than that proposed by Hirschfelder for a spherical molecule.

In the transitions from dilute gas to dense gas to liquid phase it has been shown that the mean free paths from the Sutherland and cell models are nearly equal at lower density. As the pressure is increased (approaching the liquid phase) the Sutherland model is inappropriate and a modification of the cell model has to be used. At low densities (approaching a dilute gas) the cell model will break down as the molecules are no longer in a cage. It has also been shown that the collision efficiencies $\bar{P}$ increase in the above transition due to long-range attractive forces becoming less important in the liquid phase.

Table 4.23

Pressure dependence of relaxation time in CS_2[a]

Pressure	τ_p/τ_0	
kg/cm^2	exp.	calc (eqn. 4.84)
1	1·0	1·0
500	0·87	0·83
1000	0·65	0·67

[a] T. A. Litovitz, E. Carnevale and P. Kendall, *J. Chem. Phys.*, **26**, 465 (1957).

Slie and Litovitz[134] carried out further measurements on vibrational relaxation in carbon disulphide with methyl, ethyl and propyl alcohol added as impurities up to 6 mole percent. These results were compared with similar experiments carried out in the gas phase. From these results it was deduced that the vibrational-relaxation frequency was a linear function of the impurity concentration. This is consistent with the concept of a binary-collision mechanism as being responsible for energy transfer. In addition it was shown that the ratio of the collision efficiencies $\bar{P}_{AB}/\bar{P}_{AA}$ is the same in the liquid and gas phase. In this context AA stands for pure carbon disulphide and AB stands for added impurities of alcohol. This result means that the same type of binary collision occurs for AB in both gas and liquid phases. Lastly from a comparison of the ratio $f_c(AB)/f_c(AA)$ of the liquid at $-63°C$ with that of the gas at $25°C$ it was decided that the temperature dependence of the collision efficiency for AB collisions is greater than that for AA collisions.

REFERENCES

1. K. F. Herzfeld and T. A. Litovitz, *Absorption and Dispersion of Ultrasonic Waves*, Academic Press, New York, 1959.
2. T. A. Litovitz, *J. Acoust. Soc. Am.*, **31**, 681 (1959).
3. J. E. Piercy, *Proc. I.E.E.E.*, **53**, 1346 (1965).
4. *Physical Acoustics* (Ed. Warren P. Mason), Vol. 2, Part A, Academic Press, New York, 1965, p. 281.
5. K. Tamm, *Z. Elektrochem*, **64**, 73 (1960).
6. H. J. McSkimin in *Physical Acoustics* (Ed. Warren P. Mason), Vol. 1, Part A, Academic Press, 1964, p. 272.
7. J. Karpovich, *J. Acoust. Soc. Am.*, **26**, 819 (1954).
8. L. E. Lawley and R. D. C. Reed, *Acustica*, **5**, 316 (1955).

9. C. J. Moen, *J. Acoust. Soc. Am.*, **23**, 62 (1951).
10. C. E. Mulders, *Appl. Sci. Res., Sect. B*, **1(5)**, 341 (1950).
11. G. Kurtze and K. Tamm, *Acustica*, **3**, 33 (1953).
12. M. Cerceo, R. Meister and T. A. Litovitz, *J. Acoust. Soc. Am.*, **34**, 239 (1962).
13. P. Eggers, *Acustica* (to be published); personal communication.
14. J. H. Andrae, C. Jupp and D. G. Vincent, *J. Acoustic Soc. Am.*, **32**, 406 (1960).
15. C. Eckart, *Phys. Rev.*, **73**, 68 (1948).
16. J. E. Piercy and J. Lamb, *Proc. Roy. Soc. (London), Ser. A*, **226**, 43 (1954).
17. J. E. Piercy, *J. Phys. Radium*, **17**, 405 (1956).
18. D. N. Hall and J. Lamb, *Proc. Phys. Soc. (London)*, **73**, 354 (1959).
19. J. R. Pellam and J. K. Galt, *J. Chem. Phys.*, **14**, 608 (1946).
20. P. D. Edmonds, V. F. Pearce and J. H. Andrae, *Brit. J. Appl. Phys.*, **13**, 551 (1962).
21. J. M. M. Pinkerton, *Proc. Phys. Soc. (London), Ser. B*, **62**, 286 (1949).
22. M. E. Pedinoff, *J. Chem. Phys.*, **36**, 777 (1962).
23. J. L. Hunter and H. D. Dardy, *J. Acoust. Soc. Am.*, **37**, 1914 (1964).
24. J. H. Andrae and P. L. Joyce, *Brit. J. Appl. Phys.*, **13**, 462 (1962).
25. J. H. Andrae, R. Bass, E. L. Heasell and J. Lamb, *Acustica*, **8**, 131 (1958).
26. T. A. Litovitz, T. Lyon and L. Peselnick, *J. Acoust. Soc. Am.*, **26**, 566 (1954).
27. G. Atkinson, S. Kor and R. L. Jones, *Rev. Sci. Instr.*, **35**, 1270 (1964).
28. H. F. Bommel and K. Dransfield, *Phys. Rev. Letters*, **1**, 7, 234 (1959).
29. A. A. Berdyev and N. B. Lezhnev, *Soviet Phys. Acoust. (English Transl.)*, **R127**, 211 (1966).
30. P. Debye and F. W. Sears, *Proc. Natl. Acad. Sci., U.S.*, **18**, 410 (1932).
31. R. Lucas and P. Biquard, *J. Phys. Radium*, **6**, 464 (1932).
32. C. J. Burton, *J. Acoust. Soc. Am.*, **20**, 186 (1948).
33. C. W. Willard, *J. Acoust. Soc. Am.*, **12**, 438 (1941); **19**, 235 (1947); **21**, 107 (1949).
34. G. W. Hazzard, *J. Acoust. Soc. Am.*, **22**, 29 (1950).
35. A. V. Itterbeck and L. Verhaegen, *Coll. Ultrason Triel* (Brussels, 1951) p. 220.
36. E. Hsutsung-Yuch, *J. Acoust. Soc. Am.*, **17**, 127 (1945).
37. M. C. Smith and R. T. Beyer, *J. Acoust. Soc. Am.*, **18**, 424 (1946).
38. F. L. McNamara and R. T. Beyer, *J. Acoust. Soc. Am.*, **25**, 259 (1953).
39. A. Barone and M. Nuovo, *Ric. Sci.*, **21**, 516 (1951).
40. M. Mokhtar and H. Youlsef, *J. Acoust. Soc. Am.*, **28**, 651 (1956).
41. G. B. Neergaard, *Acustica*, **17**, 143 (1966).
42. J. C. Hubbard, *Phys. Rev.*, **38**, 1011 (1931).
43. J. H. Andrae and P. D. Edmonds, *J. Sci. Instr.*, **38**, 505 (1961).
44. B. I. Hal'yenov and V. F. Nazdrev, *Soviet Phys. Acoust. (English Transl.)*, **5**, 377 (1959).
45. D. Tabuchi, *J. Chem. Phys.*, **28**, 1014 (1950).
46. V. Ilgunas and K. Paulauskas, *Soviet Phys. Acoust. (English Transl.)*, **12(2)**, 225 (1966).
47. E. L. Carstensen, *J. Acoust. Soc. Am.*, **26**, 858, 862 (1954).
48. H. Siegert, *Acustica*, **13**, 48 (1963).
49. L. Landau and G. Placzek, *Z. Phys. Sowjetunion*, **5**, 172 (1934).
50. R. G. Brewer and K. E. Rieckhoff, *Phys. Rev, Lett.*, **13**, 334 (1964).

51. E. Garmine and C. H. Townes, *Appl. Phys. Lett.*, **5**, 84 (1964).
52. R. Y. Chiao and C. H. Townes, *Phys. Rev. Lett.*, **12**, 592 (1964).
53. R. Y. Chiao and B. P. Stoicheff, *J. Opt. Soc. Am.*, **54**, 1286 (1964).
54. D. T. Mash, V. S. Sharunov and I. L. Fakelensbi, *Soviet Phys. JETP (English Transl.)*, **20**, 523 (1965).
55. D. Rank, E. Kiero, U. Fink and T. Wiggins, *J. Opt. Soc. Am.*, **55**, 925 (1965).
56. G. Benedek and T. Greytak, *Proc. I.E.E.E.*, **53**(10), 1623 (1965).
57. J. E. Piercy, *J. Acoust. Soc. Am.*, **33**, 198 (1961).
58. E. L. Heasell and J. Lamb, *Proc. Roy. Soc. (London), Ser. A*, **237**, 233 (1956).
59. R. O. Davies and J. Lamb, *Quart. Rev. (London)*, **11**, 134 (1957).
60. E. Wyn-Jones, *Ph.D. Thesis*, Wales, 1963.
61. M. S. de Groot and J. Lamb, *Proc. Roy. Soc. (London), Ser. A*, **242**, 36 (1957).
62. E. Wyn-Jones and W. J. Orville-Thomas, *Chem. Soc. (London), Spec. Publ.* **20**, 209 (1966).
63. J. Lamb and J. M. M. Pinkerton, *Proc. Roy. Soc. (London) Ser. A*, **199**, 114 (1949).
64. J. H. Andrae, P. L. Joyce and R. J. Oliver, *Proc. Phys. Soc. (London)*, **75**, 82 (1960).
65. A. V. Anantaraman, A. B. Walters, P. D. Edmonds and C. J. Prings, *J. Chem. Phys.*, **44**, 2651, 1966.
66. K. Tamm, *Dispersion and Absorption of Sound by Molecular Processes*, (Ed. D. Sette) Academic Press, N.Y., 1963, p. 175.
67. J. F. McKellar and J. M. Andrae, *Nature*, **195**, 778, 865 (1962).
68. S. Mizushima, *Structure of Molecules and Internal Rotation*, Academic Press, New York, 1954.
69. N. Sheppard, *Advan. Spectr.*, **1**, 288 (1959).
70. L. P. Melinkova, Yu A. Pentin and O. D. Ul'Yanova, *Zh. Strukt. Khim.*, **4**, 535 (1963).
71. J. H. Andrae, P. D. Edmonds and P. L. Joyce, *Proc. Third Int. Congress, Acoustics*, (Ed. L. Cremer), Elsevier, Amsterdam (1961), p. 542.
72. K. Krebs and J. Lamb, *Proc. Roy. Soc. (London), Ser. A*, **244**, 558 (1958).
73. N. I. McDevitt, A. L. Rozek, F. F. Bentley and A. O. Davison, *J. Chem. Phys.*, **42**, 1173 (1965).
74. Estimated from Raman spectrum in ref. 70.
75. W. J. Dauben and K. S. Pitzer, *Steric effects in Organic Chemistry*, (Ed. M. S. Newman), Wiley, New York, 1956, Chap. 1.
76. J. E. Piercy and S. V. Subrahmanyam, *J. Chem. Phys.*, **42**, 4011 (1965).
77. J. Karpovich, *J. Chem. Phys.*, **22**, 1767 (1954).
78. J. Lamb and J. Sherwood, *Trans. Faraday Soc.*, **51**, 1674 (1955).
79. G. S. Verma, *Proc. Phys. Soc. (London)*, **74**, 192 (1960).
80. R. P. Singh, G. S. Darbari and G. S. Verma, *J. Chem. Phys.*, **46**, 151 (1967).
81. L. W. Reeves and K. O. Stromme, *Can. J. Chem.*, **38**, 1241 (1960).
82. Yu. A. Pentin, Z. Sharipov, G. G. Kolova, A. V. Kamarnitski and A. A. Akhren, *Zh. Strukt. Khim.*, **4**, 194 (1963).
83. T. A. Litovitz and E. A. Carnevale, *J. Acoust. Soc. Am.*, **30**, 134 (195°).
84. B. A. Arbuzor and G. A. Guzhavina, *Dokl. Akad. Nauk SSSR*, **61**, 63 (1948).
85. R. A. Padmanabhan and E. L. Heasell, *Proc. Phys. Soc. (London)*, **76**, 321 (1960).

86. E. J. Williams, E. Wyn-Jones, T. H. Thomas and W. J. Orville-Thomas, unpublished data.

87. M. Tamres, S. Searles, E. M. Leighly and D. M. Mohma, *J. Am. Chem. Soc.*, **76**, 3983 (1954).

87. H. C. Brown, *J. Chem. Soc.*, 1248 (1956).
   H. C. Brown, M. D. Taylor and M. Gerstern, *J. Am. Chem. Soc.*, **66**, 434 (1944).

89. J. B. Bentley, K. B. Everard, R. J. B. Marsden and L. E. Sutton, *J. Chem. Soc.*, **1949**, 2957.

90. N. L. Owen and N. Sheppard, *Trans. Faraday Soc.*, **60**, 634 (1964).

91. N. L. Owen and N. Sheppard, *Spectrochim. Acta.*, **22**, 1101 (1966).

92. P. Biquard, *Ann. Physik.*, **6**, 195 (1936).

93. D. H. A. Huddard, *M.S. Thesis*, University of London, 1950.

94. J. M. M. Pinkerton, 'Ultrasonic Conference, Brussels,' *Mededel. Koninkl. Vlaam. Acad. Wetenshap, Belg., Kl. Welenshap.*, 1951, p. 117.

95. J. Karpovich, *J. Acoust. Soc. Am.*, **26**, 819 (1954).

96. D. Tabuchi, *J. Chem. Phys.*, **28**, 1074 (1958).

97. S. V. Subrahmanyam and J. E. Piercy, *J. Acoust. Soc. Am.*, **37**, 340 (1965).

98. D. N. Hall and J. Lamb, *Trans. Faraday Soc.*, **55**, 784 (1959).

99. U. F. Nozdrev, L. E. Belinebaza and B. A. Belinstan, Proc. 3rd I.C.A. (Ed. L. Cremer), Elsevier, Amsterdam, 1959, p. 564.

100. B. F. Kal'yanov and V. F. Nozdrev, *Soviet Phys. Acoust. (English Transl.)*, **5**, 377 (1959).

101. W. A. Slie and T. A. Litovitz, *J. Chem. Phys.*, **39**, 1538 (1963).

102. J. E. Piercy and S. V. Subrahmanyam, *J. Chem. Phys.*, **42**, 1475 (1965).

103. J. M. Riveros and E. Bright Wilson, Jr., *J. Chem. Phys.*, **46**, 4605 (1967).

105. J. Lamb and D. H. A. Huddart, *Trans. Faraday Soc.*, **46**, 540 (1950).

106. E. Freedman, *J. Chem. Phys.*, **21**, 1784 (1953).

107. J. E. Piercy and J. Lamb, *Trans. Faraday Soc.*, **52**, 930 (1956).

108. D. Tabuchi, *Z. Elektrochem.*, **64**, 141 (1960).

109. P. White and G. C. Benson, *Can. J. Chem.*, **30**, 1135 (1958).

110. C. Moriamiez, M. Moriamiez and A. Moreaux, *J. Chim. Phys.*, **63**, 615 (1966).

111. F. B. Stumpt and L. A. Crum, *J. Acoust. Soc. Am.*, **38**, 170 (1965).

112. J. E. Piercy and M. G. Seshagiri Rao, 'Meeting of the American Acoustic Society,' *J. Acoust. Soc. Am.*, **41**, 1591 (1967).

113. R. Y. Chiao, *Physics of Quantum Electronics (Conference)*, (Ed. P. L. Kelley, B. Lax and P. E. Tannenwald), McGraw-Hill, London, 1922, p. 241.

114. J. H. Andrae, E. L. Heasell and J. Lamb, *Proc. Phys. Soc. (London)*, *Ser B*, **69**, 625 (1956).

115. R. Bass and J. Lamb, *Proc. Roy. Soc. (London)*, *Ser. A*, **243**, 94 (1957).

116. J. Lamb, *Physical Acoustics*, (Ed. Warren P. Mason), Vol. 2, Part A, Academic Press, N.Y., 1965.

117. H. Z. Cummin and R. W. Gammon, *Appl. Phys. Letters*, **6**, 171(L) (1965).

118. C. L. O'Connor and J. P. Schlupf, *J. Acoust. Soc. Am.*, **40**, 663 (1966).

119. J. E. Piercy and G. R. Hanes, *J. Chem. Phys.*, **42**, 3401 (1965).

120. J. L. Hunter, H. D. Dardy and J. A. Bucaro, *Proc. Intern. Congr. Acoust*, 5th, Liege, 1965, D 26.

121. J. H. Andrae, *Proc. Phys. Soc. (London)*, *Ser. B*, **70**, 71 (1957).

123. J. L. Hunter and H. D. Dardy, *J. Chem. Phys.*, **42**, 2961 (1965).
124. N. A. Clark, C. E. Moller, J. A. Bucaro and E. F. Carome, *J. Chem. Phys.*, **44**, 2528 (1966).
125. G. R. Hanes, R. Turner and J. E. Piercy, *J. Acoust. Soc. Am.*, **38**, 1057 (1967).
126. J. L. Hunter and H. D. Dardy, *J. Chem. Phys.*, **42**, 3637 (1965).
127. J. L. Hunter, E. F. Carome, H. D. Dardy and J. A. Bucaro, *J. Acoust. Soc. Am.*, **40**, 313 (1966).
128. E. F. Carome, S. P. Singal and C. R. Kunsitis, '73rd Meeting of the Acoustical Society of America,' *J. Acoust. Soc. Am.*, **41**, 1601 (1967).
129. C. L. O'Connor, J. P. Schlupf and T. S. Stakelon, '73rd Meeting of the Acoustical Society of America,' *J. Acoust. Soc. Am.*, **41**, 1601 (1967).
130. R. D. Mountain, *J. Res. Natl. Bur. Std. A.*, **70**, 207, (1966).
131. T. A. Litovitz and C. M. Davies, *J. Chem. Phys.*, 840 (1965).
132. W. S. Gornall, G. I. A. Stegman, B. P. Stoicheff, P. H. Stolen and V. Voltena, *Phys. Rev. Letters*, **17**, 297 (1966).
133. T. A. Litovitz, *J. Chem. Phys.*, **26**, 469 (1957).
134. W. M. Slie and T. A. Litovitz, *J. Acoust. Soc. Am.*, **33**, 1412 (1961).
135. H. Eyring and J. O. Hirschfelder, *J. Phys. Chem.*, **41**, 249 (1957).
136. C. Kittel, *J. Chem. Phys.*, **14**, 614 (1946).
137. W. M. Madigosky and T. A. Litovitz, *J. Chem. Physics*, **34**, 489 (1961).
138. J. O. Hirschfelder, C. F. Curtiss and R. B. Bird, *Molecular theory of Gases and Liquids*, John Wiley, New York, 1954.
139. M. C. Henderson and L. Peselnick, *J. Acoust. Soc. Am.*, **29**, 1074 (1957).
140. M. C. Henderson and T. Z. Klose, *J. Acoust. Soc. Am.*, **31**, 29 (1958).
141. F. P. Shields, *J. Acoust. Soc. Am.*, **29**, 450 (1956).

5

Collisional Excitation and Chemical Reaction

H. O. Pritchard

5.1 INTRODUCTION

Ideas about how chemical reactions take place are subject to a continuous process of evolution, which can be traced back to the realization that the rate of a reaction must be dependent on molecular-collision processes and yet could obey first-order kinetics[1,2]. Our understanding of thermal chemical reactions has advanced since that time on two broad fronts, one relating essentially to unimolecular reactions and the other to bimolecular reactions. There is an underlying unity (shared in common with photochemistry, ion and electron-impact reactions), but in the way that the theories have developed, it has not always been apparent. This is because each field presents its own particular problems, and the facets of the theory which can be tested experimentally are different. This review is aimed towards the re-establishment of this unity, and in so doing draws together some of the problems in contemporary theoretical reaction kinetics, principally those associated with the vibrational-energy content of reacting molecules. The presentation is not in any way exhaustive; considerable areas which lie within the scope of this review have been admirably treated by Bunker[3] in his recent monograph, and my hope is that, taken together with Bunker's book, this review will prove useful in pointing the way to future advances in our understanding of reaction processes.

5.2 THE ACCUMULATION OF VIBRATIONAL ENERGY

We are concerned here with vibrational energy and the way in which the accumulation of vibrational energy by a molecule leads to reaction. Before discussing reactions themselves, we must first examine the principal features of this process of accumulation of vibrational energy. Immediately we run into difficulties, because we know relatively little about such energy-transfer phenomena. When a diatomic gas is heated, its translational degrees of freedom adjust very quickly to a Boltzmann distribution corresponding to the new temperature. It takes a little longer (say some 20–200 collisions) for the rotational degrees of freedom to come into equilibrium with the translational temperature, and usually, though not always, some millions of collisions before the vibrational-energy levels reach their new equilibrium populations.

Most of our experimental knowledge of rates of vibrational relaxation comes from ultrasonic-dispersion studies[4]. In essence, the velocity of sound in a gas depends on the specific heat of the gas. If the sound frequency is very low, then during each compression and rarefaction, the temperature of the gas has time to adjust completely, and the vibrations and rotations remain in thermal equilibrium with the translational motion throughout each cycle. However, as the sound-wave frequency rises, we reach a stage where feeding energy into and out of the vibrational degrees of freedom becomes a slow process in comparison with the frequency of the compressions and rarefactions. The gas then behaves as though it had no vibrational degrees of freedom, its effective specific heat drops, and the velocity of sound in the gas increases. At still higher frequencies, the rotations fail to equilibrate with the translational motion, and there is another rise in sound velocity, although this rotational dispersion region is much more difficult to observe experimentally.

In outline, the results of ultrasonic-dispersion studies are as follows. For very simple molecules like N_2 and CO, it takes of the order of 10^6 collisions before one quantum of vibrational energy is taken up by the molecule, but for more complex molecules, the figure may be more like 10^3 collisions. There appears to be a rough correlation between the vibration frequency and the efficiency of translation $\leftrightarrow$ vibration (V–T) conversion in the sense that the larger the vibrational quantum of energy, the less efficient is the interconversion process. There are some very marked impurity effects, however. For example, H_2O seems to be very efficient[5] in causing relaxation of the $v = 0 \leftrightarrow 1$ transition in CO. Similarly, Cl_2 is very efficient in relaxing CO, but not in relaxing N_2 which is a very similar molecule as regards mass, size and vibration frequency. This has been interpreted[6,7] as a 'chemical effect'—in other words, because Cl_2 can react with CO at higher temperatures, the approach of Cl_2 to CO seriously perturbs the electron distribution in both molecules. The result of this will be to make the encounter between Cl_2 and CO much longer and more complex than an encounter between Cl_2 and N_2, and so increase the probability of a redistribution of the relative motions. (We note in passing the recent discovery that the O atom[8] is a very efficient third body in the dissociation of O_2, and the H atom[9] in the dissociation of H_2; these can be interpreted as examples of a 'chemical effect', although alternative explanations may also be possible.) Again, for simple molecules, there is a fairly consistent trend that the transition probability increases as the temperature rises, but for more complicated molecules there are indications that the transition probability goes down as the temperature goes up[10]. This observation, if it were found to be general, would be consistent with studies on the deactivation of highly vibrationally excited molecules, where it is found that the probability of vibrational deactivation falls as the temperature rises[11-13]. This last remark brings us to the crux of the whole problem. In discussing the dissociation of a diatomic molecule we are interested in knowing how all the vibrational levels in a molecule relax, i.e. what are the probabilities of all transitions $v = i \leftrightarrow j$, but ultrasonic measurement only tells us information about the $v = 0 \leftrightarrow 1$ transition. (For a polyatomic molecule, the situation is even more difficult, because there are many vibrational modes, and *usually* it is only possible to obtain information about the $v = 0 \leftrightarrow 1$ transition of the one with the lowest vibration frequency. Such additional factors which arise, e.g. the probability of interconversion of energy between various vibrational degrees of freedom within a molecule, are on the fringe of our experimental resources at the present time[14]. It may turn out in fact that nearly all the vibrational energy enters a molecule by a V–T process through the lowest vibrational mode, and then appears in other modes by intramolecular conversion, or in V–V transfers [see below] with other similar molecules.) Recently, experimental techniques have been developed to monitor the concentrations of the first half-dozen vibrational levels in a diatomic molecule in a reacting gas[15,16], and it seems that the kind of information we want may become available for a few heteronuclear molecules before too long, for a few of the low-lying levels. When it comes to the higher vibrational levels, the experimental difficulties are greater still because the concentrations of these states in the reacting gas are much lower. However, there are some indications that transitions between the higher vibrational levels of a diatomic molecule occur more easily than those between low-lying levels. The experiments concerned suggest a faster relaxation time for the N_2 molecule when its higher vibrational levels are significantly populated, both in excitation[17] and in deexcitation[18], and a similar effect has been observed[19] for deexcitation in O_2. However, the interpretation of these observations must still be open to question. Firstly, there is the problem, which is not a trivial one[3,20], of what exactly is being measured in the particular experiment and, furthermore, we have to bear in mind the difficulty in deducing anything quantitative about individual transition probabilities from the rate of the overall relaxation process[21]. Nevertheless, such an interpretation would fit in with our supposition that the ease of translation ↔ vibration energy transfer goes up as the energy gap goes down.

Consequently, the picture of V–T energy interchange which we will use for a diatomic molecule is as follows. The diatomic molecule is an anharmonic oscillator and as such will have anything between 15 and 150 bound vibrational levels which become closer and closer together as the dissociation limit is reached. Thus, the transition probabilities between adjacent vibrational levels on collision with an inert molecule will increase as the levels approach the dissociation limit. Transitions between non-adjacent levels will be discriminated against because of the larger amount of energy which has to be interchanged with translation, but near the top of the vibrational ladder multiple-quantum jumps will still require much less energy than say the $v = 0 \rightarrow 1$ transition, and so may occur more readily. All the transition probabilities will increase with temperature, but it is not clear whether an Arrhenius or a Landau–Teller ($T^{-1/3}$) form will more nearly represent the actual behaviour[4]. Furthermore, because of detailed balancing, the probability of an upward jump from $v = i \rightarrow j$ will be less by a factor of $\exp(-\Delta_{ij}/RT)$ than the probability of the reverse jump[7] (this may be regarded as a direct consequence of the fact that all collisions can induce a downward jump, but only those collisions having more relative kinetic energy than Δ_{ij}, the separation between the two levels, can bring about an upward jump[22]). In the case of polyatomic molecules, our picture is very vague—principally because our theories of kinetics have not advanced sufficiently to be blocked completely by ignorance of these quantities—but we know that the transition probabilities are much higher, that there still seems to be a correspondence between the probability and the smallness of the vibrational quantum which has to be created out of or converted into translational energy, and that perhaps, the probabilities might sometimes decrease with increasing temperature.

So far, our discussion of vibrational-transition probabilities has been oriented towards the description of the dissociation of a molecule in a heat bath consisting only of inert-gas atoms. When we come to consider real experimental systems, there are many complicating factors. The best-documented of these seems to be a resonance-transfer effect where the colliding partner is excited in some way and the quantum of energy which it loses matches the quantum of vibrational energy gained by the diatomic molecule. For example, the transfer of vibrational energy between N_2 and the $A^2\Sigma^+$ state of NO, whose vibration frequencies match almost exactly, is considerably faster than any V–T process involving the N_2 molecule. Such exchanges are termed V–V processes and it is fairly well established now that there is a systematic variation in the rate of such processes with the difference in energy between the two vibration frequencies. This means that in an undiluted gas, such processes as

$$2A_2(v = 1) \rightleftarrows A_2(v = 0) + A_2(v = 2)$$

will play a marked rôle in the relaxation process, since the energy discrepancy will not be very large, even for quite an anharmonic oscillator. Furthermore[23], when polyatomic molecules are used to transfer vibrational energy to or from NO, the transition probabilities are even higher, and seem less severely dependent on the vibrational-frequency difference, which represents the amount of energy that has to be converted to or from translation. Perhaps this is because not all of the energy discrepancy has to be converted into translation, but can be converted into rotation instead, i.e. a V–R process. Singularly little is known about V–R processes, except that[24] molecules with low moments of inertia are more efficient in relaxing vibrational energy than those with high moments of inertia, other things being equal. Also, it has been found that *ortho*-hydrogen is less efficient than *para*-hydrogen in relaxing CO vibrations[25] and it seems much more plausible that this arises from the difference in rotational properties between *ortho*- and *para*-hydrogen, rather than any differences in long-range interaction of the nuclear spins with the CO. A little information is also available on R–R processes[26].

We must also remember not to restrict consideration to interconversion of energy only between translation, rotation and vibration, but must also include electron $\leftrightarrow$ vibration transfers. This is shown by recent work[27] on the deactivation of the 4^3P_0 state of the selenium atom by various gases. The multiplet splittings are $[4^3P_0 - 4^3P_1] = 544\ \mathrm{cm}^{-1}$, $[4^3P_1 - 4^3P_2] = 1990\ \mathrm{cm}^{-1}$ and $[4^3P_0 - 4^3P_2] = 2534\ \mathrm{cm}^{-1}$, and, just as in the V–V case, molecules having vibration frequencies near these values are some 10^3 or 10^4 times as effective as those having frequencies some 500 or 600 cm^{-1}

away. On the little evidence that is available in this work and in the deactivation of $I[5^2P_{1/2} \rightarrow 5^2P_{3/2}]$ ($\Delta E = 7600\,cm^{-1}$), it seems[28] that, unlike the V–V case, polyatomic molecules are not markedly more efficient than diatomics in E–V processes, although there may be an enhancement when the molecules are paramagnetic. Such interchange of electronic and vibrational energy will be important in some dissociation processes—for example, two very common molecules, NO and O_2, have low-lying electronic states, some of which are paramagnetic.

In this discussion of vibrational-energy transfer, the anharmonic nature of molecules has been deliberately stressed. A great deal of theoretical work has been carried out on harmonic-oscillator approximations, principally because the symmetry of the wave functions and the equal spacing of the levels reduce the mathematical complexity very considerably. However, some of the results so obtained cannot be regarded as relevant to real molecules. For example, such a treatment would suggest that if the temperature of a diatomic gas were suddenly changed in a shock wave, the vibrational populations would relax smoothly through a series of Boltzmann distributions until equilibrium is reached[29]; or alternatively, if A_2 and B_2 are diatomic harmonic oscillators[30], the rate of relaxation of A_2 is independent of the initial vibrational distribution in B_2. It seems very unlikely that real molecules will behave in such a simple fashion although one would not question the conclusion that the vibrational populations relax monotonically towards their equilibrium values. There is a satisfying elegance about these harmonic-oscillator treatments, and we should continue to explore them and modify them where necessary if they lead to unreasonable predictions. A similar situation arises for rotation, where the prediction that the approach to equilibrium may be oscillatory is in conflict with the existing experimental evidence[31].

Before concluding this section on vibrational-energy transfer, we should also make some comment about theoretical calculations of transition probabilities. Whilst satisfactory methods[32] exist for calculations on the $v = 0 \leftrightarrow 1$ transition, severe difficulties are encountered when one tries to discuss transitions amongst higher vibrational levels or transitions at high temperatures. This is because they are essentially perturbation calculations which will only be accurate if the chance of a transition occurring is very low[33]. However, it does not appear unreasonable to expect a limited numerical solution of this problem in a large computer before too long, e.g. for, say, collisions between He and H_2.

5.3 DISSOCIATION OF DIATOMIC MOLECULES DILUTED IN A LARGE EXCESS OF AN INERT GAS

The kinetics of dissociation of diatomic molecules, particularly the halogens, in shock waves have been extensively studied in recent years[34-36]. These experiments have been reviewed briefly in a recent paper where it is shown that, for all the diatomic molecules so far studied, there are two factors in common[37]. First, that the bimolecular rate constant for the dissociation

$$M + X_2 \rightarrow M + X + X$$

where M is an inert gas, has an 'activation energy' which is less than the known heat of dissociation, and second, that this discrepancy between the activation energy and the dissociation energy is dependent on the nature of the diluent gas. This strongly suggests that the rate of the reaction, and therefore its temperature coefficient, are determined by energy-transfer considerations. The model of the dissociation process which is proposed in that paper is as follows (other theoretical approaches to the problem have been reviewed recently[38]). Initially, before the shock wave arrives, the undissociated gas is at some low temperature where the translations, rotations and vibrations are all subject to their respective Boltzmann population distributions. For vibration this means that almost all the molecules are in the lowest vibrational state, and at room temperature, there are virtually no molecules in energy levels more than about 5000 cm^{-1} (say 15–20 kcal) above the ground state. In a matter of a few collisions after the arrival of the shock, the translational degrees of freedom have adjusted themselves to the new temperature, which in a typical experiment would be between 1000 and 2000°K higher than the initial temperature. Since it takes many collisions (probably of the order of hundreds at the new temperature) between the diatomic molecule and the inert gas to establish a Boltzmann equilibrium between the $v = 0$ and $v = 1$ vibrational levels, there must be a considerable time-lag before all the vibrational levels reach the new Boltzmann distribution. In the early stages, when only the lower levels are still populated, the relaxation takes place by *predominantly* single-step processes, that is

$$M + X_2(v = 0) \rightarrow M + X_2(v = 1)$$

$$M + X_2(v = 1) \rightarrow M + X_2(v = 2) \qquad \text{etc.}$$

Such processes as

$$M + X_2(v = 0) \rightarrow M + X_2(v = 2) \quad \text{etc.}$$

are *relatively* unimportant[22]. If $v = k$ denotes the topmost bound vibrational level, then as the upper levels begin to be populated, processes like (say)

$$M + X_2(v = k - 8) \rightarrow M + X_2(v = k - 5)$$

$$M + X_2(v = k - 8) \rightarrow M + X_2(v = k - 4) \quad \text{etc.}$$

become feasible. The importance of multiple jumps will vary from molecule to molecule, being much more likely for molecules with close vibrational spacings. (It should be remembered that although there is a net flow of population upwards through the vibrational ladder, any individual collision will probably result in a deactivation more often than it will result in an activation.) Eventually, dissociation begins to take place by such steps as

$$M + X_2(v = k) \rightarrow M + X + X$$

$$M + X_2(v = k - 1) \rightarrow M + X + X \quad \text{etc.}$$

For convenience we term these states capable of being dissociated as 'interesting'.

The rate constant for dissociation will depend directly on the fraction of the molecules which are in 'interesting' vibrational states. After the shock wave passes, there will be an induction period, since no dissociation can take place at all until some of these levels become populated. As time goes on, the populations of the 'interesting' levels rise, but it seems likely that the dissociating steps will be very fast and that these particular levels will never attain their equilibrium populations until the dissociation itself comes to equilibrium—in aerodynamic jargon we say that the dissociation and the vibrational relaxation are coupled. This means that the rate of dissociation will be a function of the extent of the reaction. Such questions as the length of the induction period, whether or not a steady state is ever established before equilibrium is reached, or the extent to which the observed rate constant will differ from the equilibrium value are still unsettled owing to the difficulty in solving the relevant relaxation equations[37]. If all the vibrational levels were to attain their equilibrium populations before dissociation commenced, it is easy to show that the Arrhenius temperature coefficient of the rate constant would be the heat of dissociation[22]. However, if it is assumed that the 'interesting' levels are somewhat depopulated and the depopulation becomes relatively more severe as the temperature rises, the Arrhenius temperature coefficient will be less than the dissociation energy. Since the actual populations of the 'interesting' states will depend on how efficient M is in its rôle as a supplier of energy, the 'activation energy' will not necessarily be the same for a given molecule for various M, and this is found to be the case experimentally[37]; furthermore, since the 'activation energy' is often as much as 10 kcal below the heat of dissociation, we may conclude that the extent of this depopulation of interesting levels is not trivial. Sometimes two-step processes have been postulated[39], but in fact no satisfactory scheme of consecutive *reversible* processes can account for an 'activation energy' much less than the endothermicity of the reaction. Simple kinetic treatments (particularly if one or more of the reverse steps are ignored) can lead to low 'activation energies', but if all the relevant steps are included the problem becomes that[37] upon which this presentation is based.

This picture is not seriously affected if we extend our model to include rotational energy. If we can ignore very large changes in rotational energy[22], we may rule out direct dissociation from low-lying levels by rotation. Thus, instead of having one vibrational ladder for the molecule to climb, we can imagine that there is a series of ladders, one for each value of the rotational quantum number, J, all with slightly different energy levels. So, instead of climbing one ladder, we can hop from one ladder to an adjacent or near adjacent one, and so on; any such path to the dissociation limit is acceptable (in the special case of *ortho-* or *para*-hydrogen, only jumps between all odd- or all even-numbered ladders are allowed because of the inter-relation between the nuclear-spin statistics and the allowed rotational states of the molecule). There is one complication, however, whose effect has not yet been investigated, and that is the existence of metastable rotation–vibration levels above the dissociation limit. In a rotating anharmonic oscillator, the effective potential curve is that for the rotationless case (i.e. a Morse-type curve) plus a term equal to $J(J+1)/R^2$ (given here in atomic units). Often, for high values of J, the effective potential curve can have a hump in it at large values of R; this means that there will be rotation–vibration levels of the molecule which are *above* the dissociation limit, but which cannot dissociate in the normal way. (If one of the atoms in the molecule is an H atom, there may be a possibility of tunnelling through this hump, though.) Whilst the effect of the existence of these metastable states should be looked into, they would seem to present high-energy paths to dissociation, and as such would tend to raise, rather than lower, the observed 'activation energy'. For this reason, their consideration has been deferred, although there seems no doubt[37] that such states are important in the reverse process—the recombination of two atoms. At this point, it is worth noting the problem, not yet completely settled, of whether the ratio of the rate constants for dissociation and recombination necessarily gives the equilibrium constant. For a fuller discussion the reader is referred to the relevant literature[3,37,40,41] but it seems that the magnitude of the discrepancy, if any, may not be very large.

5.4 DISSOCIATION OF UNDILUTED DIATOMIC GASES

It would be nice if we could be certain that the kind of treatment we have just given was applicable to at least one experimental situation. However, as we have pointed out before, V–V transitions are much more efficient than V–T transitions (and we know nothing about V–R transitions). Therefore, although most dissociation studies on halogens have been carried out in the presence of a hundred-fold excess of inert gas, such processes as (say)

$$X_2(v = 10) + X_2(v = 2) \rightarrow X_2(v = 15) + X_2(v = 0)$$

$$X_2(v = 10) + X_2(v = 2) \rightarrow X_2(v = 7) + X_2(v = 4) \qquad \text{etc.}$$

could be very important. If these processes were only 100 times as efficient as V–T transfers between M and X_2, they would have to be given equal consideration under the experimental conditions of a $100:1$ M:X_2 ratio. Their effect will be to accelerate the vibrational relaxation and reduce the induction period; they may or may not help in reducing the departure of the upper levels from their equilibrium populations during the dissociation process, or in setting up a steady-state condition during the reaction. The presence of these V–V (and possibly V–R) transfers whilst not introducing any further conceptual difficulties does render the kinetic scheme much harder to solve because the whole set of relaxation equations becomes non-linear. On the other hand, it may be possible for the experimental techniques to be so improved that much higher dilutions are possible, and significant advances are being made in this direction[42].

5.5 DISSOCIATION OF POLYATOMIC MOLECULES

It might seem feasible at first sight to generalize the sort of treatment given in the previous section to the polyatomic dissociation case. However, a polyatomic molecule has to be regarded as a system of coupled oscillators, each with many vibrational levels, and associated with each, a large number of rotational states. Apart from the immense task of calculating probabilities of transitions amongst all these (rather ill-defined) states, one has to remember that each rotation–vibration state we consider gives us another differential equation in the relaxation scheme, and whilst it might be possible to solve (say) a thousand simultaneous differential equations for a light diatomic molecule, it would seem to be too difficult to go much further. (Reduction in the magnitude of the problem, on the other hand, could perhaps be effected by grouping together levels of similar energy[43] or by making the [possibly reasonable] assumption that all 'non-interesting' levels have equilibrium populations[44] after the induction period is over.) To make matters more difficult still, the picture is complicated by two additional factors—the storage of vibrational energy within the molecule in modes of vibration which do not directly contribute to the reaction coordinate and the interconversion of energy between various modes. In compensation however, there is some simplification we can effect. It is known that one collision between a highly vibrationally excited molecule and another similar but unexcited molecule may remove, on average, as much as 30 kcal of vibrational energy[13]. Thus, very large amounts of internal energy can be exchanged in single collisions, and the effect of this will be to cause an extremely rapid equilibration of the internal energy. It is usual to assume that in an *undiluted* gas, internal energy is randomized after each collision—equivalent to assuming that after a time long enough for each molecule to have suffered one collision, the internal energies will have Boltzmann distributions. It is unlikely that this limiting behaviour will ever occur in practice. The chemical-activation experiments just discussed refer to molecules having something like 130 kcal excess vibrational energy, which is much more than is encountered in a normal thermal decomposition; however, studies are now being undertaken on molecules having only about 85 kcal excess energy[12] and it will be interesting to see whether it can be established that very large amounts of internal energy can be exchanged in this case also. Nevertheless the 'strong collision approximation'[3] is a reasonable one, and could be correct, as we shall see later, to within a factor of five.

The essential experimental difference between diatomic and polyatomic dissociations is that under all attainable experimental conditions, diatomic dissociations are second order, whereas for polyatomic molecules, they can be second order at low pressures and first order at high pressures[3,7,45,46]. This is a consequence of the storage and redistribution of vibrational energy within the polyatomic molecule. If the pressure is very low, and the time between collisions very long, every molecule which finds itself with more than the critical amount of internal energy after a collision will dissociate. The energy may have been accumulated in various vibrational modes, but because these modes are anharmonic, there is a continuous redistribution of the internal energy between them, and in time, the chances are that sufficient energy will assemble in the mode or combination of modes which represent the dissociation process. Since, under these conditions, every molecule which is activated by a collision decomposes, the reaction is second order. At high pressures, on the other hand, most 'interesting' molecules will be deactivated before they can react. There will then be an equilibrium population of these molecules, and in the time between the activating collision and the next one which (probably) deactivates it, the 'interesting' molecule has a certain chance that the energy will accumulate into the right mode(s) for it to dissociate. Thus, since an equilibrium population of these 'interesting' molecules is always maintained, the chance per unit time of decomposing bears a constant ratio to the number of molecules present, and the reaction is first order. The more complicated the molecule, the more complex are its vibrations and the longer is the time that it will, on average, sit around before its energy will accumulate in such a way that dissociation can occur—hence the lower will be the pressure at which the transition from second- to first-order behaviour occurs. Early theoretical treatments of this change in behaviour, associated with the names of Lindemann and Hinshelwood, assumed that all 'interesting' molecules were equally likely to react, but by the early 1930's, Kassel[47] had reviewed numerous variants of the theory, some of which allowed for the fact that the more energetic amongst the 'interesting' molecules would dissociate more quickly. Here, theory far outstripped experiment, and it was not until the last fifteen years that unambiguous illustrations of the transition from first- to second-order behaviour were found; there are now many examples[3]. An alternative approach, that of Slater[48,49], gained a great deal of impetus from the simultaneous publication of experimental data on the thermal decomposition of cyclopropane[50]. In this theory, the molecule was regarded as a collection of harmonic oscillators, with no coupling between them, and dissociation was assumed to occur when they came into phase in such a way that a certain distance within the molecule became so large that we could regard the reaction as having occurred. The agreement between theoretical prediction and the experimental behaviour of the reaction in the 'fall-off' region was remarkable; (we say 'fall-off' region, because once the pressure is below that which maintains first-order kinetics, the population of 'interesting' molecules is less than its equilibrium value, so that the rate of reaction, expressed as a fraction of those molecules present, falls). The theory lacks what we now must regard as an essential ingredient—that of rapid redistribution of energy amongst modes of vibration for all but the simplest molecules—and for this reason does not adequately describe certain observations. For example, a very symmetric molecule will have lots of degenerate modes of vibration, and the removal of this degeneracy by substitution of one isotopically different atom would, according to the theory, make a large difference; intuitively one would expect, and in fact it is found[51,52], that there is only a small effect. We have therefore fallen back on the earlier theory, which has been developed further by Marcus, and which is now referred to as the RRKM theory (Rice, Ramsperger, Kassel and Marcus): a sufficiently comprehensive review of the application of this theory to unimolecular reactions can be found in Bunker's book[3]. The basic facts are that thermal decomposition reactions are usually first order, but the simpler the molecule, the higher is the pressure at which the unimolecular rate constant begins to fall off. Fall-off behaviour can be fairly easily studied for molecules less complex than about cyclobutane, but one has to go to much simpler molecules before the complete transition from first-order to second-order kinetics can be followed. However, if one goes to too simple a molecule, for example N_2O, not only does it have very little capacity for storage and redistribution of energy, but the strong-collision approximation also becomes a poor one, and the RRKM theory, which can give a reasonable representation for more complex molecules, becomes inadequate[3]. (It is worth noting in passing that the reverse of a dissociation, a combination reaction, shows analogous behaviour. In the recombination of two atoms, the reaction is always third order, since unless a third body comes along and removes some energy from the colliding pair of atoms, they will redissociate at the end of the collision [we are assuming that radiative stabilization is unimportant]. On the other hand, a pair of complex radicals combine according to second-order kinetics because the excess vibrational energy can flow out into the various modes of the newly-formed molecule, and it will take so long for the energy to reassemble in the dissociating mode(s) that stabilization by collision will happen first. The recombination of two CH_3 radicals, or of two CF_3 radicals, falls between these two extremes[53].) In principle, although it has not been observed yet, when we enter this weak-collision regime, the characteristics of the change in order of the reaction with pressure should depend on temperature, since energy-transfer efficiencies are temperature dependent, and the magnitude and direction of the effect will depend on the magnitude and sign of the temperature coefficient of the energy-transfer efficiency. This region in between the strong-collision and weak-collision regimes will obviously be an important experimental field at some time in the future.

We have said earlier that the strong-collision assumption was probably good to within a factor of five for molecules of reasonable complexity. Apart from the fact that the strong-collision assumption seems to give an adequate description of unimolecular reactions within the framework of the RRKM theory[57], there is also some supporting experimental evidence. When a reaction is being carried out in the fall-off region, where there are not enough collisions to maintain an equilibrium population of 'interesting' molecules, the addition of an inert gas causes the rate of the reaction to rise because the inert gas can help to repopulate the 'interesting' levels by collisional-energy transfer. A comparison of the increase in rate constant for the addition of some inert gas against the addition of an equivalent amount[3] of the reactant, tells us the relative efficiency of the additive and of the reactant in transferring 'interesting' amounts of energy to and from reactant molecules. It is usually found that molecules of the reactant gas are the most effective (the reason for this is not very clear) but there are a few molecules which are slightly more efficient—for example, mesitylene would appear to be about 25 % more effective than cyclobutane in the thermal decomposition of cyclobutane[54]. Clearly then, the strong-collision assumption is not absolutely correct, although it could still be very nearly so. However, some experiments on the thermal decomposition of cyclopropane at very low pressures suggest that the assumption is not quite as good as this[55]. If the pressure is reduced sufficiently, the situation arises when the mean free path of the molecules becomes the same as the diameter of the reaction vessel, and we must include collisions with the reaction-vessel walls in our energization scheme. Once energization becomes principally a wall process, instead of a gas-phase one, the transition from first-order towards second-order behaviour with falling pressure ceases, and the reaction becomes first order again at the lowest pressures. If wall $\leftrightarrow$ cyclopropane and cyclopropane $\leftrightarrow$ cyclopropane collisions were equally effective in the energization process, the changeover in behaviour would occur when the reaction-vessel size equalled the mean free path—however, it was observed that the changeover took place at slightly higher pressures, consistent with a wall efficiency some 3 to 5 times greater than that of cyclopropane. Clearly then, it must take at least 3 to 5 cyclopropane–cyclopropane collisions to randomize the internal energy completely, and until more experiments have been done, the possibility still remains that it could take more. Note also that the best fit of RRKM theory to the experimental data occurs[57] for cyclo-propane with a value of 1 collision in every 4. Another approach to this problem is currently being tested[56] where one has a low-pressure flow system in which the molecules suffer known numbers of collisions with the hot walls of the reaction vessel before the products are analysed with a mass-spectrometer.

There are three other aspects of polyatomic-molecule decompositions which warrant further investigation, both experimental and theoretical—the phenomena of alternative decomposition paths, simultaneous breaking of two or more bonds and the problem of abnormal Arrhenius parameters. The phenomenon of alternative reaction paths does not present us with much conceptual difficulty: for example, we consider the molecule CCl_3CH_3 which could conceivably split up in the following ways

1. $CCl_3CH_3 \rightarrow CCl_3CH_2 + H$ $E \approx 100 \, kcal$
2. $CCl_3CH_3 \rightarrow CCl_3 + CH_3$ $E \approx 85 \, kcal$
3. $CCl_3CH_3 \rightarrow CCl_2CH_3 + Cl$ $E \approx 70 \, kcal$
4. $CCl_3CH_3 \rightarrow CCl_2CH_2 + HCl$ $E \approx 50 \, kcal$
5. $CCl_3CH_3 \rightarrow CCl_2 + CH_3Cl$ $E \approx ?$
6. $CCl_3CH_3 \rightarrow CCl_3H + CH_2$ $E \approx ?$

Clearly the step with the lowest activation energy (reaction 4) will dominate in the thermal decomposition although there is no reason why some or all of the others might not be observed in chemical activation studies[12,58-60]. However, it often happens that two alternative paths have virtually the same activation energy, as for example in the thermal decomposition of tert-butylamine which can eliminate NH_3 or CH_4 with about equal ease[61], and there are many examples of such behaviour in the cycloalkane series[62]. Clearly, if both alternative processes required identical activation energy, the relative rates of decomposition would be governed by a statistical factor which determined the relative chances of assembling the internal energy into the correct sets of vibrational modes. A study of the fall-off characteristics of such pairs of reactions might be quite interesting, particularly at very low pressures.

The problem of simultaneous dissociation arises when a molecule can break into three or more fragments. In the thermal decomposition of acetone, for example, the activation energy[63] corresponds to the heat of dissociation of one C–C bond in the molecule, and it is fairly safe to assume that the overall reaction

$$CH_3COCH_3 \rightarrow CO + 2CH_3$$

takes place by two consecutive reactions

$$CH_3COCH_3 \rightarrow CH_3CO + CH_3$$

$$CH_3CO \rightarrow CO + CH_3$$

the radical CH_3CO is known to exist at lower temperatures, and there seems no problem in assuming that it has a short but finite lifetime at the dissociation temperature. (Basically, the difference between this reaction scheme and that for the diatomic dissociation is that this one is made deliberately irreversible by adding a radical trap like toluene, or in any case the products disappear by other processes which are not the reverse of the dissociation reactions.) On the other hand, in the thermal decomposition of azomethane, the activation energy E is some 20–30 kcal more than the endothermicity D of the reaction[45,64,65]

$$CH_3N{=}NCH_3 \rightarrow N_2 + 2CH_3$$

If we postulate a stepwise process, we must conclude that the $CH_3{-}N{=}N$. radical is vibrationally excited to the extent of $(E{-}D)$ kcal and will decompose very rapidly indeed. The question is whether in the departure of the two methyl radicals from the N_2 molecule, the trajectories are in phase or not, and if they are in phase, will this have any observable effect? Possible situations where simultaneous dissociation could occur might be for molecules like the mercury dialkyls or di-*tert*-butylperoxide, although there is no conclusive evidence on this matter at the present time; photochemical dissociations could also occur in this way[66].

The occurrence of high frequency factors in unimolecular decompositions does however present a serious unsolved problem. Frequency factors of around 10^{13} sec^{-1} are usually regarded as normal[45,46], based on the argument[67] that the vibrations which lead to decomposition have cyclic frequencies of the order 10^{12}–10^{14} sec^{-1}. However, frequency factors well outside this range do occur, and whilst those which are very low seem well understood[45,46], there is still a problem about those which are very high. The thermal decomposition of di-*tert*-butylperoxide has probably been subjected to more studies than any other reaction of this type, and whilst frequency factors and activation energies are known to be pressure dependent[45,65,68], there seems no doubt that for this reaction the activation energy is about 39 kcal and the frequency factor is about 10^{17} sec^{-1} (despite some small side reactions which are not fully understood[58,69]). It is easy to manipulate a Kassel-type theory to give high-frequency factors[46,70,71]. All one needs to say is that the activation energy needed for the dissociation process does not have to be concentrated in one vibration (or squared term)[45,47], but may be distributed amongst several degrees of freedom; we have left the door open for this assumption throughout this review. Rotation in the activated state is one idea which has been suggested with reference to the high-frequency factor for the dissociation

$$C_2H_6 \rightarrow CH_3 + CH_3$$

(If one takes the measured Arrhenius parameters for the reverse recombination reaction, and assumes that the reaction is reversible with $E_f = D$ and $E_r = 0$, one finds[45,46] $A_f = 10^{17}$ sec^{-1} at room temperature: one has of course to assume that a reaction which takes place in 10^{40} years, and its reverse which takes place every collision, can be regarded as reversible; the actual decomposition of ethane at higher temperatures is not fully understood[85], but the initiation step does appear to have a high frequency factor. If this reaction were to take place, there is no doubt that the two methyl groups would have very considerable rotational motion before dissociation occurred. However, it is hard to see what contribution rotation about the internuclear axis could really make to the process, since it will be essentially a free rotation; on the other hand, rotations about the other two axes would lead to dissociation, but it is hard then to see that any consideration of the vibrations within the groups would have much relevance to the process. It has also been suggested that a similar manipulation of this argument to describe simultaneous rupture of two or more bonds can account for high-frequency factors[46,70]. However, we are awaiting a decisive experiment or a new advance in the theory, but the possibility[22] that perhaps some non-equilibrium effect may be involved should not be forgotten.

5.6 BIMOLECULAR REACTIONS

Very little is known about whether vibrational energy of one or both molecules can contribute to the activation process in a bimolecular-exchange reaction. It had been suggested that such a possibility could account for the high-frequency factors found in certain bromination reactions[72], but it was later found that the experiments were wrong[73]. However, the possibility remains that internal energy can be used as energy of activation[72]. Some recent papers[74] on isotopic exchange between simple molecules in shock waves do point in this direction and one can hope that one day molecular-beam experiments will provide further information in this area[75,76]. Of course, we cannot discuss the simplest bimolecular reaction, a diatomic dissociation, without considering the vibrational energy of the reacting molecules.

5.7 CONSERVATION OF SPIN IN CHEMICAL REACTIONS

The thermal dissociation of the CO_2 molecule has been studied a number of times in shock waves. In one study[42], where the concentration of CO_2 in argon was 0·04%, thus minimizing (although perhaps not eliminating) V–V transitions, the temperature coefficient of the dissociation rate-constant corresponds to an activation energy of 99 kcal/mole. The CO_2 molecule may dissociate into CO and O in many ways, the lowest energy paths being

$$CO_2 \rightarrow CO + O(^3P) \qquad D \approx 125 \text{ kcal}$$
$$CO_2 \rightarrow CO + O(^1D) \qquad D \approx 171 \text{ kcal}$$

In the latter process spin is conserved, but the observed activation energy suggests, in the light of the ideas presented earlier in this paper, that really the former process is the one which is taking place, and that allowing for depopulation of the upper levels of the CO_2 molecule, the value of 99 kcal is an approximation to 125 rather than 171 kcal. If this is the case, spin is not conserved in this process. The thermal decomposition of N_2O is another example[46] of dissociation in which spin is obviously not conserved. Also, there are several ion reactions in which spin is not conserved[77]. Kineticists have usually tended to think that kinetic processes take place more or less in accordance with spectroscopic selection rules[7], although there is no a priori reason why they should when we consider the very severe perturbations that exist in a reaction process. The simplest way to discuss spin conservation is to look at the reverse of a dissociation process, the recombination of two H atoms. According to stationary-state quantum mechanics, as two H atoms are brought together, there is one chance in four that the interaction will be attractive ($^1\Sigma$ state) and three chances in four that it will be repulsive ($^3\Sigma$ state). In the absence of emission processes, one then argues that only one collision in four can lead to recombination, since collision complexes which are in the $^3\Sigma$ states cannot be collisionally deactivated into the $^1\Sigma$ ground state. The unsatisfactory consequences of such an assumption when applied to more complex atoms have been noted by Bunker[3]. However, it might be that as these two H atoms approach each other with their spins parallel, the electromagnetic field caused by one moving collection of spins and charges (i.e. a H atom) could cause the spin of the other H atom to flip, as in an e.s.r. experiment. Thus, two atoms which initially started off on a repulsive trajectory ($^3\Sigma$) might suddenly find themselves on an attractive collision trajectory ($^1\Sigma$) and could then be deactivated collisionally to one of the bound vibrational levels in the normal way. Of course, both energy and angular momentum would have to be conserved in the process, so that the loss in spin angular momentum in the system would have to appear as rotational angular momentum of the collision complex. This amount of rotational angular momentum would correspond, for any given configuration of the complex, to a certain amount of extra rotational energy, so that energy will be conserved when the separation between the repulsive and attractive curves corresponds to this amount of energy; thus the internuclear separation at which such a transfer could occur would vary with the initial-impact parameters. It might even be that such a transfer would be very fast, in which case we would have the satisfying situation that all collisions could lead to recombination, given a suitable deactivating collision with a third body. We may make a crude estimate of the likelihood of such transitions by looking at the spin–orbit interaction constants for the atoms concerned. The separation between $I(5^2P_{1/2})$ and $I(5^2P_{3/2})$ is 7600 cm^{-1}, which is equivalent to an interconversion frequency of about 2.5×10^{14} cycles/sec, whereas the separation between $H(2^2P_{1/2})$ and $H(2^2P_{3/2})$ is 0.36 cm^{-1}, corresponding to about 10^{10} cycles/sec. Since a repulsive collision would last of the order of 10^{-12}–10^{-13} seconds, it would seem that a spin-change process for I atoms would be quite likely, but maybe not for two H atoms. Some recent calculations on a slightly different problem, that of the collision between two ground-state H atoms in which the two atoms eventually separate, suggests that the cross-section for a change in spin is of the order of gas-kinetic collision cross-sections[78].

Experimental evidence on such combination processes is unfortunately not very helpful. Detailed studies of emission in Cl atom and Br atom recombinations have been made[79-82], but have thrown no light on this particular aspect of the problem.

An interesting application of this spin-change phenomenon, if it happens easily, would be to the collisional deactivation of excited electronic states of molecules. Collisional deactivation of excited states of atoms is inefficient[28], but for molecules it can be quite an efficient process[11,83,84]. A spin ↔ rotation interconversion cannot occur for an atom, but it could occur in an excited molecule either intramolecularly or on collision, and would provide a convenient explanation of the inter-system crossing processes that are known to occur so readily and of these rapid collisional deactivation processes.

REFERENCES[86]

1. F. A. Lindemann, *Trans. Faraday Soc.*, **17**, 598 (1922).
2. J. A. Christiansen and H. A. Kramers, *Z. Physik. Chem.*, **104**, 451 (1923).
3. D. L. Bunker, *Theory of Elementary Gas Reaction Rates*, Pergamon, London, 1966.
4. T. L. Cottrell and J. C. McCoubrey, *Molecular Energy Transfer in Gases*, Butterworths, London, 1961.
5. M. W. Windsor, N. R. Davidson and R. L. Taylor, *Symp. Combust.*, *7th*, London–Oxford, 1958, p. 80; see also R. L. Taylor, M. Camac and R. M. Feinberg, *Symp. Combust.*, *11th*, Berkeley, 1966, p. 49.
6. A. Euken and R. Becker, *Z. Physik. Chem.*, **B27**, 235 (1934).
7. V. N. Kondratiev, *Chemical Kinetics of Gas Reactions*, Pergamon, London, 1964.
8. J. H. Kiefer and R. W. Lutz, *Symp. Combust.*, *11th*, Berkeley, 1966, p. 67.
9. I. R. Hurle, *Symp. Combust.*, *11th*, Berkeley, 1966, p. 827.
10. A. W. Read, *Progr. Reaction Kinetics*, **3**, 705 (1965).
11. D. H. Shaw and H. O. Pritchard, *J. Phys. Chem.*, **70**, 1230 (1966).
12. G. O. Pritchard, M. Venugopalan and T. F. Graham, *J. Phys. Chem.*, **68**, 1786 (1964).
13. M. C. Flowers and B. S. Rabinovitch, *Quart. Rev. (London)*, **18**, 122 (1964).
14. T. L. Cottrell, I. M. Macfarlane, A. W. Read and A. H. Young, *Trans. Faraday Soc.*, **62**, 2655 (1966).
15. P. E. Charters and J. C. Polanyi, *Discussions Faraday Soc.*, **33**, 107 (1962).
16. C. C. Chow and E. F. Greene, *J. Chem. Phys.*, **43**, 324 (1965).
17. B. P. Levitt and D. B. Sheen, *Chem. Soc. (London) Spec. Publ.* **20**, 269 (1966).
18. I. R. Hurle, *Chem. Soc. (London) Spec. Publ.* **20**, 276 (1966).

19. I. I. Glass, unpublished data.
20. H. O. Pritchard, *J. Phys. Chem.*, **66**, 2111 (1962).
21. T. Carrington, *J. Chem. Phys.*, **35**, 807 (1961).
22. H. O. Pritchard, *J. Phys. Chem.*, **65**, 504 (1961).
23. See A. B. Callear and G. J. Williams, *Trans. Faraday Soc.*, **62**, 2030 (1966) for references to this work.
24. T. L. Cottrell, *Dynamic Aspects of Molecular Energy States*, Oliver and Boyd, Edinburgh, 1965.
25. R. C. Millikan, *Chem. Soc. (London) Spec. Publ.* **20**, 219 (1966).
26. R. Campargue, *Chem. Soc. (London) Spec. Publ.* **20**, 287 (1966).
27. A. B. Callear and W. J. R. Tyerman, *Trans. Faraday Soc.*, **62**, 2313 (1966).
28. R. J. Donovan and D. Husain, *Trans. Faraday Soc.*, **62**, 2023 (1966).
29. E. W. Montroll and K. E. Shuler, *J. Chem. Phys.*, **26**, 454 (1957).
30. K. E. Shuler and G. H. Weiss, *J. Chem. Phys.*, **45**, 1105 (1966).
31. R. G. Gordon, *J. Chem. Phys.*, **44**, 1830 (1966); M. L. Unland and W. H. Flygare, *J. Chem. Phys.*, **45**, 2421 (1966); see also T. A. Bak, reference 40.
32. K. F. Herzfeld and T. A. Litovitz, *Absorption and Dispersion of Ultrasonic Waves*, Academic Press, New York, 1959; see also K. Takayanagi, *Advan. At. Mol. Phys.*, **1**, 149 (1965).
33. D. Rapp and T. E. Sharp, *Symp. Combust.*, *11th*, Berkeley, 1966, p. 77.
34. H. O. Pritchard, *Quart. Rev. (London)*, **14**, 46 (1960).
35. E. F. Greene and J. P. Toennies, *Chemical Reactions in Shock Waves*, Academic Press, New York, 1964.
36. J. N. Bradley, *Shock Waves in Chemistry and Physics*, Methuen, London, 1962.
37. D. G. Rush and H. O. Pritchard, *Symp. Combust.*, *11th*, Berkeley, 1966, p. 13.
38. K. N. C. Bray and N. H. Pratt, *Symp. Combust.*, *11th*, Berkeley, 1966, p. 23.
39. E. S. Fishburne, K. R. Bilwakesh and R. Edse, *J. Chem. Phys.*, **45**, 160 (1966).
40. B. Widom, *Science*, **148**, 1555 (1965); see also T. A. Bak, *Contributions to the theory of chemical kinetics*, W. A. Benjamin, New York, 1963, for a discussion of non-equilibrium effects and oscillations.
41. R. M. Krupka, H. Kaplan and K. J. Laidler, *Trans. Faraday Soc.*, **62**, 2754 (1966).
42. H. A. Olschewski, J. Troe and H. G. Wagner, *Symp. Combust.*, *11th*, Berkeley, 1966, p. 155.
43. P. C. Haarhoff, *Mol. Phys.*, **7**, 101 (1963).
44. W. G. Valance and E. W. Schlag, *J. Chem. Phys.*, **45**, 216 (1966).
45. A. F. Trotman-Dickenson, *Gas Kinetics*, Butterworths, London, 1955.
46. K. J. Laidler, *Chemical Kinetics*, 2nd ed., McGraw-Hill, New York, 1965.
47. L. S. Kassel, *Kinetics of Homogeneous Gas Reactions*, Chemical Catalog Co., New York, 1932.
48. N. B. Slater, *Proc. Roy. Soc. (London)*, *Ser. A*, **218**, 224 (1953).
49. N. B. Slater, *Theory of Unimolecular Reactions*, Cornell University Press, Ithaca, 1959.
50. H. O. Pritchard, R. G. Sowden and A. F. Trotman-Dickenson, *Proc. Roy. Soc. (London)*, *Ser. A*, **217**, 563 (1953).
51. E. W. Schlag and B. S. Rabinovitch, *J. Am. Chem. Soc.*, **82**, 5996 (1960).

52. see F. J. Fletcher, B. S. Rabinovitch, K. W. Watkins and D. J. Locker, *J. Phys. Chem.*, **70**, 2823 (1966) and B. H. Mahan, *Ann. Revs. Phys. Chem.*, **17**, 173 (1966) for further references.
53. K. J. Ivin and E. W. R. Steacie, *Proc. Roy. Soc. (London), Ser. A*, **208**, 25 (1951); N. Arthur and T. Bell, *Chem. Comm.*, **1965**, 166.
54. H. O. Pritchard, R. G. Sowden and A. F. Trotman-Dickenson, *Proc. Roy. Soc. (London), Ser. A*, **218**, 416 (1953); see also reference 52.
55. A. D. Kennedy and H. O. Pritchard, *J. Phys. Chem.*, **67**, 161 (1963).
56. S. W. Benson and G. N. Spokes, *Symp. Combust., 11th*, Berkeley, 1966, p. 95.
57. G. M. Wieder and R. A. Marcus, *J. Chem. Phys.*, **37**, 1835 (1962).
58. D. M. Tomkinson and H. O. Pritchard, *J. Phys. Chem.*, **70**, 1579 (1966).
59. J. C. Hassler and D. W. Setser, *J. Chem. Phys.*, **45**, 3237, 3246 (1966).
60. G. O. Pritchard, J. L. Bryant and R. C. Thommarson, *J. Phys. Chem.*, **69**, 2804 (1965).
61. H. O. Pritchard, R. G. Sowden and A. F. Trotman-Dickenson, *J. Chem. Soc.*, **1954**, 546.
62. R. J. Ellis and H. M. Frey, *J. Chem. Soc.*, **1964**, 4184.
63. D. Clark and H. O. Pritchard, *J. Chem. Soc.*, **1956**, 2136.
64. M. Page, H. O. Pritchard and A. F. Trotman-Dickenson, *J. Chem. Soc.*, **1953**, 3878.
65. C. Steel and A. F. Trotman-Dickenson, *J. Chem. Soc.*, **1959**, 975.
66. P. G. Bowers and G. B. Porter, *J. Phys. Chem.*, **70**, 1622 (1966).
67. M. Polanyi and E. Wigner, *Z. Physik. Chem.*, **A139**, 439 (1928).
68. B. M. H. Billinge and B. G. Gowenlock, *Trans. Faraday Soc.*, **59**, 690 (1963).
69. S. W. Benson, *J. Chem. Phys.*, **40**, 1007 (1964).
70. H. O. Pritchard, *J. Chem. Phys.*, **25**, 267 (1956).
71. C. Steel, *J. Chem. Phys.*, **31**, 899 (1959).
72. H. O. Pritchard, *Rec. Trav. Chim. Pays Bas.*, **74**, 779 (1955).
73. G. C. Fettis, J. H. Knox and A. F. Trotman-Dickenson, *J. Chem. Soc.*, **1960**, 4177.
74. S. H. Bauer, *Symp. Combust., 11th*, Berkeley, 1966, p. 105.
75. H. O. Pritchard, *Discussions Faraday Soc.*, **33**, 278 (1962).
76. D. R. Hershbach, *Discussions Faraday Soc.*, **33**, 278 (1962).
77. J. B. Hasted, *Physics of Atomic Collisions*, Butterworths, London, 1964.
78. F. J. Smith, *Planet Space Sci.*, **14**, 929 (1966).
79. H. B. Palmer, *J. Chem. Phys.*, **26**, 648 (1957).
80. L. W. Bader and E. A. Ogryzlo, *J. Chem. Phys.*, **41**, 2926 (1964).
81. D. B. Gibbs and E. A. Ogryzlo, *Can. J. Chem.*, **43**, 1905 (1965).
82. E. Hutton and M. Wright, *Trans. Faraday Soc.*, **61**, 78 (1965).
83. P. Seybold and M. Gouterman, *Chem. Reviews*, **65**, 413 (1965).
84. S. K. Laver and M. A. El-Sayed, *Chem. Reviews*, **66**, 199 (1966).
85. M. C. Lin and M. H. Back, *Can. J. Chem.*, **44**, 505 (1966).
86. Literature survey completed November, 1966.

Author Index

The numbers in square brackets refer to the reference numbers under which the Author's work is quoted in full at the end of the chapter.

Akhren, A. A. 323, 334[82]
Albaceete, L. M. 118[24]
Allen, R. T. 69, 110[36], 131[36], 156[36]
Allport, J. J. 221[99]
Alterman, E. B. 88
Alyamovskii, V. N. 219[82]
Amme, R. C. 105, 109, 112, 114, 115, 159
Anantaraman, A. V. 310[65]
Andrae, J. H. 285[14], 286, 290[20, 24, 25], 293[43], 306, 312, 327, 349[114], 350 [114], 352, 353[114], 354[121], 357 [114]
Andreiss, J. H. 362
Angona, F. A. 38, 362
Antripov, E. T. 219[82]
Appleton, J. P. 109, 221, 234, 242[102], 245, 258
Arbuzor, B. A. 337[84]
Arnold, J. W. 235[126], 255[175]
Arthur, N. 381[53]
Atkinson, G. 290[27]
Aybar, S. 126, 128

Bader, L. W. 387[80]
Bair, E. J. 233
Bak, T. A. 377[40]
Barone, A. 291[39]
Bartlett, M. S. 69
Basco, N. 153[253], 156[253], 242[144], 252, 253
Bass, R. 122[189], 290[25], 349[115], 350 [115], 351[115], 352, 353, 356[115], 359, 360

Bauer, E. 71, 82
Bauer, H. J. 4[7], 7, 8, 17, 21, 22, 23, 24[14], 25, 26, 42[44], 45[7], 90, 96, 97, 99, 105, 108, 113, 116, 119, 142, 143, 144[100], 145, 153, 168[171], 242[145], 249, 252, 253
Bauer, S. H. 104, 180, 152, 385[74]
Becker, R. 114[154], 115, 369[6]
Belinebaza, L. E. 343[99]
Belinstan, 343[99]
Bell, T. 381[53]
Benedek, G. 296[56]
Bennewitz, H. J. 169
Benson, G. C. 346, 348[109]
Benson, S. W. 63, 72[46], 105, 107, 109, 111, 112, 114, 166, 167, 229, 235, 237, 239, 240, 382[56], 384[69]
Bentley, F. F. 328[73], 329
Bentley, J. B. 340[89]
Berdyev, A. A. 290[29]
Berend, G. C. 63, 72[46], 105, 107, 109, 111, 112, 114, 166, 167, 229, 235, 237, 239, 240
Berlincourt, D. A. 28
Berne, B. J. 59[269]
Besse, A. L. 214
Bethe, H. A. 228[109]
Beyer, R. T. 18, 291[37], 291[38]
Bhangu, J. K. 185, 220[14], 242[14], 255, 256
Billinge, B. M. H. 384[68]
Bilwakesh, K. R. 376[39]
Biquard, P. 290

391

Bird, R. B. 74, 95[50], 106[50], 107[50], 124[50], 127[50], 155[50], 357
Blackman, V. H. 213[61], 235, 236, 237, 245, 247, 255, 256
Blitz, J. 31
Blythe, A. R. 28, 29, 30, 110, 131
Blythe, P. A. 117, 185[15], 216, 230, 256
Boade, R. R. 134, 139[225], 159, 160, 230
de Boer, P. C. T. 216, 256
Bommel, H. F. 290[28]
Borrell, P. 60[270], 103, 112, 180[5], 219 [80], 220[91, 96], 234, 240[91], 241, 242 [19, 86], 254, 256[91], 257
Boudart, M. 44, 120, 146, 147, 249[164]
Bowers, P. G. 383[66]
Bradley, J. N. 193[43], 206, 209[43], 374 [36]
Bray, K. N. C. 374[38]
Breeze, J. C. 220[92]
Brewer, R. G. 296[50]
Brickl, D. 255, 256
Bridgman, K. B. 117, 118[165], 119
Bristow, M. 214
Brown, H. C. 335[88]
Brown, R. L. 192[40], 234
Brugger, R. M. 137
Bryant, J. L. 382[60]
Bucaro, J. A. 132, 353[120, 127], 354[120, 124, 127], 355
Bugrim, E. D. 191
Bunker, D. L. 368, 371[3], 377[3], 378[3], 379[3], 380, 381[3], 386[3]
Burks, J. A. 152
Burnett, D. 19
Burton, C. J. 291[32]
Busala, A. 25[41], 31, 131, 133, 134
Byron, S. 246[155]

Callear, A. B. 6, 45[102, 103], 46, 47[10, 103], 91, 113[147], 153, 155, 156[253], 191, 219, 233, 242[144], 252, 254, 372 [23, 27]
Calvert, J. B. 105, 109, 112, 114, 147, 159
Camac, M. 153[287], 183, 242[148], 256
Campargue, R. 372[26]
Campbell, G. S. 206

Carey, C. 219[83]
Carnevale, E. A. 219[83], 335, 363
Carome, E. F. 132, 353[127, 128], 354, 355, 362[128]
Carrington, T. 371[21]
Carstensen, E. L. 294
Cary, B. 109, 111, 112, 182[17], 237, 246 [17, 254]
Cerceo, M. 284[12]
Chamberlain, J. W. 154[256]
Chandler, T. R. D. 117, 118[165], 119
Chareyre, R. 246
Chen, J. H. 318, 323, 325
Cheo, P. K. 258
Chiao, R. Y. 296[52, 53], 349[113]
Chow, C. C. 112, 221, 222, 238, 242[101], 234, 254, 258, 370[16]
Christiansen, J. A. 386[2]
Clark, A. E. 325
Clark, A. V. 42, 120, 146, 149, 249[163]
Clark, D. 383[63]
Clark, N. A. 132, 354[124]
Clarke, J. F. 60, 69[9], 99[9]
Cole, K. S. 23
Corran, P. G. 122[184]
Cottrell, T. L. 6, 7, 10, 18, 28, 30, 31, 50, 51, 52, 60, 68, 72, 86[4], 101, 102[4], 110, 112[4], 116, 117, 119, 120, 123[4], 124, 125, 126, 127, 128, 130, 131, 133, 138, 151, 152, 164, 165, 168[200], 189, 191 [31], 228, 229, 240[129], 256, 257, 259, 369[4], 370[14], 371[4], 372[24]
Coulson, C. A. 8
Cramer, K. H. 169
Cross, P. C. 66, 80, 169[60]
Crum, L. A. 348[111]
Cummin, H. Z. 353[117]
Cummings, F. W. 82
Curran, D. R. 28
Curtiss, C. F. 74, 95[50], 106[50], 107[50], 124[50], 127[50], 155[50], 357

Daen, J. 215, 256
Darbari, G. S. 333[80]
Dardy, H. D. 132, 133, 290, 352, 353[120, 127], 354[120, 123, 126, 127], 355

Davies, C. M. 356[131]
Davies, R. O. 18, 90, 303[59], 326
Davison, A. D. 328[73], 329
Davidson, N. R. 219, 220[84, 85], 233, 249, 242[84], 369[5]
Dauben, W. J. 329
Day, M. A. 28, 30, 119, 120[172], 126, 128, 151, 240[129], 259
Debye, P. 290
Decius, J. C. 43, 49, 80, 125, 126[195], 169[60], 220, 232, 255[176]
Devonshire, A. F. 75
Dickens, P. G. 74[51], 76[53], 105, 114, 115, 121, 122, 123, 124, 132, 219
Dobbie, R. C. 125, 126[197], 165, 229 [110]
Donnelly, G. J. 42
Donovan, R. J. 373[28]
Doyennette, L. 49, 109, 110, 112[127], 183[9], 250
Dransfield, K. 290[28]
Ducuing, J. 53, 119, 193, 242[42], 243, 244

Eckart, C. 287[15]
Edmonds, P. D. 39, 40[83], 126, 128, 290 [20], 293[43], 310[65], 312[20], 327[71]
Edse, R. 376[39]
Edwards, A. J. 25, 26[45], 30[45], 138 [233], 141[233], 155, 160, 161
Eggers, P. 285
Ehlers, J. G. 206[52]
Ellis, R. J. 382[62]
El-Sayed, M. A. 387[84]
Englander-Golden, P. 86, 153[68]
Eucken, A. 114[154], 115, 126, 128, 152, 255, 369[6]
Everard, K. B. 340[89]
Eyring, H. 357, 358

Faizulhov, F. S. 219[82]
Fakelensbi, I. L. 296[54]
Ferguson, M. G. 49, 50, 109, 110, 112, 183[10, 12], 190, 192[10], 250, 254
Feinberg, R. M. 153[287]
Ferriso, C. C. 220[92]
Fettis, G. C. 385[73]
Feuer, P. 69, 110[36], 131[36], 156[36]

Fink, U. 296[55]
Fishburne, E. S. 376[39]
Fitzsimmons, R. V. 233
Fletcher, F. J. 380[52]
Flowers, M. C. 370[13], 378[13]
Flygare, W. H. 373[31]
Flynn, G. W. 53, 117, 192[39], 258[39]
Fox, J. N. 207, 210
Frank, N. H. 64[19]
Freedman, E. 347, 348
Frey, H. M. 382[62]
Fritsche, L. 39, 42
Fujii, Y. 120, 121

Galt, J. K. 288
Gammon, R. W. 353[117]
Garmine, E. 296[51]
Gasilevich, E. S. 116
Gaydon, A. G. 109, 110, 193[44], 206, 209 [44], 213[44], 216, 217, 218, 237, 242 [140], 243, 249, 257
Geide, K. 121
Generalov, N. A. 105, 108, 242[136], 247, 249
Gerrard, J. H. 117, 185[15], 230[15], 256, 257[15], 258[16]
Gerstern, M. 335[88]
Gibbs, D. B. 387[81]
Glass, I. I. 193[45], 206, 371[19]
Gordon, R. 59[269]
Gordon, R. G. 373[31]
Gordon, S. 206[52]
Gorelik, G. 49
Gornall, W. S. 356[132]
Gouterman, M. 387[83]
Gowenlock, B. G. 384[68]
Graham, T. F. 370[12], 378[12] 382[12]
Gravitt, J. C. 126, 128, 130
Green, J. A. 45[102]
Greene, F. E. 112, 193[46], 206, 209[46], 221, 222, 234, 238, 242[101], 254, 258, 370[16], 374[35]
Greenspan, M. 19, 21, 36, 37
Greenspan, W. D. 256
Greytak, T. 296[56]
Griffith, W. 44, 255, 256

Grigor, A. F. 130
de Groot, M. S. 303[61], 338, 339, 340, 342
Guenoche, H. 246
Gutteridge, R. 219[86], 220[86], 234, 242 [86], 254
Guzahvina, G. A. 337[84]

Haarhoof, P. C. 378[43]
Haebel, E. U. 134
Hageseth, G. T. 132
Hall, D. N. 188[18], 245, 287, 288, 343
Hal'yenov, B. I. 293[44]
Hames, G. R. 353[119], 354[119, 125]
Hanes, J. R. 132
Harlow, R. G. 42, 146
Harris, E. L. 149, 188[24]
Hasted, J. B. 385[77]
Hassler, H. 32[64], 382[58, 59]
Hartmann, B. 89
Hayashi, M. 329
Hazzard, G. W. 291[34]
Heasell, E. L. 290[25], 303[58], 306, 334, 335, 336, 337, 349[114], 350[114], 351 [58], 352, 353[114], 357[114], 362
Heimel, S. 206[52]
Henderson, M. C. 42, 105, 108, 110, 111, 120, 122[190], 146, 147, 149, 183, 249 [163], 357[139, 140], 358, 360, 362
Henry, L. 49, 109, 110, 112[127], 183[9], 250
Herman, R. 4, 226
Hershbach, D. R. 385[76]
Herzberg, G. 220[90], 224[90]
Herzfeld, K. F. 1, 6, 7, 8, 11, 13, 17, 18, 21, 26, 60, 61[10], 63[10], 66, 68[3], 71, 72, 74[3], 75[3, 10], 76[3], 77[3], 78, 81, 83, 90[3], 90[57], 93, 103, 107, 114, 118, 121[61], 146, 147, 149, 152, 189, 228, 229, 266[1], 350[1], 397[1], 361, 373[32]
Hess, H.-D. 126
Higgs, R. W. 151[246], 152
Hinsch, H. 19, 21, 32, 38[31], 126, 133, 134
Hirschfelder, J. O. 74, 95[50], 106[50], 107[50], 110, 124[50], 127[50], 155[50], 357, 358
Hobson, R. M. 207, 210

Hocker, L. O. 53, 117, 192, 258
Hodkinson, T. B. 138[284], 165[284]
Holbeche, T. A. 188, 189, 245
Holmes, J. G. 245
Holmes, R. 19, 24, 32, 36[68], 37, 40, 105, 107, 108, 112, 119, 134, 136, 137, 138, 248
Hooker, W. J. 109, 110, 153, 219[87], 220[87], 233, 242[87], 246[87], 249, 250
Houghton, J. T. 53, 117
Hu, B. L. 117, 119[164], 110[164], 152 [164], 171[164]
Hubbard, J. C. 131, 133, 134, 293[42]
Huber, P. W. 44, 237, 245[150], 246[150]
Huddard, D. H. 343[93], 346
Hudson, G. H. 136
Hunter, J. L. 132, 133, 290, 352, 354[120, 123, 126, 127], 355
Hunter, T. F. 51[117], 127
Hurle, I. R. 109, 110, 111, 188, 193[44], 206, 209[44], 213[44], 216, 217, 218, 236, 237, 242[67, 140, 141], 243, 245, 246, 256, 257, 370[9]
Husain, D. 373[28]
Hutton, E. 387[82]
Hsutsung-Yuch, E. 291[36]

Ilgunas, V. 293
Isaev, I. L. 219[82]
van Itterbeek, A. 9, 291[35]
Ivin, K. J. 381[53]

Jackson, J. M. 62, 68[29], 75
Jaffe, H. 28
Jameson, A. K. 83
Javan, A. 53, 117, 192[39], 258[39]
Jenkins, D. R. 218[77]
Johannesen, N. H. 117, 185, 213[16], 216, 230, 231, 232, 256, 257, 258
Johnson, A. C. J. 24
Johnson, R. B. 86, 87, 113
Johnson, R. L. 382[58]
Jones, D. G. 19, 20, 24, 77, 96, 99, 112[64], 120, 121, 124, 125[98], 127, 128, 142, 143[98], 144[98], 145[98], 146, 147, 148, 149, 150[98], 151, 153[98], 154, 155, 156, 248

Jones, G. R. 19, 136, 137, 138
Jones, R. L. 290[27]
Jortner, J. 59[269]
Joyce, P. L. 290[24], 306[64], 312[24], 327 [71], 352
Jupp, C. 285[14], 286[14]

Kaiser, R. 190[30]
Kal'yanov, B. F. 343[100]
Kamac, M. 242[135]
Kamarnitski, A. V. 332, 334[82]
Kantrovitz, A. J. 44, 237, 245[150], 246 [150]
Kaplan, H. 377[41]
Karpovich, J. 248[7], 331[77], 334, 343 [95]
Kelley, J. D. 64, 66, 69, 72[47], 83, 87, 88, 167, 214
Kindall, P. 363
Kennedy, A. D. 381[55]
Kern, R. A. 21
Kerner, E. H. 69
Kessel, L. S. 379, 384[47]
Keeson, W. H. 9
Kiefer, J. H. 104, 105, 108, 215, 230, 232, 242[67, 68], 244, 247, 248, 370[8]
Kiero, E. 296[55]
Kirchhoff, G. 18
Kishimoto, T. 71[40]
Kitaeva, V. F. 219[82]
Kitching, R. 42
Kittel, C. 357
Klebs, K. 318
Klemperer, W. 192, 234
Klimas, P. C. 206
Klose, T. Z. 357[140]
Kneser, H. O. 6, 11, 13, 39, 60, 98[5], 112 [5], 113[148], 116, 119, 120[5], 149, 242 [145], 252[145], 253[171]
Knotzel, H. 39
Knox, J. H. 385[73]
Kohlmaier, J. H. 180[6]
Kolova, G. G. 334[82]
Kompaneets, A. S. 193[49]
Kondratiev, V. N. 369[7], 379[7], 385[7]
Konowalow, D. D. 110

Kor, S. 290[27]
Korobkin, I. 89
Kosche, H. 113[148], 253[171]
Kotova, G. G. 332
Kovacs, M. A. 53, 117, 192[39], 258[38]
Kramers, H. A. 368[2]
Krauss, F. 82
Krebs, K. 325, 327, 336, 337
Kruger, H. 193[47]
Krupka, R. M. 377[41]
Kuhl, W. 32
Kunsitis, C. R. 353[128], 355[128], 362 [128]
Kuratani, K. 180[1], 218[79]
Kurtze, G. 284[11]

Lagemann, R. T. 38, 126, 128, 130
La Graff, J. E. 246
Laider, K. J. 377[41], 379[46], 383[46], 384[46], 385[46]
Lamb, J. 18, 39, 40[83], 90, 122[189], 126, 128, 287, 288, 290[25], 298, 303[58, 59, 61], 306, 318, 325, 326, 327, 331[76, 78], 334, 335, 336, 337, 338, 339, 340, 343, 345, 346, 347, 348, 349[114, 115], 350 [114, 115, 116], 351[58, 115], 352, 353, 356[115], 357[114], 359, 360, 362
Lambert, B. 6
Lambert, J. D. 18, 25[40, 42], 26[45], 30 [45], 95[96], 120, 121, 122[187], 137, 139, 140, 141[233], 146, 148, 149, 151, 154, 155, 157, 158, 160, 161, 163, 230, 241[131], 249
Landau, L. 3, 79, 80, 224, 228, 296
Lapp, M. 220[94]
Larson, G. 219[83]
Laver, S. K. 387[84]
Lavercombe, B. J. 52, 53, 119
Lawley, L. E. 284[8]
Lawrence, R. 137, 138
L'Ecuyer, J. 154[256]
Lee, K. P. 38, 107, 108, 112, 151, 249
Legvold, S. 25[43], 28[51], 31, 133, 134, 137, 155, 158, 159, 160
Leighly, E. M. 335[87]
Leroy, G. E. 137

Levitt, B. P. 153[288], 221, 246, 370[17]
Lewis, J. W. L. 119, 151, 180[5]
Lezhnev, N. B. 290[29]
Lifschitz, A. 180[2]
Lifschitz, C. 180[2]
Light, J. C. 100, 235, 246
Lindemann, F. A. 368[1]
Lindsay, R. B. 18, 120
Linnett, J. W. 122, 219
Lintz, P. R. 42, 120, 146, 149, 249[163]
Lipscomb, F. J. 45
Liska, E. 26, 116, 119
Litovitz, T. A. 7, 8, 13, 17, 18, 21, 26, 60,
   68[3], 72, 74[3], 75[3], 76[3], 77[3],
   78[3], 90[3], 103, 114, 189, 228, 229,
   266[1], 267, 284[12], 290[26], 297[1],
   325, 335, 343, 344, 350[1], 356, 357[2,
   133], 359, 360, 362, 363, 373[32]
Liutyi, A. I. 191[37]
Locker, D. J. 380[52]
Longuet-Higgins, H. C. 72
Losev, S. A. 138, 242[136], 257
Lucas, R. 290
Lukasik, S. J. 4[6], 42, 109, 111, 237, 245
   [151]
Lutz, R. W. 104, 105, 108, 215, 230, 233,
   242[67, 68, 120], 244, 247, 248, 370[8]
Lyon, T. 290[26]

McBride, B. J. 206[52]
McChesney, M. 60, 69[9], 99[9], 214
McCoubrey, J. C. 6, 7, 18, 50, 60, 68, 72,
   86[4], 101, 102[4], 112[4], 116, 119, 122
   [185], 123[4], 123, 131, 133, 136, 138,
   189, 228, 235[120], 255[175], 369[4],
   371[4]
McDevitt, N. I. 328[73], 329
McElroy, M. B. 154[256]
Macfarlane, I. M. 10, 51, 52[116], 116,
   117[157], 119[157], 120[175], 125[175],
   127, 152, 191[31], 256[31], 257[31], 307
   [14]
McKellar, J. F. 312[67]
McLain, J. 125, 126[197], 165, 229[110]
McLaren, T. I. 207, 210
McNamara, F. L. 291[38]

McSkimin, H. J. 31, 283[6]
Madigosky, W. M. 59[2], 126, 357, 358,
   359, 360, 362
Mahan, B. H. 89, 380[52]
Mandl, F. 70
Markham, J. J. 18
Marriott, R. 86, 87, 88, 100, 109, 110, 118,
   121, 152
Marsden, R. J. B. 340[89]
Martin, P. E. 126, 128
Martinez, J. V. 43, 255[176]
de Martini, F. 53, 104, 119, 193, 242[42],
   243, 244
Mash, D. T. 296[54]
Mason, E. A. 28, 73, 74, 121[49], 122, 131,
   272[4], 297[4]
Massey, H. S. W. 68, 88[32]
Matheson, A. J. 31, 125, 126, 128, 130,
   164, 165, 168[200], 229[111, 112]
Matthews, D. L. 109, 110, 242[142], 249
Meister, R. 284[12]
Meixner, J. 4, 7, 90, 96
Melinkova, L. P. 322[70], 325
Meyer, E. 32, 37, 132
Mies, F. H. 66, 82, 83, 87, 88
Miller, A. R. 10, 134
Millikan, R. C. 47[105], 48, 49, 60, 105,
   108, 109, 110, 112, 113, 138, 139[7], 153,
   155, 164, 165[7], 167, 183[11], 192, 206,
   219[87, 88], 220[87], 229, 230, 233, 236,
   237, 238, 239, 240, 241, 242[87, 88, 137,
   138, 143], 244, 245, 246, 247, 248, 249,
   250, 252[88, 168], 372[25]
Milward, R. C. 122[185]
Mirels, H. 207, 210
Miyama, H. 246[156]
Mizushima, S. 314[68], 317, 319, 321, 324
   [68], 328[68], 329
Modica, A. P. 246
Moeller, C. E. 132, 345[124]
Moen, C. J. 284[9]
Mohma, D. M. 335[87]
Mokhtar, M. 291[40]
Monchick, L. 73, 74, 112, 121[49], 131
Monson, P. R. 229[99]
Montroll, E. W. 90[91], 99[91], 373[29]

Moore, C. P. 52, 53, 117, 118, 119, 120 [164, 168], 127, 128, 129, 130, 151[210], 152[164], 153, 165, 168[264], 170[264], 171[164, 168], 192, 239
Moreaux, A. 346, 348[110]
Moreno, J. B. 104, 216, 242[72]
Morgan, E. J. 21
Moriamez, M. 346, 348[110]
Moriamez, C. 346, 348[110]
Mott, N. F. 67, 68[29], 75, 88[32]
Mountain, R. D. 355
Moyal, J. E. 69
Mulders, C. E. 284[10]

Nakamura, K. 329
Neergaard, G. B. 291[41]
Nikitin, E. E. 89, 113, 252, 254
Nolan, M. E. 146
Norrish, R. G. W. 45, 153[253], 156[253], 191[32], 218[78], 242[144], 252
North, A. M. 138[284], 165[284]
Nozdrev, V. F. 293[44], 343
Numann, E. 152
Numann, Z. 255
Nuovo, M. 291[39]

O'Connor, C. L. 353[118], 355
Ogryzlo, E. A. 387[80, 81]
Oliver, R. J. 306[64], 352
Olschewski, H. A. 377[42], 385[42]
Olsen, J. R. 134
Oppenheim, I. 90[272]
Orville-Thomas, W. J. 303, 325, 327, 336, 337[86]
Osberg, L. A. 164, 165[263], 167, 239
Osipov, A. I. 138
Ossa, E. 180[3]
Owen, N. L. 342[90, 91]

Padmanabhan, R. A. 318, 320, 325, 336, 337
Page, M. 383[64]
Palmer, H. D. 387[79]
Parbrook, H. D. 119
Parker, J. G. 40, 42, 64, 68, 104, 105, 107, 108, 109, 111, 114, 126, 127, 128, 151, 244, 246[149], 248

Parks-Smith, D. J. 38[71], 95[96], 122, 124, 137[96], 138[96], 154[255], 155, 157, 158, 161, 163
Partington, J. R. 9[16]
Paulauskas, K. 294
Pauly, H. 169
Pearce, V. F. 290[20], 312[20]
Pedinoff, M. E. 290[33], 332, 333
Pellam, J. R. 288
Pemberton, D. 25, 26[45], 30[45], 122 [188], 138[233], 141[233], 155, 160, 161
Penner, S. S. 58[1]
Pentin, Yu A. 322[70], 325, 332, 334[82]
Peselnick, L. 290[26], 357[129], 358, 360
Petersen, O. 43
Pethrick, R. A. 338
Petrauskas, A. A. 318, 322, 325
Phillips, C. G. S. 6
Phinney, R. 230
Pielemeir, W. H. 152
Piercy, J. E. 132, 270[3], 271, 287[16, 17], 288, 301, 302, 318, 323, 331[76], 331, 333, 343, 344, 345, 347, 348, 349[112], 353[119], 354[119, 125]
Pinkerton, J. M. M. 290[21], 306, 343, 344, 345, 346
Pitzer, K. S. 329
Placzek, G. 296
Polanyi, J. C. 370[15]
Polanyi, M. 384[67]
Porter, G. 191[33]
Porter, G. B. 383[66]
Potapov, A. V. 219[82]
Pratt, N. H. 374[38]
Prings, C. J. 310[65]
Pritchard, G. O. 370[12], 378[12], 382[60]
Pritchard, H. O. 370[11], 371[20, 22], 374 [34, 37], 375[22, 37], 377[37], 379[50], 381[54, 55], 382[57, 61], 383[63, 64], 384[22, 70, 75], 385[75], 387[11]
Pusat, N. 19, 136, 137

Rabinovitch, B. S. 180, 370[13], 378[13], 380[51, 52]
Rank, D. 296[55]
Rankin, C. C. 100, 235, 246

Rapp, D. 64, 65, 66, 68, 83, 84, 85, 86, 87, 88, 118, 153, 170, 373[33]
Read, A. W. 10, 49, 50, 51, 52[116], 53, 60, 109, 110, 112, 113, 116, 117[157], 119, 120[157, 175], 125, 126, 127, 128, 131, 133, 134, 152, 183[10, 12], 190, 191 [31], 192[10], 229[110], 241, 242[132], 245[132], 250, 254, 255[132], 256[31], 257[31], 258[132], 370[10, 14]
Ream, N. 127
Reed, R. D. C. 284[8]
Reeves, L. W. 332, 344[81]
Rehm, R. G. 100, 171, 235, 246
Resler, E. L. 180[4]
Rhodes, C. K. 53, 117, 192[39], 258[39]
Rice, F. O. 1, 11
Rich, J. M. 100, 171, 235, 246
Richards, L. W. 126, 127, 128, 221, 242 [100], 258
Rieckhoff, K. E. 296[50]
Ripamonti, A. 74[51], 105, 114, 115, 121, 124
Riveros, J. M. 345
Roach, K. F. 117
Robben, F. 221, 252, 253
Roberts, G. A. H. 122[187]
Roberts, J. K. 10
Roesler, H. 25, 40, 41, 42[44, 86], 99, 105, 108, 116, 120, 142, 143, 144[100], 145, 153, 249
Rontgen, W. C. 48
Rossilkhin, V. S. 191[37]
Rossing, T. D. 28[51], 133
Roth, W. 91, 221, 242[95, 147], 254, 255 [147]
Rowlinson, J. S. 101, 122[187]
Rozek, A. L. 328[73], 329
Rubin, R. J. 99, 100
Rubin, N. J. 226
Rush, D. G. 374[37], 375[37], 377[37]
Russo, A. L. 109[133], 188, 219, 245, 246

Sahm, K. F. 21, 22, 113, 120, 252[169]
Salkoff, M. 71
Salter, R. 25[40, 42], 30[54], 121, 122, 137, 139, 230

Saxton, H. L. 152
Schapiro, A. H. 204
Schlag, E. W. 378[44], 380[51]
Schlupf, J. P. 353[118], 355
Schnauss, V. E. 42, 114[153], 115, 149, 249
Schodder, G. R. 32
Schofield, D. 77[56], 93, 118, 119, 132
Schroder, F.-K. 32
Schwartz, R. N. 71, 72[45], 78, 90[57], 229, 360, 361
Sears, F. W. 290
Searles, S. 335[87]
Sebacher, B. I. 188[25]
Secrest, D. 87, 113
Sell, H. 32
Seshagiri Rao, M. G. 318, 323, 348[112], 349[112]
Seshagiri Rao, T. 158
Sessler, G. 32, 37
Setser, D. W. 180[7, 8], 382[58, 59]
Sette, D. 25[41], 31, 131, 133, 134
Seybold, P. 387[83]
Sharipov, Z. 332, 334[82]
Sharp, T. E. 66, 83, 84, 85, 86, 87, 88, 118, 170, 373[33]
Sharunov, V. S. 296[54]
Shaw, D. H. 370[11], 387[11]
Sheen, D. B. 221, 246, 370[17]
Sheppard, N. 314[69], 317, 324[69], 325, 328[69], 342[90, 91]
Sherwood, J. 331[78], 334
Shields, F. D. 38, 105, 107, 108, 112, 114, 115, 117, 119, 122[186], 123, 126, 127, 128, 152, 221, 242[100], 248, 255, 258, 357[141], 362
Shimanouci, T. 329
Shin, H. K. 89, 167
Shuler, K. E. 4, 90, 99, 100, 373[29, 30]
Siegert, H. 294
Simons, J. W. 180[7, 8]
Simpson, C. J. S. M. 117, 118, 119, 220 [93], 257
Singh, R. P. 333[80]
Singal, S. P. 353[128], 355[128], 362[128]
Sittig, E. 114, 115, 242[145], 252[145]

Slater, J. C. 64[19], 379
Slawsky, Z. I. 71, 74, 78, 78[57], 89, 90[57], 229
Slie, W. A. 343, 344, 356[134], 363
Slobodskaia, P. V. 116, 258
Smiley, E. F. 114, 115, 213[62], 242[62], 255, 256
Smith, D. H. 42
Smith, F. A. 40, 105, 107, 108, 112
Smith, F. G. 248
Smith, F. J. 387[78]
Smith, I. W. M. 91, 153[93]
Smith, M. C. 291[37]
Smith, W. M. 117, 191[35], 233, 254
Soboleve, N. N. 219[82]
Sovers, O. 76[53], 219
Sowden, R. G. 379[50], 381[54], 382[61]
Spokes, G. N. 382[56]
Srinivasachari, E. 158
Stakelon, T. S. 355
Steacie, E. W. R. 381[53]
Steel, C. 383[65], 384[71]
Steele, W. A. 130
Stegman, G. I. A. 356[132]
Steinberg, M. 221, 234, 242[102], 245, 258
Steinveld, J. I. 170[268], 192[41], 234
Stewart, E. S. 30[55], 31
Stewart, J. L. 28[52], 30[55], 31
Stoicheff, B. P. 296[53], 356[132]
Stokes, G. C. 18
Stolen, P. H. 356[132]
Stott, M. A. 24, 37, 138[283]
Strømme, K. O. 332, 334[81]
Strong, A. A. 137
Stretton, J. L. 18, 19, 20, 25, 26[45], 30 [45], 32, 33, 35, 36[128], 38[28], 77[55], 80, 81, 86, 93[55], 94[62], 95[62, 96], 102[55], 105[62], 106, 110, 120, 123, 124, 127, 128, 129[55], 130, 131, 134, 135, 137[96], 138, 141[233], 146, 148, 149, 151, 154[255], 155, 157, 158, 160, 161, 163, 164, 168[55], 170[55]
Stumpt, F. B. 348[111]
Subrahmanyan, S. V. 331[76], 332, 333, 343, 345
Sutton, L. E. 340[89]

Swope, R. H. 42, 105, 107, 108, 126, 127, 128

Tabuchi, D. 293[45], 343
Takayanagi, K. 60, 63, 64, 65, 66, 68, 71, 72, 88[76], 167, 228
Tamm, K. 283[5], 284[11], 312
Tamres, M. 335[87]
Tanczos, F. I. 18, 78, 80, 81, 90[58], 91 [58], 93, 94, 95[58], 96, 118, 127, 131, 133, 134
Taylor, M. D. 335[88], 369[5]
Taylor, R. 219, 220[84, 85], 233, 242[84], 249
Taylor, R. L. 153[287]
Telfair, D. 152
Teller, E. 3, 79, 80, 224, 228
Tempest, W. 32, 37, 40, 105, 107, 108, 112, 119, 134, 136, 248
Tetevski, U. M. 325
Thiele, E. 170
Thomas, T. H. 323, 325, 336, 337[86]
Thommarson, R. C. 382[60]
Thompson, M. C., Jr. 36, 37
Thrush, B. A. 45
Toborg, R. H. 151[246], 152
Toennies, J. P. 169, 193[46], 206, 209[46], 374[35]
Tomkinson, D. M. 382[57], 384[57]
Townes, C. H. 296[51, 52]
Treanor, C. E. 69, 84, 171
Troe, J. 377[42], 385[42]
Trotman-Dickenson, A. F. 379[45, 50], 381[54], 382[61], 383[45, 64, 65], 384 [45], 385[73]
Truesdell, C. 21
Tschuikow-Roux, E. 89[78]
Tsuchiya, S. 109, 111, 218[79], 237
Tuesday, C. S. 44, 120, 146, 147, 249[164]
Turner, R. 132, 354[125]
Turner, R. E. 65
Tyerman, W. J. R. 372[27]
Tyndall, J. 48

Ubbelohde, A. R. 122[185], 136, 235[126], 255[175]

Ul'Yanova, O. D. 322[70], 325
Unland, M. L. 373[31]
Urushihara, K. 120

Valance, W. G. 378[44]
Valley, L. M. 25[43], 31, 137, 158
Venugopalan, M. 370[12], 378[12], 382
  [12]
Verma, G. S. 333
Verhaegen, L. 291[35]
Vincent, D. G. 285[14], 286[14]
Vincenti, W. G. 193[47]
Voitonis, V. V. 31[57]
Voltena, V. 356[132]

Wagner, H. G. 377[42], 385[42]
Walker, R. A. 28[51]
Walters, A. B. 310[65]
Wang, M. L. 159[262]
Warburton, B. 122[184]
Wares, G. W. 219[83]
Watkins, K. W. 350, 380[52]
Watt, W. S. 180[5]
Weaner, D. 117
Weare, J. 170
Weiss, G. H. 90[92, 272], 100, 137, 373[30]
Werner, H. W. 152[247]
de Wette, F. W. 71, 74
White, D. R. 105, 108, 112, 113, 138, 149,
  151, 153, 213[63], 219[88], 229, 230,
  236, 237, 238, 240, 241, 242[63, 88, 130,
  134, 137, 138, 139], 243, 244, 245, 246,
  247, 248, 249, 250, 252[88], 259, 346,
  348[109]
Whetstone, C. N. 126, 128, 130
Whitteman, W. J. 152[247], 215
Whitten, G. L. 89[78]
Widom, B. 89, 152, 235, 377[40]
Wiggins, T. 296[55]
Wigner, E. 384[67]
Wiley, W. J. 38
Wilkinson, V. J. 122[187]
Willard, C. W. 291[33]
Williams, E. J. 336, 337
Williams, G. J. 45[102, 103], 47[103], 113
  [147], 146[147], 153, 155, 191[36], 254
  [36], 372[23]

Wilson, D. J. 88
Wilson, E. B. 80, 169[60]
Windsor, M. H. 219, 220[84, 85], 233, 242
  [84], 249, 369[5]
Winkler, E. H. 114, 115, 213[62], 242[62],
  255, 256
Winter, T. G. 168[265]
Witteman, W. J. 68, 100, 118, 242[70], 256
Wolfsberg, M. 64, 69, 72[47], 83, 87, 88,
  167
Wolnik, S. 219[83]
Wood, J. L. 137[231]
Wood, R. E. 117, 119[164], 120[164],
  152[164], 171[164]
Woodley, J. G. 188, 189[27], 245
Woodmansee, W. E. 49
Wray, K. L. 89, 98[80], 113, 242[146], 246
  [155], 252, 253
Wright, J. K. 193[48]
Wright, M. 387[82]
Wright, W. M. 32
Wu, J. C. 63
Wyn-Jones, E. 303, 318, 320, 323, 325,
  327, 336, 337[86], 338

Yakolev, V. F. 31[57]
Yamada, K. 120, 121
Yardley, J. T. 52, 53, 117, 118, 119[164,
  168], 120[164, 168], 127, 128, 129, 130,
  151[210], 152[164], 153, 171[164, 168],
  192
Youlsef, H. 291[40]
Young, A. H. 116, 117[157], 119[157], 120
  [157], 127, 191[31], 256[31], 257[31],
  370[14]
Young, J. E. 42, 51, 52[116], 110, 111, 237,
  245[151]

Zartman, I. F. 30
Zeldovitch, I. B. 193[49]
Zener, C. 66, 228
Zienkiewicz, H. K. 117, 185, 213, 215,
  230, 231, 232, 256, 257[15], 258[15, 16]
Zuecrow, M. J. 204

Subject Index

Absolute calculations 228
Absorption 2, 3, 18, 36, 37, 38, 39, 40, 306
  coefficient 7, 266, 269, 274, 284, 289,
    291, 292, 296, 347, 350
  equation 14, 15
  height 23
  maximum 23, 25
  measurements 26, 31
  of sound 295
  spectroscopy 6, 45
  /wave length 266, 269
Accuracy 26
Acetaldehyde 38
Acetic acid 345, 346
Acetone 383
Acetylene 24, 239, 242
Acetylenes, substituted 164
Acoustic approximation 92
  chamber 28, 33, 34, 43
  data 271
  experiments 94, 101, 105, 109, 116, 141,
    168, 169
  instruments 27
  interferometer 27, 114
  measurements 5
  methods 5, 6, 155
  path 286
  -resonator technique 142
  relaxation times 90
  streaming 286
  waves 269
Acrolein 338, 340

Activation energy 349
  enthalpy 303
  entropy 303
Active vibrations 205
Adiabatic compressibility 15, 273, 277
  curve 198, 199
  principle 61, 63, 78
  process 61
  relaxation time 11, 12, 15, 18, 94
Aerodynamic methods 5, 43
Affinities 97
Air 249
Aldehydes 338
Alphatron 38
Ammonia 43, 125, 167, 169, 239
Amplitude spectrophone 49
  density functions 86
  of vibration 64
Angular frequency 7
  momentum 62, 70
Anharmonicity 118, 171
Anti-Stokes components 295
Argon 87, 112, 113, 123, 130, 166, 239,
  242, 251
Arrhenius parameters 328
Arsine 165, 239
Atoms 76
  recombination 377
Attenuation 266
Attractive forces 65, 74, 77
Average molecular weight 6
Azomethane 383

401

Band-reversal techniques 216, 219
Barrier heights 315, 320, 322, 323, 329
Bauer method 96
Benzene 354
Bimolecular collisions 4, 59
  reactions 385
Binary collisions 90
  mixtures 60, 112, 140
Boltzmann distribution 45, 373, 374
Born approximation 67
Boron trifluoride 123
Boundary layer 207, 210, 211
Bragg equation 295
'Breathing sphere' model 80, 167
Brillouin spectroscopy 295, 349
  scattering 132
de Broglie wave length 69, 75, 77
Bromine 88, 114
Bromocyclohexane 332
Bromoethylene 239
2-bromohexane 323, 325
2-bromopentane 323
3-bromopentane 323, 325
Butane 137
n-butane 318
Butyric acid 346
n-butyraldehyde 339, 342

Calculation, shock tube parameters 210
Capacitance manometer 38
Carbon dioxide 1, 26, 38, 52, 53, 90, 100,
  116, 118, 120, 151, 153, 168, 220, 230,
  240, 242, 256, 259, 339, 352, 353, 358,
  362, 385
  disulphide 38, 116, 352, 353, 362
  monoxide 25, 47, 49, 50, 104, 109, 112,
  113, 145, 153, 155, 164, 183, 220, 233,
  235, 240, 242, 245, 249, 259
  oxysulphide 116
  tetrachloride 133, 170, 356
  tetrafluoride 25, 26, 133
Carboxylic acids 345
Cartesian displacement 80
Cell model 360
Centrifugal energy 71
Chemical reaction 4, 60, 97, 368

Centrifugal barrier 76
Chlorine 114, 230, 242, 255
1-chloro-2-bromopropane 320, 325
Chlorocyclohexane 332
Chlorobromo methane 133
Chloroethylene 239
2-chloroethyl vinyl ether 342
Chlorofluoro methane 133
Chloroform 134
2-chloropentane 323
3-chloropentane 325
Cinnamaldehyde 338, 340
cis-1,2-dichloroethylene 326
Classical absorption 18, 19, 33
  calculation 60, 63, 68, 79, 84, 109, 110,
  114
  limit 68, 69
  mechanics 58
  perturbation 81
  turning point 71, 74
Cole–Cole plot 24, 116
Collision 4
  cross section 60, 387
  life time 44, 47, 52, 58
  number 60, 139, 155, 162, 184
  pair 63
  process 58
  rate, 58, 59, 75
  trajectories 63, 64, 69, 75
  velocity 74
Collisional deactivation 387
  excitation 368
Comparison measurements 294
Complex exchange 161
  process 58, 60, 77, 79, 102, 103, 118,
  133, 141, 152, 159
  transfer 113
  transitions 92
  velocity 14
Compressibility 272
  adiabatic 15
  , isothermal 15
Compression 7, 14
Condenser microphone 38, 39, 50
  transducers 33, 34, 37, 42, 49
Conservation equations 196, 201, 205

Continuous waves 32, 36
Coordinate systems 195
Coriolis coupling 118
Coupled modes 141
Critical frequency 266
Cross section 185
  talk, electrical 36, 37
  talk, mechanical 36
Crotonaldehyde 338, 340
Crystal impedance 30
Cyanide radical 221, 242, 255
Cyanogen 123
Cyclobutane 381
Cyclohexane 330
  derivatives 329
Cyclohexanol 334
Cyclohexylamine 334
Cyclopentane 137
Cyclopropane 168, 356, 382

Degeneracy 81
Detailed balancing, principle of 78, 91
Deuterated molecules 140, 164
Deuterium 104, 119, 139, 168
  oxide 120, 155
  sulphide 120
Difluorochloromethane 160
Difluorodichloromethane 160
Diatomic gases 141, 168
  molecules 4, 59, 64, 76, 81, 82, 86, 90,
    103, 111, 167
  oscillator 63
1,2-dibromobutane 308, 323, 325
d-1,2,3-dibromobutane 320, 325
1,2-dibromoethane 318, 326
Dibromomethane 352
1,2-dibromo-2-methylpropane 304, 308,
  316, 318, 325, 329
1,4-dibromopentane 323, 325
1,2-dibromopropane 320, 325
1,2-dibromopentane 323, 325
1,2-dibromo-1,1,2,2-tetrafluoroethane
  318, 325
2,3-dichlorobutane 312
d-1,2,3-dichlorobutane 320, 325
1,2-dichloroethane 318, 326
1,2-dichloroethylene 326

dichloromethane 352
1,2-dichloro-2-methylpropane 318, 325,
  329
1,2-dichloropropane 320, 325
dicyclohexylamine 334
dideuteroacetylene 123
3,N-diethylaminobutane-2-one 336
1-diethylamino-2-chloropropane 336
2-diethylaminoethylamine 336
3,N-diethylaminopropiononitrile 336
3,N-diethylaminopropylamine 336
N,N-diethylcyclohexylamine 336
Diffraction angle 291
Diffusion effects 168
Difluoromethane 239
Dihalogenated ethane 327
2,3-dimethylbutane 318, 325
1,4-dimethylcyclohexane 331
Dimethyl ether 155
Dinitrogen tetroxide 97, 120
Dipolar attraction 73
  interactions 59, 89, 146
  moments 112
3-N-dipropylaminopropiononitrile 336
Dispersion 1, 5, 26, 43, 266
  measurements 40
  equation 14, 15, 17
Dissociation 180, 374, 377, 384
  , polyatomic molecules 378
  process 386
Distorted waves 67, 71, 80
  wave approximation 71
  wave method 88
Distribution of velocities 79
Debye waves 295
Decay time 284
Dense media 59
Densitometry techniques 214
Density 7, 272
  ratio 201
Deuterium 240, 242, 243, 244
Deutero acetylene 19
Double quantum jumps 80
  relaxation 26, 122, 131, 133, 136, 137,
    141, 160, 170
  transducer instruments 31
Dynamic viscosity 270

Effective compressibility 15
  frequency 4
  specific heat 11, 14
  velocity of approach 71
Ehrenfest's principle 61, 63, 78
Eigen-frequencies 39, 40
Einstein transition probability 183
Elastic collision 59
  scattering 67, 68
  cross section 75, 77
Electrical cross talk 36
Electronic energy 21
  energy levels 89
  relaxation 21
  states 153
Empirical correlations 91, 138
  suggestions 229
Energy box 267
Energy of liquids 267
  relaxation 91
  , two states 315
  transfer 58, 63
Enthalpy difference 314, 324
  equation 7, 17
Entropy 273
  change 203, 204
  difference 327
  increase 201
Equation of continuity 7
  of motion 7, 17, 63
Equations of state 7, 9
Equilibrium constant 276
  populations 377
  region 196
Esters 342
Ethane 25, 134, 137, 158, 239, 384
  derivatives 302, 303
Ethyl acetate 343
  formate 343
Ethylene 134, 188, 242
  oxide 168, 239
  oxide mixtures 158
Ethylenes, substituted 164
2-ethyl-3-propyl acrolein 339, 341
Ethyl vinyl ethers 342
Excited oxygen 233
Expansions 188, 234

Expansion coefficient 278, 297
External degrees of freedom 5

Fall-off region 380
First approximation 68, 71
  order perturbation theory 83
  order solution 66
Fixed wall model 361
Flash photolysis 47, 191
  spectroscopy 45, 53
  technique 152
Fluorescence 52, 53
  technique 44, 47, 49, 117, 119, 153
Fluorine 38, 114
Fluoroethylene 239
Formic acid 346, 348
Forward rate constant 276
Free energy 297
Fresnel region 289
Frozen gas 196
Furacrolein 338, 341
Furfural acetone 339, 341

Gas-density microbalance 6
  dynamics 60
  mixtures 6, 90, 92, 98
  purity 6, 37
Group V hydrides 123, 125
Germanium–copper detectors 220
  –gold detectors 220
  tetrachloride 134

Half-band width principle 42
Halogens 114
Halomethanes 134, 155, 159
Hamiltonian, unperturbed oscillator 65
Harmonic oscillator 62, 64, 80
Heat conduction 11
Helium 26, 82, 113, 120, 168, 242, 251
Heteromolecular collisions 127, 140, 155
  force constants 95
Hexane 137, 323
Hexyl cinnamic aldehyde 339, 341

High energy collisions 83
  effective frequencies 31
  frequency factors 383
  frequency specific heat 15
  vibrational levels 232
Hinshelwood theory 379
Homomolecular collisions 140
Hugoniot relations 196, 197, 198
Hydrides 123, 125, 164
Hydrocarbons 21, 134, 164, 167
Hydrogen 49, 53, 69, 82, 104, 112, 119,
  139, 164, 168, 242, 243, 251
  atoms 81
  bonding 265
  bromide 166, 239, 240, 242, 254
  chloride 87, 220, 234, 239, 242, 254
  cyanide 116, 120
  halides 82, 112, 139, 168, 238, 254
  iodide 112, 221, 234, 238, 239, 242, 254
  sulphide 120, 155
Hypersonic velocity 353, 354

Ideal gases 7
Impact tube 5, 43, 146
Impurities 6, 105
Indium antimonide detectors 220
Inelastic collision 59, 60
  cross section 75, 76, 86, 87
  partial wave 75
  scattering 67
Inert gases 119
Infrared active modes 89, 116, 127
  fluorescence 192, 219
Instantaneous temperature 90, 93
  vibrational temperature 11
Interferometer 213, 292
Intermolecular collisions 4
  exchange 118, 130
  potential 60, 62, 68, 70, 72, 75, 82, 112,
  138, 169, 170, 229
  repulsion forces 4
Internal degrees of freedom 5
  rotation 334, 338, 348
Intersystem crossing 387
Intramolecular exchange 118, 130
Inversion 125

Iodine 114, 170, 234
2-iodobutane 312
Iodoethylene 239
Ionization 4, 60
Irreversible thermodynamics 90, 96
Isentropic process 43
Isobaric process 11
Isobutyl bromide 316, 318, 325, 329
  chloride 318, 325, 329
  iodide 318
Isochoric process 11
3-isomer system 321
Isopropyl formate 343
  ketone 341
Isothermal compressibility 15, 277
  relaxation 11
  relaxation time 11, 13, 15, 18, 92, 99,
  101

Kassel theory 379
Ketones 338
Kinetic energy 4
k matrix 94, 96, 99, 123, 169
Krypton 242, 251

Lambert–Salter plot 39, 168
Landau–Teller model 80, 91, 92, 99, 184,
  224, 228, 230, 235, 258, 371
Laplace's formula 9
Lasers 44, 52, 53, 119, 129
Lead sulphide detectors 220
  zirconate 28
Lennard-Jones potential 72, 229
Lifetime, collisional 58
  , radiative 58
Lindemann theory 379
Linear molecules 81, 116, 123
Line-reversal technique 216
Liquids 59
London dispersion forces 59, 73
Long chain molecules 323
Low frequency techniques 38
  -lying vibrational states 4
Lycot coronagraph 43

Mach number 202
Martian atmosphere 154
Maxwellian distribution of velocities 76
Mean free path 356
Mercury manometer 31
Mesityl oxide 341
Methacrolein 338, 341
Methane 52, 80, 125, 129, 130, 134, 149,
   151, 153, 155, 168, 221, 239, 242, 258
Methyl acetate 343
5-methyl-2-acetyl furane 341
Methyl bromide 130
2-methylbutane 318, 325
α-methyl-cinnamaldehyde 338
Methyl chloride 130, 356
  cyclohexane 302, 332, 334
  cyclohexanone 334
Methylene bromide 132
  chloride 25, 131, 136
  fluoride 131, 159, 160
  halides 125, 131, 354
Methyl fluoride 81, 130
  formate 343
  halides 130, 164
  isopropenyl ketone 339
2-methylpentane 323, 325
3-methylpentane 323, 325
Methyl radicals 381
  vinyl ethers 342
Methylvinyl ketone 339, 341
Microphone calibration 51
Mixtures 60, 145, 151, 226, 240
 , halo-methanes 159
  polyatomic molecules 155
Modified wave number 71
Modulus 272
  of a liquid 270
Molar volume 9
Molecular acoustics 265, 266
  beam 169
  -collision processes 368
  collision rate 19
  diameter 357
  oscillator 58, 65, 86
  rotation 125
  weight 26

Moment of inertia 164, 166
Monosilanes 130
Morse potential 68, 74, 75, 105, 229
Movable reflector 27
  seal 28
Moving coil microphone 49
  coil transducer 40
Multi-jump transitions 88
Multiple relaxation 25, 119, 121, 124, 125,
   130, 131, 162, 168
Multi-quantum jumps 82, 130, 171

N-th order perturbation approximation
  83
Napier number 185
  probability 185
  time 181, 183, 185, 228, 258
Navier–Stokes equation 19
Neon 87, 130, 251
Neo-pentane 137
Nitric oxide 21, 45, 48, 89, 113, 153, 155,
   221, 233, 238, 242, 252
Nitrogen 42, 82, 83, 85, 87, 99, 104, 109,
   111, 112, 114, 116, 142, 153, 221, 230,
   234, 235, 236, 237, 242, 245
Nitrogen dioxide 120
Nitrous oxide 42, 116, 220, 242, 255, 356
Nomenclature 17, 18
Non-adiabatic process 61
Non-ideality in a second virial coefficient
  10
Non-linear molecules 125
  triatomic molecules 120
Normal coordinate of vibration 79
Number of collisions 184

Observation time 206
  time, compression of 206
Octanoic acid 348
One-dimensional collision 62
Optic–acoustic effect 6, 44, 48, 51, 62
Optical activation 104
  diffraction gratings 43
  methods 6, 44, 152, 169, 290
Orientation 76
Orthohydrogen 240

Oscillator eigenfunction 65, 66, 70
Oxygen 6, 25, 38, 40, 61, 71, 81, 82, 99,
    104, 105, 112, 113, 114, 142, 145, 146,
    149, 151, 153, 166, 168, 230, 231, 235,
    240, 242, 247
    mixtures 38, 44

Parahydrogen 240
Parallel excitation 100, 123, 159
Partial waves 70, 71, 76
Penta-atomic molecules 125
Pentane 137, 323
Perfect gas 9
Perturbation treatment 69, 78
Perturbed oscillator 58
    stationary states 88
Piezoelectric crystals 28, 43
Planck–Einstein formula 350
Polar molecules 73
Polyatomic molecules 4, 5, 59, 76, 81, 92,
    94, 139, 378
    gas 10, 168
12-6-3 potential 113
Potential barrier 340
    energy 62
Phase lag 49, 51, 53
    velocity 266
    spectrophone 49, 50, 51, 53
Phosphine 165, 239
Pressure amplitude 266
Probability of energy transfer 184
Propagation of sound 7, 21, 60
    of sound at high frequencies 21
Propane 137, 239
Propionaldehyde 339, 342
Propionic acid 346
Propylene 136
n-propylbromide 318, 325
Pulses 33, 34
Pulse methods 288
Purity of gases 6

Q factor 284, 286
Quantum mechanical calculation 60, 66,
    75, 78, 86, 167

Quartz 28
    crystal 30, 36, 37

Radiation 4, 58
Radiative lifetime 44, 47, 52, 58, 112, 183
Radiation pressure 291
Raman method 53, 192
Rankine–Hugoniot relations 196, 197
Rate equation 90, 280
Reabsorption of fluorescence 49
Real gases 209
    phase velocity 7
Receiver 27
Reciprocity technique 51
Reduced mass 63, 102, 166
Reech's theorem 9
Reflected shock 94
Reflector 30
Refraction fan 194
Relative kinetic energy 63
    velocity 63
Relaxation 3
    constants 95
Relaxational coupling 268, 269
Relaxation equation 4, 7, 59, 81
    frequency 275, 283
    parameters 271
    strength 17, 23, 95, 351
    time 3, 4, 17, 59, 86, 89, 98, 120, 124,
        129, 130, 134, 182, 268, 269, 278, 302
    time, adiabatic 11, 12, 15
    time, binary collision 356
    time, isothermal 11, 13, 15
    time, translational 19
Relaxing gas 7, 11
Repulsive potential 65, 68, 73, 74, 79, 170
Resonance band width 39, 40
Resonance-exchange 77, 86, 113, 150, 234
    fluorescence 6
    -reverberation methods 284
Resonant complex process 59
Resonator 39, 40, 42, 285
Reverberation techniques 39, 40, 126
Reverse rate constant 276
Reversible process 376

Ribbon transducer 38
RRKM theory 380
Rotation 62, 89, 164, 170
Rotational absorption 136
  energy 5, 21, 81, 376
  -energy exchange 167
  isomerism 302, 303, 314, 327, 342
  relaxation 19, 21, 120
  specific heat 168
  states 164
  temperature 48
  transitions 81
Rotation–vibration coupling 82

Schlieren technique 104, 214
Schrödinger equation 66, 86
sec-butyl bromide 320, 325, 329
  chloride 320, 325, 329
  halide 321
  iodide 320, 325
Second virial coefficient 74
Semiclassical calculation 60, 65, 68, 78,
  81, 86, 105, 110, 114, 153
Series excitation 100, 102, 123
Shear viscosity 349
Shock techniques 5, 43, 103, 114, 126,
  168, 169, 180, 185, 193, 210, 212
  wave 100, 193, 374
  wave profiles 104
Schottky's function 310, 341
Silane 164, 239
Silicon tetrachloride 134
Simple deactivation 136
  process 58, 60, 140, 141
Sonic absorption 43
  velocity 1, 28, 31, 34, 200, 292, 294, 295
Sound propagation 7, 18
  wave 1, 92, 266
Specific heat 1, 5, 277, 280, 297
  , effective 14
  ratio 297
Spectrophone 6, 44, 48, 52, 109, 112, 116,
  190
Spin change 386
  conservation 385

SSH–acoustical approximations 91
  calculations 114, 119, 124, 131, 142,
  154, 155, 162, 170, 229, 240, 258
  –Tanczos calculation 75, 81, 105, 112
Stagnation temperature 44
Stannic tetrachloride 134
Standing wave 27, 30, 32
Stationary perturbation theory 66
Static molar specific heat 92
Steric factor 77, 79, 80, 81, 83, 118, 167,
  171
Stokes component 295
Strain 270
Stress 270, 272
Strong-collision assumption 381
Strong coupling 88
  shocks 203
Structural energy 269
Substituted ethanes 314
Sulphur dioxide 25, 120, 122, 123, 136,
  221, 352, 353
  hexafluoride 19, 94, 138, 155, 160, 165,
  356
Sutherland model 360
Surface atoms 80, 81
Symmetrization 66

Temperature dependence 110, 114, 118,
  120, 136, 137, 165, 168, 170
  dependence relaxation time 5, 107
  ratio 202
  relaxation equation 91
Tertiary amines 334
Tetra-atomic molecules 123
1,1,2,2-tetrabromoethane 318, 325
Tetradeuteroethylene 134, 239
Tetradeuteromethane 164, 166, 239
Tetradeuterosilane 130, 239
Tetrafluoroethylene 25, 26, 138, 155
Tetrafluoromethane 133
Theoretical transition probabilities 61
Thermal conductivity 17, 18
  losses 38
  relaxation 4
  velocity 61
Thiophene-2-aldehyde 339, 341

Three-dimensional collision 69
Threshold energy 76
  level 68
Time-dependent perturbation 65, 77
  -dependent wave equation 65
Torsional mode 137, 158
trans-1,2-dichloroethylene 326
  -1,2-dimethylcyclohexane 334
  -1,4-dimethylcyclohexane 334
Transducer 284, 286, 287
Transition probability 59, 60, 67, 68, 72,
  82, 89, 93, 100, 168, 371, 373
Translational dispersion 19, 137
  energy 1, 164, 267
  relaxation 19, 120
  temperature 2, 11, 48, 59, 92, 100, 117
Transport properties 72, 74, 75, 112
Transverse modes 28, 31
Triallylamine 336
Triatomic molecules 100, 116
Tribromoethane 318, 325
Tributylamine 336
Trichloroethane 318, 325, 382
Trideutero ammonia 125, 238
Trideuteroarsine 239
Trideuteromethyl fluoride 94
Trideuterophosphine 239
Triethylamine 312, 336
Trihexylamine 336
Trifluoromethyl radicals 381
Trimethylamine 336
Tripentylamine 336
Triisopentylamine 336
Triple-quantum jump 137
Tripropylamine 336
Tube method 38
Tungsten hexafluoride 165
Tuning fork 285
Two-state approximation 83, 87, 184,
  223, 229

Ultrasonic dispersion 369
  frequencies 1
  interferometer 292
  measurements 126, 189
  waves 5

Ultraviolet absorption 221
  fluorescence 192, 221
Unimolecular decompositions 383
Units 185
Unsaturated molecules 338
Unsteady expansion 189

Valeric acid 346
Velocity 38, 266
  of approach 74
  dispersion 266, 275, 282, 350, 354
  of light 295
  measurements 36, 212
  , sound 34, 295, 306
Vibrational coordinates 79
  deactivation 370
  degrees of freedom 1, 181
  energy 1, 4, 164, 267, 269, 368, 378
  energy exchange 4, 58, 60, 72, 113, 167,
    372, 373
  factors 67, 79, 80, 154
  levels 181
  matrix element 82
  relaxation 19, 97, 99, 180, 185, 265, 334,
    348, 369, 375
  relaxation times 4, 6, 17, 21
  –rotation model 166
Vibrational specific heat 104, 116, 168,
  349
  states 65, 181
  temperatures 2, 11, 48
  -translational energy transfer 48, 89
Vibration–vibration transfer 103, 159,
  162, 227
Vinyl ethers 342
Virial coefficient 10, 26
  data 75
  expansion 9
Virtual transitions 83, 87
Viscous losses 38
  restoring force 8
Viscosity 17, 18, 74, 270, 287
Viscothermal effects 168
Volume 2

Wall corrections 42
  energization 381
  losses 38, 40, 284
Water 49, 120, 167, 169, 151
  vapour 6, 44, 105, 146
Wave-mechanical calculation 68

Weak coupling 88

Xenon 88, 242

Zero approximation 66, 68, 71

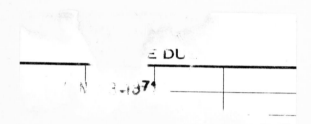